System Level Control and Optimisation of Microgrids

Other related titles:

You may also like

- PBRN0060 | S. Chowdhury, S.P. Chowdhury and P. Crossley | Microgrids and Active Distribution Networks | 2009
- PBPO0860 | Federico Milano | Advances in Power System Modelling, Control and Stability Analysis | 2016
- PBPO1300 | Rupp Carriveau and David S-K Ting | Wind and Solar Based Energy Systems for Communities | 2018
- PBPO0900 | Shin'ya Obara and Jorge Morel | Clean Energy Microgrids | 2017

We also publish a wide range of books on the following topics:
Computing and Networks
Control, Robotics and Sensors
Electrical Regulations
Electromagnetics and Radar
Energy Engineering
Healthcare Technologies
History and Management of Technology
IET Codes and Guidance
Materials, Circuits and Devices
Model Forms
Nanomaterials and Nanotechnologies
Optics, Photonics and Lasers
Production, Design and Manufacturing
Security
Telecommunications
Transportation

All books are available in print via https://shop.theiet.org or as eBooks via our Digital Library https://digital-library.theiet.org.

IET ENERGY ENGINEERING SERIES 149

System Level Control and Optimisation of Microgrids

Edited by
Alessandra Parisio, Johannes Schiffer and
Christian A. Hans

The Institution of Engineering and Technology

About the IET

This book is published by the Institution of Engineering and Technology (The IET).

We inspire, inform, and influence the global engineering community to engineer a better world. As a diverse home across engineering and technology, we share knowledge that helps make better sense of the world, accelerate innovation, and solve the global challenges that matter.

The IET is a not-for-profit organisation. The surplus we make from our books is used to support activities and products for the engineering community and promote the positive role of science, engineering, and technology in the world. This includes education resources and outreach, scholarships and awards, events and courses, publications, professional development and mentoring, and advocacy for governments.

To discover more about the IET, please visit https://www.theiet.org/.

About IET books

The IET publishes books across many engineering and technology disciplines. Our authors and editors offer fresh perspectives from universities and industry. Within our subject areas, we have several book series steered by editorial boards made up of leading subject experts.

We peer review each book at the proposal stage to ensure the quality and relevance of our publications.

Get involved

If you are interested in becoming an author, editor, series advisor, or peer reviewer please visit https://www.theiet.org/publishing/publishing-with-iet-books/ or contact author_support@theiet.org.

Discovering our electronic content

All of our books are available online via the IET's Digital Library. Our Digital Library is the home of technical documents, eBooks, conference publications, real-life case studies, and journal articles. To find out more, please visit https://digital-library.theiet.org.

In collaboration with the United Nations and the International Publishers Association, the IET is a Signatory member of the Sustainable Development Goal (SDG) Publishers Compact. The Compact aims to accelerate progress to achieve the SDGs by 2030. Signatories aspire to develop sustainable practices and act as champions of the SDGs during the Decade of Action (2020–30), publishing books and journals that will help inform, develop, and inspire action in that direction.

In line with our sustainable goals, our UK printing partner has FSC accreditation, which is reducing our environmental impact on the planet. We use a print-on-demand model to further reduce our carbon footprint.

British Library Cataloguing in Publication Data

A catalogue record for this product is available from the British Library.

ISBN 978-1-78561-875-8 (hardback)
ISBN 978-1-78561-876-5 (PDF)

Typeset in India by MPS Limited

Cover image credit: Justin Paget/DigitalVision via Getty Images

Contents

About the editors

Alessandra Parisio is a professor of control of sustainable energy networks at the University of Manchester, UK. She has been a principal or co-investigator on research projects supported by e.g. Innovate UK, EC H2020, and industrial partners. She serves as an IEEE senior member and co-chair of the IEEE RAS Technical Committee on Smart Buildings and as editor of the IEEE Transactions on Control of Network Systems, and the European Journal of Control and Applied Energy. She received the IEEE PES Outstanding Engineer Award in January 2021. Her research interests include distributed optimisation and control, power systems, and optimisation and control of multi-energy networks. She received the IEEE PES Outstanding Engineer Award in January 2021 and the Energy and Buildings Best Paper Award (for a 10-year period between 2008 and 2017) in January 2019. Her main research interests span the areas of control engineering, in particular Model Predictive Control, distributed optimisation and control, stochastic constrained control, and power systems, with energy management systems under uncertainty, optimisation and control of multi-energy networks, and distributed flexibility.

Johannes Schiffer is a professor of control systems and network control technology at the Brandenburg University of Technology Cottbus-Senftenberg, Germany, in a joint appointment with the Fraunhofer Research Institution for Energy Infrastructures and Geothermal Systems (IEG), where he heads the Business Area 'System Integration, Automation and Operation Management'. Prior to that, he held appointments as a lecturer (assistant professor) at the School of Electronic and Electrical Engineering, University of Leeds, UK, and as a research associate in the Control Systems Group and as the Chair of Sustainable Electric Networks and Sources of Energy both at TU Berlin. Prof. Schiffer and his co-workers received the Automatica Paper Prize over the years 2014–16 and the at – Automatisierungstechnik Best Paper Award 2022. He currently also serves as co-coordinator of the Energy Innovation Center Cottbus (EIZ) and as coordinator of the EU MSCA Doctoral Network Dependable Smart Energy Systems (DENSE).

Christian A. Hans is a professor (W1 with tenure track) of Automation and Sensorics in Networked Systems at the University of Kassel, Germany. He earned a PhD degree with distinction from TU Berlin in 2021. Before joining TU Berlin, he worked as an engineering expert at Younicos AG (now Aggreko plc), Germany, developing control algorithms for low-inertia microgrids. His research combines automatic control theory and applications in power systems with intermittent decentralized renewable generation.

Introduction and overview

The need to build a more sustainable urban environment and smarter energy systems is widely acknowledged. In this context, the microgrid concept is regarded as one of the most promising enablers of the energy transformation to a resilient, efficient, and sustainable low carbon future. Evidence for this is the wide attention that microgrids have received over the past years, both from academia and industry. As highlighted in Chapter 3, the concept of microgrid traces back to the beginning of this century [1]. The IEEE standard 2030.7 [2] provides a comprehensive definition of a microgrid, as given below.

Microgrid definition [2]

A microgrid is a group of interconnected loads and distributed energy resources, both controllable and uncontrollable, with clearly defined electrical boundaries that act as a single controllable entity with respect to the grid and can connect and disconnect from the grid to enable it to operate in both grid-connected or island modes.

Typical distributed energy resources in microgrids comprise distributed generators, renewable energy sources, and energy storage devices. At times, the term *distributed energy resources* is also used to mainly refer to distributed generating units.

In [2], the role of the microgrid energy management system is also elucidated. In fact, a key element of microgrid operation is the microgrid energy management system. It includes the control functions that define the microgrid as a system that can manage itself, operate autonomously or grid connected, and seamlessly connect to and disconnect from the main distribution grid for the exchange of power and the supply of ancillary services.

Microgrids are envisioned to be capable of managing and coordinating distributed generation, storage devices and loads in a distributed and flexible manner, hence reducing the need for conventional rigid centralized coordination, supporting the integration of renewable energy sources, decreasing transmission losses, and ensuring a continuous power supply by operating independently of the main grid during emergencies. The increased control over modern communication network and digitization of energy systems renders microgrids increasingly active, i.e., they can generate, sense, compute, communicate, and actuate. This will also enable a more proactive role of consumers.

Prompted by the need to decarbonize not only the electric, but also other energy sectors, such as heat and transport, the microgrid concept is being extended to *multi-energy* microgrids. Multi-energy microgrids, also referred to as *integrated energy systems* (at a local, microgrid scale) in some of the chapters of this book, extend the traditional microgrid benefits by integrating various forms of energy—such as electricity, heat, and gas (hydrogen)—optimizing overall energy efficiency and flexibility and enhancing both energy security and environmental benefits by providing a more robust solution to meet varied energy demands.

It is evident that, to fully utilize the new possibilities offered by microgrids, advanced optimization, and control algorithms are indispensable. Microgrid optimization and control need to address several challenging aspects, span several time scales, and meet different control requirements, ranging from fast electrical control of frequency, on time scales of seconds or less, to unit commitment and economic dispatch of generating units and storages, on longer time scales (e.g., several minutes or hours). To meet these requirements, a hierarchical control approach is typically adopted to managing and operating a microgrid and combining fast, local responses with microgrid-wide optimization and coordination. Typically, three control layers are considered: (i) *primary control*, usually decentralized, operates at the fastest timescale and is responsible for voltage and frequency regulation in real-time; (ii) *secondary control* addresses slower timescale issues and fine-tunes the voltage and frequency settings, ensuring the synchronization of the microgrid with the main grid, if needed; and (iii) *tertiary control* operates at the slowest timescale and focuses on the economic and optimal operation of the microgrid, the energy management to optimize operational costs, as well as the interaction with the main grid.

In practice, variations of the control structure above are adopted. This is, for instance, reflected in Chapter 3, where economic dispatch is part of the secondary control layer and the tertiary control layer is considered as a functionality of the utility, since it addresses the interaction of the microgrid with the main grid and the neighboring microgrids. In general, it is expected that microgrid energy management systems will have an increasingly significant role in the tertiary control layer. Besides, given the growing complexity and rising number of resources, decentralized and distributed techniques aid the tractability of the microgrid optimization and control problems, since they decompose it into a set of smaller sub-problems, which can be solved in parallel.

Within this context, the main aim of this book is to provide a comprehensive guide on the most promising advanced techniques for microgrid optimization and control available. It brings together studies and contributions from leading researchers to provide a unifying overview of the current relevant state of the art, ranging from advanced control methods for frequency control to economic dispatch, optimized operation, and energy management. This exposition is complemented by two real-world case studies, which are discussed in Chapters 9 and 10, providing practical insights and lessons learned, demonstrating the application of theoretical concepts in real settings, and offering valuable knowledge on problem-solving and decision-making processes. We hope that this book, covering a broad spectrum of topics, from theoretical frameworks to practical implementations, will be a valuable resource for researchers, engineers, and practitioners engaged in the development and management of microgrids.

I.1 Structure of the book

The book is structured into five parts, each comprising two chapters, as outlined next. Each chapter may have its own list of acronyms and notation, when appropriate.

I.2 Part I. Dynamics and modeling

This part encompasses Chapter 1, *Modeling of integrated energy systems*, and Chapter 2, *Energy-based modeling and dissipativity analysis of district heating systems*. The focus of this part is on the modeling aspects of microgrids, including both steady-state and dynamic operating conditions, as well as on different energy systems, such as power, gas, heat, and hydrogen systems. Dynamic models related to fast time scales and low-level control, as well as to slow time scales and high-level control, are illustrated. Particular emphasis is put on the heating sector and innovative control of district heating systems, because of the importance of the decarbonization and electrification of the heating sector to the energy transition to Net-Zero. This part also provides an overview of optimization methods for operating integrated energy systems, including centralized and decentralized methods.

I.3 Part II. Frequency control

The focus if this part is on faster time scales and frequency control in microgrids. It comprises Chapter 3, *Dynamics and control of grid-connected microgrids*, and Chapter 4, *Fully distributed and economic frequency regulation solutions for autonomous microgrids*.

Therein, the impact of grid-connected microgrids on power system dynamics is analyzed, as well as their ability to provide efficient frequency support, while maximizing their revenue. Fully distributed control approaches for autonomous microgrids, which both stabilize frequency and minimize operating costs, are also discussed.

I.4 Part III. Electric energy management

This part includes Chapter 5, *Distributed optimization for energy grids: a tutorial on ADMM and ALADIN*, and Chapter 6, *Integrating distributed energy resources in real-world sector-coupled microgrids: challenges, strategies, and experimental insights*.

The main focus is on slower time scales and energy management in electric microgrids. A comprehensive overview of the state of the art of microgrid energy management systems is provided, which includes practical applications in real-world case studies. Efficient distributed optimization algorithms applied to optimal power flow and energy management in microgrids are also presented and analyzed.

I.5   Part IV. Multi-energy management

This part is dedicated to multi-energy microgrids and consists of Chapter 7, *Risk-averse transactive energy management for a multi-energy microgrid*, and Chapter 8, *Operation of multi-energy microgrids with laboratory validation*. Novel energy management systems for multi-energy microgrids are examined, based on transactive energy and game theory. The benefits of coordinating the operation of district heating and cooling systems together with electric distribution systems are explored and validated in a laboratory infrastructure.

I.6   Part V. Case studies

This part comprises Chapter 9, *Electricity market-oriented control of the EUREF Energy Workshop*, and Chapter 10, *A fog computing-based architecture for the decentralized energy management of microgrids*. Examples of real-world implementation and experimental results are provided and discussed, in particular the EUREF Energy Workshop, located in Berlin, Germany, which incorporates Germany's first combined power-to-heat/power-to-cold system, as well as test-bed facilities at the University of Genova Savona Campus, in Italy, which include a Smart Polygeneration Microgrid and a Sustainable Energy Building connected to the microgrid.

In Chapter 9, the development and implementation of a model predictive control framework to interface the EUREF Energy Workshop with the day-ahead, intraday, and experimental congestion management markets are presented and analyzed.

A novel cloud-fog-based decentralized approach for the day-ahead optimal power scheduling and its application to polygenerative microgrids are explored in Chapter 10.

Acknowledgments

The editors would like to thank all the authors who contributed to the chapters of this book. The latter is truly a collaborative effort. We gratefully acknowledge the professional support from the Institution of Engineering and Technology (IET) and its staff, particularly the Senior Commissioning Editor, Christoph von Friedeburg, the Assistant Editors, Olivia Wilkins and Megan McGill, and the Editorial Assistant, Brittany Insull, for their valuable and unwavering support.

References

[1]   Lasseter B. Microgrids [distributed power generation]. In: *2001 IEEE Power Engineering Society Winter Meeting. Conference Proceedings* (Cat. No. 01CH37194). vol. 1; 2001. p. 146–149.
[2]   IEEE Standard for the Specification of Microgrid Controllers. IEEE Std 20307-2017; 2018. p. 1–43.

Part I

Dynamics and modeling

Chapter 1
Modeling of integrated energy systems

Hongjie Jia[1], Xiandong Xu[1], Xiaolong Jin[1] and Jianzhong Wu[2]

Traditionally, modeling of different energy systems, such as power, gas, heat, and hydrogen systems, is conducted independently. For integrated energy systems, which include multi-energy microgrids, different models need to be combined to reflect interactions and coordination potentials between various energy systems. This chapter introduces the current modeling and operating methods of integrated energy systems, including energy networks, coupling components, energy storage, and multi-energy loads. Steady-state models, dynamic models, and modeling tools are taken into account along with typical cases under different scenarios. Optimization methods for operating integrated energy systems are also summarized, including centralized and decentralized methods. The methods and approaches described in this chapter can be adopted to model multi-energy microgrids.

1.1 Introduction

1.1.1 Transformations of energy systems

Technologies of demand-side management (DSM), renewable generation, and information and communication technologies (ICT) are drawing more and more attention due to the growing concerns over energy depletion and environmental issues all over the world [1]. These new technologies are transforming modern electric systems, especially electric systems, by creating new opportunities and challenges to their generations, networks, and demands. The transformations take place on the generation, network, and consumption sides of the electric systems, which is summarized in Figure 1.1. First, generation is changing due to growth in renewable generation. Second, for networks, multi-vector energy networks (e.g., electricity networks, natural gas networks, and heating networks) are integrated together and the ICT enables more advanced network control and management. Third, for demands, ICT technologies offer the possibility to manage demand-side resources, which enables

[1]The Key Laboratory of Smart Grid of Ministry of Education, Tianjin University, China
[2]School of Engineering, Cardiff University, United Kingdom

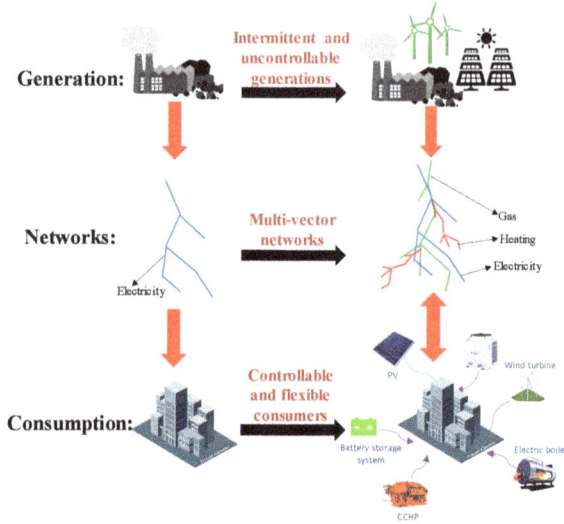

Figure 1.1   Diagram of transformations of modern electric systems (adopted from [2])

local energy trading and local flexibility trading. The specific transformations are introduced from the following aspects:

(1)   Intermittent and uncontrollable generation

In order to cope with the increasing level of pollution and depletion of fossil fuels, countries around the world are promoting the development of renewable generation [3]. As pointed out in [4], renewable generation's share of the worldwide electricity production is around 25%. In Europe, a series of regulatory targets are set up for the development of renewable generation, e.g., 20% of EU energy from renewable generation by 2020 in "2020 climate & energy package" [5] and at least 32% share for renewable generation by 2030 in "2030 climate & energy framework" [6]. Denmark plans to satisfy 100% of Denmark's energy demand with renewable generation by 2050 [7]. Following this target, oil for heating purposes and coal are to be phased out by 2030 [7]. Germany also decides its goal to cover 80% of the electricity demand with renewable generation by 2050 [8].

(2)   Controllable and flexible consumers

The recent years experienced a fast increase of distributed energy resources (DERs), which are normally connected to residential houses, smart buildings, microgrids, and distribution systems [9]. DERs mainly include distributed generators (DGs), energy storages, and flexible loads (e.g., electric vehicles (EVs), heat pumps). The DERs can be potential providers of flexibility with demand response (DR) technologies [10]. With the increasing connections of DERs and the fast developments in ICT at the demand side, traditional consumers are transforming into prosumers [11], who can not only consume energy but also share excessive energy with other prosumers and/or the distribution system operator (DSO). Therefore, the prosumers can participate in the energy generation/consumption process and provide flexible services by managing their DERs and communicating with other prosumers and the DSO [12]. In other words, the prosumers can trade energy and flexibility among themselves and the DSO locally based on the information exchange and local market platforms [13].

Take a smart building as an example, as a prosumer with ICT infrastructure, the smart building can manage its resources in an economically efficient way (as shown in Figure 1.2), and trade energy and flexibility with other smart buildings or DSO locally [14,15] (as shown in Figure 1.3). As pointed out by the International Energy Agency, buildings' energy consumption occupies around 40% of the worldwide energy usage, and their heating, ventilation, and air conditioning (HVAC) systems are responsible for almost half of the buildings' energy consumption [16]. With the rapid progress of urbanization and the transformation of industrial structures around the world, this proportion will be increasing continually [15]. Therefore, buildings are potential prosumers that can provide flexibility locally [15].

(3) Integrated energy systems

Traditionally, different energy systems, e.g., gas networks, electricity networks, and heating networks, are planned and operated independently. Actually, these energy systems are tightly coupled via different coupling technologies and devices, including combined cycle gas turbines (CCGT),

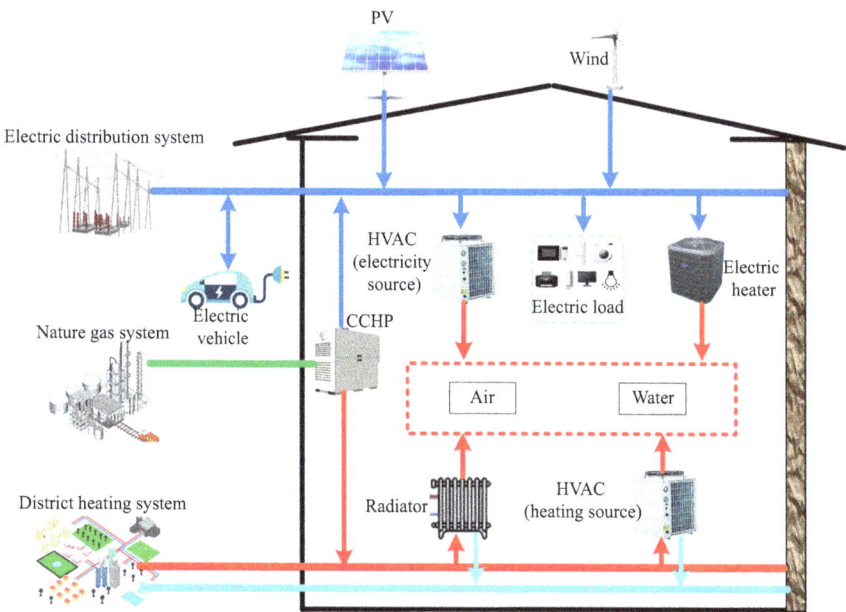

*Figure 1.2   Schematic diagram of a smart building with connection of DERs
(adopted from [2])*

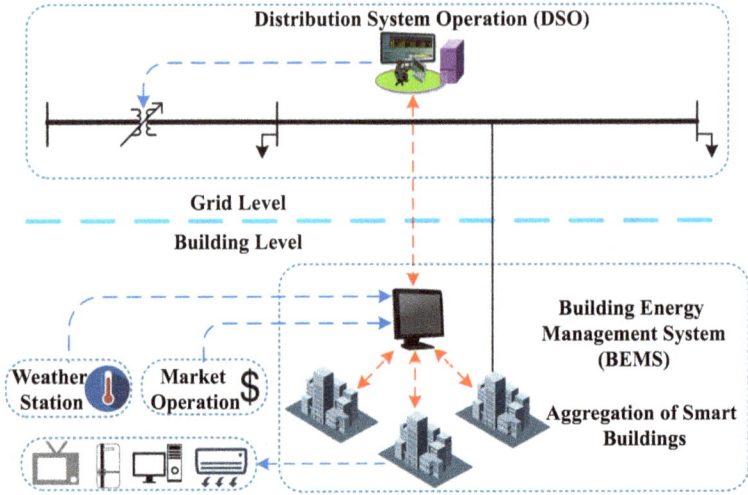

Figure 1.3   Schematic diagram of energy trading among the smart buildings and the DSO (adopted from [15])

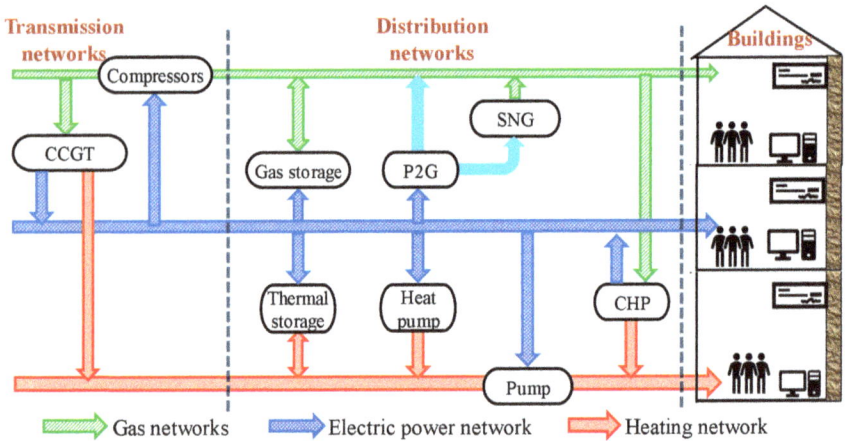

Figure 1.4   Schematic diagram of a multi-energy system from transmission, distribution scale down to building level (adopted from [17])

combined heat and power units (CHPs), power to gas (P2G) equipment, heat pumps, and HVACs [17]. Consequently, it leads to the multi-energy systems. Figure 1.4 shows the schematic diagram of multi-energy systems with interactions among different energy networks at various scales, i.e., from transmission, distribution scale down to building level [17]. In this context, the synergies among electricity and other energy vectors can be used to provide cost-efficient flexibility required.

To sum up, the massive integration of renewable generation, the integration of multi-vector energy systems, and the controllable and flexible consumers are changing the landscape of modern electric systems.

1.1.2   Drivers for integrated energy systems

This sub-section summarizes the driving forces of integrated energy system development from the perspectives of environment and economy.

(1)   *Environment*

The environmental issue has become a global challenge constraining human sustainable development. Governments worldwide are under immense pressure to conserve energy, enhance efficiency, reduce emissions, and embrace renewable energy. Taking the European Union (EU) as an example, it has formulated a series of climate and energy development goals and plans covering the entire EU and spanning decades, with greenhouse gas emissions reduction, renewable energy development, and energy efficiency improvement as core objectives. In 2022, the European Union unveiled the "REPowerEU" plan, which aims to focus on three key areas: enhancing energy efficiency, accelerating the adoption of clean energy, and diversifying energy sources. In 2024, the European Commission recommended a new target to reduce net greenhouse gas emissions by 90% by 2040 compared to 1990 levels. This target acts as a crucial step between the 2030 target and the ultimate goal of climate neutrality by 2050. The plan emphasizes emissions reductions, carbon capture, and carbon dioxide removal technologies. Furthermore, to address environmental challenges, China has committed to peaking CO_2 emissions around 2030, and multiple studies have outlined technological roadmaps for China to achieve over 80% renewable energy in electricity generation by 2050. China has published a guideline for the construction of a standardized system for carbon peaking and carbon neutrality as summarized in Table 1.1. Countries worldwide are facing unprecedented environmental pressures, actively seeking solutions, and thus driving the development of IESs.

(2)   *Economy*

The energy industry, as a traditional pillar of the national economy, has a significant impact on economic development and industrial upgrading. Taking China as an example, in 2014, the top 500 energy companies in China had a total operating income of 19.58 trillion yuan, accounting for 34.43% of the total GDP. However, the current situation of high pollution, high energy consumption, and low energy efficiency in the traditional energy industry restricts its sustainable development, thereby indirectly constraining the rapid growth of China's economy. Furthermore, after over 30 years of rapid development, China's economy has entered a new normality, facing the important task of transitioning its economic development mode. There is an urgent need to cultivate new energy industries to promote industrial upgrading. For instance, in the 2015 government work report, the Chinese government proposed implementing major strategic projects such as new energy and gas turbines, aiming

Table 1.1   Guidelines for building a carbon peak carbon neutral standards system

Items	Details
Basic General	✓ Classification of terms and carbon information disclosures ✓ Carbon monitoring and accounting ✓ Low carbon management and evaluation
Reduce carbon emissions	✓ Energy-saving ✓ Non-fossil energy ✓ New electricity power systems ✓ Fossil energy cleaner utilization ✓ Emission reduction in production and service processes ✓ Resource recycling
Carbon elimination	✓ Carbon sequestration and enhancements of ecosystem ✓ Carbon capture, utilization and storage (CCUS) ✓ Direct air carbon capture and storage (DACS)
Market-based mechanisms	✓ Green financial market ✓ Carbon emission trade ✓ Ecological product value realization

to foster a group of emerging energy industries into leading industries and support the development of strategic emerging industries such as new electric vehicles. Therefore, as a typical representative of the new energy industry, research on IESs has been vigorously promoted.

(3)   *Technology advancement*

The development of clean technology in the generation makes it intermittent and uncontrollable. This poses a great challenge to the grid for balancing the power imbalance and controlling the consequent frequency fluctuations. Thus, sufficient flexibility is needed by the grid to cope with these issues. On the other side, the increasing number of DERs connected to the demand side makes the electric distribution systems more complex (i.e., the bi-directional and decentralized systems). It causes reverse power flows, voltage violations, congestion, and line losses. The safe and secure operation of the distribution systems would be challenging. The emergence of these technologies and challenges is also contributing to the development of technology on IESs. The multi-energy systems can offer great opportunities for the electric systems to get cost-effective flexibility by capitalizing on synergies and complementary advantages of the various energy systems at various scales.

(4)   *Society*

As society progresses, the landscape of energy production and consumption is undergoing a significant transformation. The rapid expansion of renewable energy has somewhat mitigated geopolitical competition over energy resources. However, competition surrounding renewable energy technologies is expected to intensify. Integrated energy systems (IESs) are a crucial initiative to ensure energy security and prevent energy shortages.

1.2 Concepts of integrated energy systems

1.2.1 Definition

As shown in Figure 1.5, an IES is defined as a system that integrates multiple various energy systems, namely power, gas, heating, hydrogen, etc. The key technologies for the integration include combined heat and power, gas furnace, P2G, heat pump, which support the energy conversion between different energy systems. In certain cases, the IES is also combined with other infrastructures, such as building, transportation, water, and communication systems. On the one hand, the IES provides energy to these infrastructures. On the other hand, these infrastructures support the operation of the IES via demand response schemes.

1.2.2 Benefits

Compared with traditional energy systems, which were designed and operated independently of each other, the IES holds tremendous benefits, especially in today's rapidly evolving energy landscape.

Firstly, the IES could enhance overall energy efficiency by optimizing the use of various energy resources, which not only reduces energy wastage but also enhances

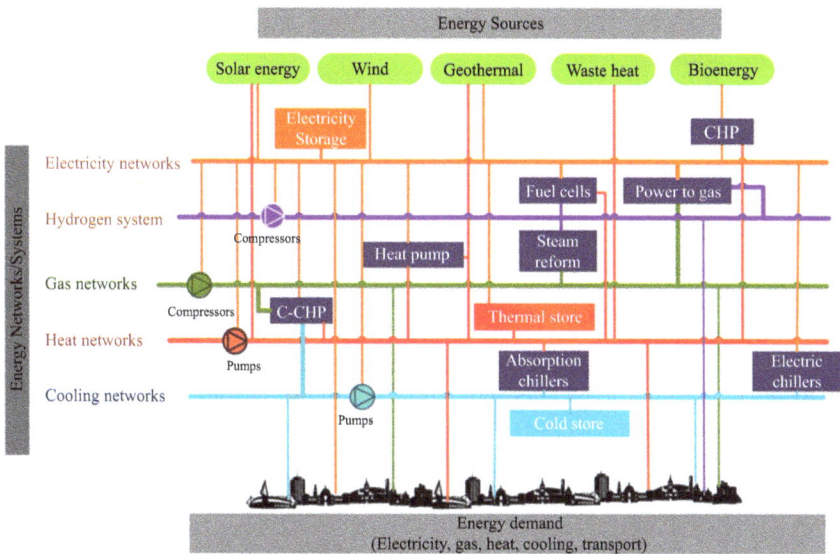

Figure 1.5   A schematic diagram of the IES

energy supply reliability and resilience. By integrating renewable energy such as solar and wind, the IES also contributes significantly to environmental sustainability.

Furthermore, the IES improves energy supply security by diversifying energy sources. This diversification reduces dependence on any single energy resource and thus minimizes the risk of energy supply interruptions. Moreover, the IES is highly flexible and adaptable. It can easily scale up or down based on energy service requirements, ensuring a constant and reliable energy supply even during peak hours. This flexibility is crucial in today's rapidly changing power systems.

Lastly, the IES often incorporates cutting-edge technologies like smart meters, sensors, automated control systems, and artificial technologies. These technologies enable real-time monitoring and optimization of energy usage, improving the overall efficiency and reliability of the grid.

In conclusion, the IES represents a significant leap forward in energy technology. The IES will be a vital component of our future energy infrastructure by combining multiple energy sources efficiently, enhancing energy security, providing flexibility, and incorporating advanced technologies.

1.3   Modeling of the IES

This section introduces the modeling mechanism of the IES to help readers understand the principles behind the different models. Then key points in obtaining the models are introduced to help readers select the appropriate models and modeling methods under different scenarios. Finally, a numerical implementation of an IES model is presented, including technical framework, numerical algorithm, and existing tools.

1.3.1   Mechanism models

This chapter introduces the steady-state and dynamic models of the IES with electric power system, natural gas system, heating/cooling systems, and coupling units presented separately.

1.3.1.1   Electric power system

Steady-state model

A power system consists of various electrical devices interconnected to produce, transmit, distribute, and consume electrical energy. The primary components necessary for calculating steady-state power flow in power systems include node power and power transmitted along lines. For a power system containing n nodes, the node power equations are

$$P_i^{\text{inj}} = V_i \sum_{j \in \mathcal{J}^{\text{bus}}} V_j \left(G_{ij} \cos \theta_{ij} + B_{ij} \sin \theta_{ij} \right) \quad \forall i \in \mathcal{J}^{\text{EPS-bus}} \tag{1.1}$$

$$Q_i^{\text{inj}} = V_i \sum_{j \in \mathcal{J}^{\text{EPs.bus}}} V_j \left(G_{ij} \sin \theta_{ij} - B_{ij} \cos \theta_{ij} \right) \quad \forall i \in \mathcal{J}^{\text{EPS-bus}} \tag{1.2}$$

where P_i^{inj} and Q_i^{inj} represent the active power and reactive power injections at node i, respectively. These are calculated from the difference between the active/reactive output (P_i^G and Q_i^G) and the active/reactive load (P_i^L and Q_i^L) of the node, respectively, which is $P_i^{\text{inj}} = P_i^G - Q_i^L$, $Q_i^{\text{inj}} = Q_i^G - Q_i^L$. G_{ij} and B_{ij} are the conductance and susceptance components of the nodal admittance matrix, respectively. θ_{ij} is the phase angle difference between nodes ($\theta_{ij} = \theta_i - \theta_j$). $\mathscr{I}^{\text{EPS-bus}}$ represents the set of all node indices within the power system.

At any given time t, for the branch connecting nodes i and j, the line transmission power equations are

$$P_{ij} = V_i^2 g_{ij} - V_i V_j \left(g_{ij} \cos \theta_{ij} + b_{ij} \sin \theta_{ij} \right) \quad \forall i, j \in \mathscr{I}^{\text{EPS-bus}} \tag{1.3}$$

$$Q_{ij} = -V_i^2 \left(b_{\text{si}} + b_{ij} \right) - V_i V_j \left(g_{ij} \sin \theta_{ij} + b_{ij} \cos \theta_{ij} \right) \quad \forall i, j \in \mathscr{I}^{\text{EPS-bus}} \tag{1.4}$$

where P_{ij} and Q_{ij} represent the active and reactive power flow between nodes i and j, respectively. g_{ij} and b_{ij} are the conductance and conductivity of the branch.

Dynamic model

The power grid model includes models for power branches and nodes. The power branch model refers to an equivalent circuit model represented by resistance, reactance, conductance, and susceptance. Assuming that the transmission line medium is uniformly distributed, its classical mathematical model is

$$\begin{cases} C_0 \frac{\partial U}{\partial t} + \frac{\partial I}{\partial x} + G_0 U = 0 \\ L_0 \frac{\partial I}{\partial t} + \frac{\partial U}{\partial x} + R_0 I = 0 \end{cases} \tag{1.5}$$

where C_0, G_0, R_0, and L_0 are per unit length branch capacitance, conductance, resistance, and inductance, respectively. U and I are branch voltage and current. t and x are time and spatial variables, respectively.

Modeling each parameter such as resistance, reactance, conductance, and susceptance per kilometer accurately for power lines that extend for tens to hundreds of kilometers is quite complex. It is usually feasible to classify the analysis based on the branch length. The long line model accounts for the distributed parameter characteristics of the branch, transforming (1.5) into an ordinary differential equation (ODE) in the phasor space for characterization and analysis. On this basis, the medium-length line model disregards the distributed parameter characteristics, establishing two approximate equivalent circuits, namely II-type and T-type, based on the voltage–current relationship at the beginning and end of the branch. The short-line model further neglects the impact of line admittance and derives the implications accordingly.

1.3.1.2 Natural gas systems

Steady-state model

For studies on flow calculation and optimization in natural gas systems, we often use a set of nonlinear algebraic equations to represent the steady-state operation of the gas network (i.e., the flow parameters of the gas in the pipeline network are independent of time). These equations primarily derive from the internal physical relationships in the pipelines and the constraints between various pipelines in the network.

Gas flow in natural gas systems can be expressed using one-dimensional compressible flow equations. For the purpose of facilitating steady-state analysis in this paper, we adopt the following assumptions: (1) there are no temperature changes inside the pipeline, and the gas flows isothermally; (2) changes in kinetic energy within the pipeline are negligible; (3) the friction factor along the pipeline direction is constant; (4) the compressibility factor of the gas is constant throughout the entire length of the pipeline; (5) the pipeline is always maintained horizontally without any inclination.

For gas transmission pipelines, there is a certain pressure difference between the inlet and the outlet, with the gas flowing from areas of high pressure to areas of low pressure. The relationship between the natural gas flow inside the pipeline and the pressures at the inlet and outlet can be represented by various equations, such as the Lacey equation, Panhandle "A" equation, Weymouth equation, etc. In this case, we use the Weymouth equation:

$$F_b^{\mathrm{P}} = \mathrm{sgn}\left(p_b^{\mathrm{in}} - p_b^{\mathrm{out}}\right) C_b^{\mathrm{P}} \sqrt{\left|\left(p_b^{\mathrm{in}}\right)^2 - \left(p_b^{\mathrm{out}}\right)^2\right|} \quad \forall b \in \mathscr{I}^{\mathrm{NGN-P}} \tag{1.6}$$

where $\mathrm{sgn}(x)$ is defined as 1 when $x > 0$ and -1 otherwise. F_b^{P} represents the gas flow rate inside pipeline b in kilograms per second (kg/s). p_b^{in} and p_b^{out}, respectively, represent the nodal pressures at the inlet and outlet of pipeline b; C_b^{P} represents the pipeline constant, typically calculated using $C_b^{\mathrm{P}} = \frac{1}{389640} \frac{l_b T^{\mathrm{A}} G Z a}{D_b^{16/3}}$ (where l_b, T^{A}, G, Z_a and D_b represent the length of the pipeline, the average temperature of the natural gas, the specific gravity of the natural gas, the average compressibility factor, and the diameter of the pipeline, respectively).

Assuming there are no gas leaks within the pipeline network, the sum of the gas flows entering any node must equal the sum of the gas flows leaving that node. This node flow continuity constraint is expressed by the following formula:

$$\sum_{b \in S_i^{\mathrm{NGN-pipe+}}} F_b^{\mathrm{P}} - \sum_{b \in S_i^{\mathrm{NGN-pipe-}}} F_b^{\mathrm{P}} = F_i^{\mathrm{S}} - F_i^{\mathrm{LD}} \quad \forall i \in \mathscr{I}^{\mathrm{NGN-ND}} \tag{1.7}$$

where F_i^{S} and F_i^{LD}, respectively, represent the supply and demand of gas at node i. $\mathscr{I}^{\mathrm{NGN-ND}}$ represents the set of node numbers in the natural gas system. $S_i^{\mathrm{NGN-pipe+}}$ and $S_i^{\mathrm{NGN-pipe-}}$, respectively, represent the sets of pipeline numbers flowing into and out of node i.

Similar to Kirchhoff's Voltage Law in electrical circuits, when a natural gas network includes a ring structure, the algebraic sum of the squares of the pressure drops across each segment of the ring, starting and ending at the same point, is always zero. Assuming the direction of gas flow within each segment of the ring is known, this constraint can be expressed as follows:

$$\sum_{b \in S_i^{\mathrm{NGN-ND-L}}} B_{l,b}\left(F_{b,t}^{\mathrm{P}}/C_b^{\mathrm{P}}\right)^2 = 0 \quad \forall l \in \mathscr{I}^{\mathrm{NGN-L}} \tag{1.8}$$

where $B_{l,b}$ represents the ring-pipeline connectivity matrix. $S^{\text{NGN-L}}$ represents the set of ring numbers in the natural gas network.

Due to the internal surfaces of pipelines not being perfectly smooth, gas generates friction as it is transported, causing pressure to decrease progressively along the pipeline. To ensure that the gas pressure meets operational requirements, natural gas networks commonly include compressor stations to increase gas pressure. Compressors can be driven by electric motors, steam turbines, or gas turbines. Compressors are akin to phase shifters in power systems. The relationship between the power consumption of a commonly used centrifugal compressor and the gas flow passing through the compressor can be expressed as:

$$P_i^{\text{C}} = K_i^1 + K_i^2 \left[\frac{p_i^{\text{out}}}{p_i^{\text{in}}} \right]^{\alpha^{\text{C}}} F_i^{\text{COMP}} \quad \forall i \in \mathscr{I}^{\text{NGN-C}} \tag{1.9}$$

where $P_{i,t}^{\text{C}}$ represents the horsepower required by the compressor i. $F_{i,t}^{\text{COMP}}$ represents the natural gas flow rate through compressor i. p_i^{in} and p_i^{out}, respectively, represent the inlet and outlet pressures of compressor i. K_i^1, K_i^2, and α^{C}, respectively, represent coefficients related to the horsepower calculations of the compressor i. $\mathscr{I}^{\text{NGN-C}}$ represents the set of compressor numbers in the natural gas system.

Some compressors consume electricity, while others are driven by natural gas. The formula for calculating the gas consumption of compressors is

$$F_i^{\text{Csupply}} = a_i^{\text{comp}} + b_i^{\text{comp}} P_i^{\text{C1}} + c_i^{\text{comp}} \left(P_i^{\text{C1}} \right)^2 \quad \forall i \in \mathscr{I}^{\text{NGN-C}} \tag{1.10}$$

where F^{Csupply} represents the gas consumption of the compressor. a_i^{comp}, b_i^{comp}, and c_i^{comp}, respectively, represent coefficients for calculating the gas consumption of the compressor. Compressors can proportionally increase the outlet pressure.

The relationship between the compression ratio and the inlet and outlet pressures of the compressor is given by

$$R_i^{\text{C}} = \max \left(p_i^{\text{in}}, p_i^{\text{out}} \right) / \min \left(p_i^{\text{in}}, p_i^{\text{out}} \right) \quad \forall i \in \mathscr{I}^{\text{NGN-C}} \tag{1.11}$$

where R_i represents the compression ratio of compressor i.

Dynamic model

For natural gas systems, especially large-scale, long-distance gas transmission systems, the speed of gas transmission is limited. Changes in pressure and flow at any point cannot be transmitted to distant nodes in a very short time. This results in a slow transition to a new steady state when the operational conditions of the gas system change, requiring a certain amount of time to stabilize. While steady-state models assume that the pipeline is in a stable condition and are not suitable for situations where the gas in the pipeline is unstable and various parameters change over time, dynamic models of natural gas are sometimes necessary for analyzing and optimizing the operating conditions of the gas system (Figure 1.6).

The dynamic mathematical model of gas in a pipeline can be represented by a system of partial differential equations (PDEs). As shown in Figure 1.5, the PDEs for

Gas flow direction

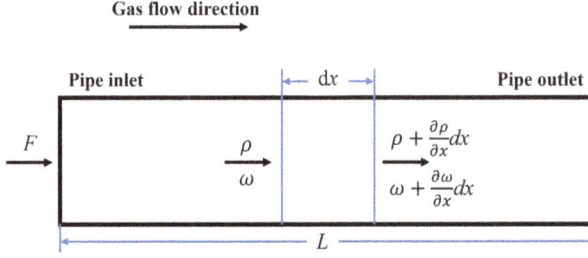

Figure 1.6   *Gas dynamics in a pipeline*

the pipeline can be expressed by the continuity equation and the momentum equation, which are written as follows:

$$\frac{\partial \rho}{\partial t} + \frac{\partial \rho \omega}{\partial x} = 0 \tag{1.12}$$

$$\frac{\partial \rho \omega}{\partial t} + \frac{\partial \rho \omega^2}{\partial t} + \frac{\partial p}{\partial x} + \frac{\lambda \rho \omega^2}{2D} + \rho g \sin \theta^G = 0 \tag{1.13}$$

where ρ, p, and ω, respectively, represent the gas density, pressure, and velocity. λ, D, g, and θ^G, respectively, represent the pipe's friction coefficient, diameter, gravitational acceleration, and the pipe's inclination angle. Identifiers for pipeline/node numbers and time subscripts are omitted here for simplicity.

Considering that gas transport is in a constant temperature state, there exists a relationship in each segment of the pipeline between the pressure and its square: $p = c^2 \rho$, where c represents a factor related to the gas. The natural gas flow rate can be calculated using the formula $F = \rho \omega A$ (where F represents the gas flow rate in the pipeline, and A represents the cross-sectional area of the pipeline). In addition, it is assumed that the pipelines within the network are laid horizontally; another relationship exists in the gas: $p = \rho ZRT$ (where Z represents the gas compressibility factor, R represents the ideal gas constant, and T represents the gas temperature). Thus, the original system of equations can be rewritten as follows

$$\frac{\partial p}{\partial t} = -\frac{c^2}{A} \frac{\partial F}{\partial x} \tag{1.14}$$

$$\frac{\partial F}{\partial t} = -A \frac{\partial p}{\partial x} - \frac{\lambda c^2}{2AD} \frac{F^2}{p} \tag{1.15}$$

For the system of PDEs, boundary conditions can be set according to actual operational requirements, choosing from $p(t, 0) = p0(t)$, $F(t, 0) = F0(t)$, $p(t, L) = pL(t)$, and $F(t, L) = FL(t)$ (where the subscripts 0 and L, respectively, represent the inlet and outlet of the pipeline). Two of these conditions are selected to define the boundaries.

For a natural gas system composed of multiple pipelines, the external constraints of the system are similar to those of a steady-state model. These include the need to satisfy node flow balance constraints (1.7), ring network pressure drop conservation constraints (1.8), and compressor boosting constraints (1.11), among others.

1.3.1.3 Heating/cooling systems

The modeling of a district heating system encompasses three primary components: heat source, heat user, and heat network. The primary focus for modeling heat sources and users involves thermal calculations, whereas the heat network is divided into hydraulic (pressure and water flow) and thermal (temperature and heat output) models. The description of each part of the model is shown below.

In a district heating system, various entities such as industrial waste heat, geothermal sources, solar energy, heat pumps, boilers, and cogeneration units are considered heat sources. The heat output of a source is calculated based on the water flow rate and the temperature differential between the supply and return flows. The formula for heat output is

$$\Phi_i^S = C_p m_i^S (TS_i^S - TR_i^S) \quad \forall i \in \mathscr{I}^{\text{DHN·HS}} \tag{1.16}$$

where Φ_i^S is the heating output of the heat source i. m_i^S is the flow rate of water supply at heat source i. C_p is the specific heat capacity of water. TS_i^S is the temperature of the water outlet at heat source i. TR_i^S is the temperature of the return outlet at heat source i. $\mathscr{I}^{\text{DHN·HS}}$ is the set of heat source node numbers for the district heating system.

The heat load supplied to the user can be calculated from the water flow rate and the temperature difference between inflow and outflow. The specific load calculation formula is

$$\Phi_i^{\text{LD}} = C_p m_i^{\text{LD}} (TS_i^{\text{LD}} - TR_i^{\text{LD}}) \quad \forall i \in \mathscr{I}^{\text{DHN·LD}} \tag{1.17}$$

where Φ_i^{LD} is the heat load at load i. m_i^{LD} is the water supply flow at load i. TS_i^{LD} is the temperature of the water inlet at load i. TS_i^{LD} is the temperature at the outlet at load i. $\mathscr{I}^{\text{DHN·LD}}$ is a collection of heat load numbers for the district heating system. For a primary heating network, the load side of the network is not the local user, but the heat energy is transferred to the heat exchanger station of the secondary heating network, and its transferred energy can also be calculated by this formula.

Similar to the continuity equation of airflow at nodes in the same gas network, the sum of water flows entering any node should be equal to the sum of water flows flowing out of that node. This constraint can be expressed using the following formula:

$$\sum_{b \in S_i^{\text{DHN-pipe+}}} M_b^P - \sum_{b \in S_i^{\text{DHN-pipe-}}} M_b^P = M_i^S + M_i^{\text{LD}} \quad \forall i \in \mathscr{I}^{\text{DHN-ND}} \tag{1.18}$$

where M_b^P is the water flow rate in pipe b, M_i^S is the injected water flow supplied at node i, M_i^{LD} is the injected water flow that consumes heat at node i, $S_i^{\text{DHN-pipe+}}$ is the set of pipe numbers flowing into node i, $S_i^{\text{DHN-pipe-}}$ is the set of pipe numbers of outgoing node i, $\mathscr{I}^{\text{DHN-ND}}$ is the set of node numbers of the district heating system.

Since the inner wall of the heat network pipe is not completely smooth, the friction between the hot water during transmission and the pipe wall leads to a gradual decrease in water pressure. The pressure drop inside the pipe is calculated by the equation

$$ps_b^{\text{in}} - ps_b^{\text{out}} = K^f M_b^{\text{P}} \left| M_b^{\text{P}} \right| \quad \forall b \in \mathscr{I}^{\text{DHN}-\text{P}} \tag{1.19}$$

where ps_b^{in} is the pressure at the inlet of pipe b, ps_b^{out} is the pressure at the outlet of pipe b, K^f is the pressure drop coefficient of pipe b, $\mathscr{I}^{\text{DHN}-\text{P}}$ is the collection of pipe numbers for district heating systems.

Assuming that there is no loss of heat from the hot water at each node, the sum of the heat energy contained in the water flow from multiple pipes when they converge at the same node should be equal to the heat energy contained in the hot water flowing out of that node. This constraint can be expressed by the following equation.

$$\sum_{b \in S_i^{\text{DHN}-\text{pipe}+}} (M_b^{\text{P}} T_b^{\text{P,out}}) = T_i^{\text{ND}} \sum_{b \in S_i^{\text{DHN}-\text{pipe}=}} M_b^{\text{P}} \quad \forall i \in \mathscr{I}^{\text{DHN}-\text{ND}} \tag{1.20}$$

where $T_b^{\text{P,\overline{out}}}$ is the water temperature at the outlet of pipe b, T_i^{ND} is the water temperature at node i. The water temperature at node i is the temperature of the water at the inlet of all the pipes starting at node i.

The constraint can be expressed by the following equation:

$$T_b^{\text{P,in}} = T_i^{\text{ND}} \quad \forall b \in S_i^{\text{DHN}-\text{pipe}-} \tag{1.21}$$

where $T_b^{\text{P,in}}$ is the water temperature at the inlet of pipe b.

During hot water transfer, there is a temperature difference between the water flow and the external environment, which results in some loss of heat from the water flow. The heat loss equation is

$$T_b^{\text{P,out}} = (T_b^{\text{P,in}} - T^{\text{A}}) \exp\left(-\frac{\lambda_b L_b}{C^{\text{P}} m_b}\right) + T^{\text{A}} \quad \forall b \in \mathscr{I}^{\text{DHN}-\text{P}} \tag{1.22}$$

where T^{A} is the ambient temperature, λ_b is the heat transfer coefficient of pipe b, L_b is the length of pipe b.

1.3.1.4   Coupling units

Gas turbine model

When considering an electrically coupled network wherein the gas turbine serves dual roles—as a consumer of natural gas and as a generator of electric power—the formula to ascertain gas consumption is

$$F_k^{\text{GT}} = \alpha^{\text{GT}} (P_i^{\text{GT}})^2 + \beta^{\text{GT}} P_i^{\text{GT}} + \gamma^{\text{GT}} \quad \forall i \in \mathscr{I}^{\text{IES}-\text{GT}} \tag{1.23}$$

where F_k^{GT} symbolizes the gas consumption of the gas turbine, while P_k^{GT} denotes its power generation; α^{GT}, β^{GT} and γ^{GT} are the respective coefficients utilized in the calculation of gas consumption. The set $\mathscr{I}^{\text{IES}-\text{GT}}$ enumerates the gas turbines within the integrated energy system.

Cogeneration unit model

Within the framework of the integrated energy system, this study adopts the back-pressure cogeneration unit, which orchestrates the simultaneous generation of electricity and thermal energy. The relationship between thermoelectric power output is expressed by

$$c_i^{\text{CHP}} = \frac{\Phi_i^{\text{CHP}}}{P_i^{\text{CHP}}} \quad \forall i \in \mathscr{I}^{\text{IES}-\text{CHP}} \tag{1.24}$$

where c_i^{CHP} represents the thermal-to-electrical energy conversion ratio of the cogeneration unit i; Φ_i^{CHP} and $\P_i^{\text{CHP}}$ being the thermal and electrical outputs, respectively. The set $\mathscr{I}^{\text{IES}-\text{CHP}}$ indexes the cogeneration units within the IES.

The gas consumption of a CHP unit is deduced as follows:

$$F_i^{\text{CHP}} = \frac{\Phi_i^{\text{CHP}}}{c_i^{\text{CHP}}\eta^{\text{CHP}}} \quad \forall i \in \mathscr{I}^{\text{IES}-\text{CHP}} \tag{1.25}$$

where F_i^{CHP} indicates the CHP unit's gas consumption, while η^{CHP} is the thermal efficiency of the cogeneration process.

Compressor model

The electrical compressors used in the natural gas system derive their energy from the electrical grid, and their power consumption is referenced from (1.9).

Water pump model

In the context of district heating systems, to maintain adequate water pressure for operational integrity, electrically driven pumps are often deployed. The computation for the power consumption of these pumps is articulated as follows:

$$P_i^{\text{pump}} = \frac{m_i^{\text{pump}} \cdot (ps_i^{\text{pump,s}} - ps_i^{\text{pump,r}})}{\rho\eta^{\text{pump}}} \quad \forall i \in \mathscr{I}^{\text{DNH}-\text{PUMP}} \tag{1.26}$$

1.3.2 Key points

1.3.2.1 Making reasonable assumptions

Existing studies often make many assumptions about the IES characteristics and operating states to reduce model complexity. Yet, the applicability of these assumptions and the resulting errors may not be acceptable under certain scenarios. For example, when the load fluctuation is relatively smooth, and the gas flow velocity is stable, a linear model can be adopted to reduce the complexity and instability of the solution. When the gas pressure level is high, the acceleration term in the momentum conservation equation can even be ignored to further reduce complexity. However, dramatic fluctuations in gas load can cause significant variations in gas flow velocity along the pipelines. A linear gas network model based on average flow velocity may lead to a mismatch between the obtained gas pressure and flow rate, as the linear model cannot accurately depict the temporal distribution of gas pressure and flow rate in such scenarios. Therefore, a nonlinear momentum equation is necessary to ensure modeling accuracy in this scenario. Moreover, The IES couples multiple energy systems, resulting in a high-dimensional model with a large number of variables and diverse operating scenarios. Modeling errors and complexity on any subsystem side may lead to the infeasibility of joint analysis.

One of the challenges in IES modeling is how to make reasonable assumptions based on operating scenarios (such as system scale, load characteristics, etc.) under general rules and select appropriate models.

1.3.2.2   Contingency analysis

Multi-energy coupling increases the risk of bidirectional fault propagation within sub-networks. For example, ice blockage in gas pipelines can cause power outages due to the interruption of GT power supply, ultimately leading to large-scale blackouts. Additionally, failures in cables can disrupt the heating cycle of electric boilers and thermal power plants, and may further result in widespread heating outages. Furthermore, three-phase short circuits in the power grid can cause power loss and temporary shutdown of pressurization stations, leading to downtime events due to insufficient gas supply for downstream gas turbines. Existing research primarily focuses on modeling the IES under normal conditions. These studies mainly refer to power system analysis, establishing network models under fault conditions by adding virtual branches and nodes. Yet, the disconnection of cables or overhead lines is a fast dynamic process, whereas the leakage and complete disconnection of gas and heating pipelines is a slow dynamic process. Thus, describing pipeline failures solely through the connection and disconnection of branches is not entirely reasonable. In addition to pipeline failures, node (i.e., equipment) failures within the system are also a typical operating condition that urgently needs to be characterized and analyzed.

Therefore, establishing a typical fault event library for the IES and characterizing the impact of fault events on the network model under normal conditions is another challenge in modeling the IES.

1.3.2.3   Multi-time scale characterization

Multi-energy flow transmission exhibits multi-time scale characteristics. Among them, electrical energy propagates at the speed of light, and its time scale under steady-state conditions ranges from seconds to minutes. Thermal power and gas propagate at fluid flow speeds, with a time scale typically ranging from tens of minutes to hours, exhibiting a noticeable delay characteristic. Given control commands, the fast dynamic characteristics of the power grid allow the electrical state to quickly stabilize within the command interval. However, the slow dynamic characteristics of the gas and heat networks cause the gas and heat state variables to change over time within the command interval. Nevertheless, practical applications often require discretizing PDEs and assume that gas and heat state variables remain constant within the command interval, which does not truly reflect the multi-time scale characteristics in the IES analysis. By using the command interval as the simulation period and the control commands as the simulation boundaries, the multi-time scale characteristics of the multi-energy network are addressed through alternating solutions of long-time scale control and small time-step simulations. On the one hand, characterizing the continuous variation of state variables within the command interval through discretization inevitably introduces approximation errors. On the other hand, both the energy-based interactive interface based on weighted averages and the embedded small time-step simulations reflect the multi-time scale characteristics by increasing additional computational complexity, making the joint analysis of IES more complicated.

Therefore, the issue of reflecting multi-time scale characteristics in IES joint modeling has not been fundamentally solved, posing a challenge for related research.

1.3.2.4   Parameter identification

Compared to electric power systems, the energy management of gas and heat systems is relatively crude, and measurement equipment is mainly deployed at important units and loads, making it difficult to obtain the whole system parameters needed for modeling. Moreover, equipment parameters will deviate from the calibrated values as time goes by, causing a mismatch between the first principle model and the actual operating conditions. Additionally, from the perspective of joint analysis, the electric power, gas, and heating/cooling systems in an IES typically belong to multiple operating entities. Due to privacy concerns, the internal parameters of different systems are usually not shared or not fully shared. The analytical premise of complete sharing of network topologies and equipment parameters among different systems does not align with the actual operational model. Therefore, how to identify model parameters is another challenge for the IES modeling.

1.3.3   Modeling tools

1.3.3.1   Technical framework

Considering the different physical characteristics of electricity, heat, and gas networks, a hybrid modeling approach is commonly used for the IES simulation. The main idea is to construct a specific model interface to connect simulation software from different energy fields to achieve joint modeling. However, due to different modeling mechanisms, inconsistent data interface forms can arise from different modeling methods and integration sequences, leading to potential data interaction errors at the energy system interface.

Another framework is the integrated modeling approach, which describes the physical characteristics of the whole system through general expressions and develops the models within the same framework. This kind of approach, such as bond graph, analyzes the operation and optimization of gas and heating systems using equivalent electrical circuit models. This enables the application of power network analysis methods to guide the analysis of other energy networks. These efforts extend the theory of power grid modeling to heating and natural gas networks, helping to break down knowledge barriers between different disciplines.

1.3.3.2   Numerical algorithm

Steady-state simulation
Generally, the steady-state energy flow calculation of an IES involves solving a set of nonlinear algebraic equations with an iterative form of the Newton–Raphson method.

$$0 = f(y) \tag{1.27}$$

$$y^{k+1} = y^k - J^{-1}(y^k)f(y^k) \tag{1.28}$$

where k denotes the kth iteration. J represents the Jacobian matrix. y includes state variables such as voltage amplitude, phase angle, mass flow rate, and temperature. f denotes the steady-state energy flow equation.

In practice, the composition of J and f can be adjusted based on the coupling relationships between various subsystems to adapt to different analysis scenarios. In addition to the classical Newton–Raphson method, improved versions of the Newton-Raphson method and other iterative methods have also been applied in steady-state energy flow calculations.

To overcome the convergence challenges of iterative methods, non-iterative approaches, represented by the holomorphic embedding method, have also found widespread application in IES energy flow calculations. The concept of the holomorphic embedding method originates from the field of power system load flow calculations. "Holomorphic" refers to the analyticity of complex functions. In this regard, the holomorphic embedding method has been applied to the grid portion of the electrical and thermal energy flow calculations. The holomorphic embedding method avoids iterations and requires only one matrix decomposition operation. Apart from the univariate holomorphic embedding method, the multivariate holomorphic embedding algorithm is also used in the IES model by constructing corresponding multivariate embedded systems. It has been shown that the holomorphic embedding method demonstrates a significant online efficiency advantage compared to the Newton-Raphson iterative method.

Dynamic simulation

Compared to traditional power systems, the dynamic model of the IES model exhibits a high degree of complexity. The IES encompasses diverse heterogeneous systems such as electricity, natural gas, and heating. These systems facilitate simultaneous production, consumption, and mutual conversion of multiple types of energy. Consequently, more variables and operational constraints must be considered, particularly the synergy among different energy sources. In addition, the IES includes various operational time scales. Due to the varying energy transmission speeds within pipelines, such as those in natural gas and heating systems, the subsystems within the IES operate on different time scales. Therefore, modeling and subsequent research must account for these distinct operational time scales and their corresponding physical characteristics.

To reflect interactions and interdependencies of different energy systems within the IES, there exists three typical methods:

- Numerical method
  Common numerical methods for dynamic equations in the IES include the finite difference method, the finite volume method, and the method of lines (MOL). A special form of the MOL is the method of characteristic (MOC), which constructs a discrete grid along the characteristic lines. This method is only applicable to dealing with first-order quasi-linear hyperbolic PDEs. The commonly used gas-thermal system model in the IES analysis just happens to exhibit

a first-order quasi-linear hyperbolic form, making the MOC an ideal method for the IES.

- Semi-analytical method
  Unlike numerical methods, which solve discrete system states using the mechanism model equations of the IES, semi-analytical method aims to obtain approximate expressions of the original model using Taylor approximation. Compared with traditional numerical methods such as backward Euler method, the semi-analytical method does not require to solve nonlinear algebra equations. Moreover, the complexity of the model is adjustable by regulating the order of Taylor approximation.

- Analytical method
  Different from numerical and semi-analytical methods, the analytical method does not rely on model discretization and approximation. Instead, it converts the original model into models that are easier to solve, using methods such as Laplace transform or Fourier transform. It should be noted that due to the discontinuous and non-periodic behavior of the IES, certain approximation is still required during the model-solving process even though the analytical method is used.

1.3.3.3   Software

After decades of development, some software has been specifically developed for the IES analysis, such as RETScreen, EnergyPlan, HOMER, SAInt, EnergySim, CloudPSS-IESLab, and IES-SIM. This kind of software usually has a library with equipment models that are commonly used in the IES. The software has been used by researchers and industries for supporting the IES planning and operation. Another type of tools that are commonly used for the IES simulation are general modeling tools such as Modelica and MATLAB/Simulink. Although these tools were not originally developed for the IES simulation, the fundamental libraries and their extensions enable these tools to mimic the behaviors of the IES under different scenarios. Moreover, researchers could combine their own models with the existing library elements supported by the powerful numerical solvers embedded in the tools.

Apart from the single software-based modeling tools, there are also some tools that conducts the IES simulation by linking different tools. A typical example is the Large-Scale Infrastructure Co-Simulation (HELICS) developed by NREL. HELICS links multiple off-the-shelf simulation tools to act as a single unified model, exchanging data at each time step. One of the important advantages is its scalability. HELICS has been tested in bringing together multiple energy systems simulations along with buildings, grid, transportation, and water system simulation. However, how to ensure numerical stability during the iterative processes is a big challenge, especially for large-scale systems.

1.3.3.4   Hardware-in-the-loop (HIL) simulation

Similar to other fields, IES also requires the HIL simulation to obtain more accurate results to support equipment transient analysis such as controller design. Some works have been conducted by researchers to combine HIL simulation tool with IES model library. The main direction is to extend the function of traditional HIL tools such as RTLAB and LabVIEW. Although some tools have been developed, there is currently no widely accepted or recognized HIL simulation tool for the IES that has passed testing in the industry. The main challenge lies in how to solve the conflict between high computational burden and high calculation speed requirement, which still requires further studies in the future.

1.4   Operational methods of integrated energy systems

Extensive centralized optimization methods have been proposed for the optimal operation of IES. However, different energy systems regularly belong to different entities. In this case, the centralized optimization model may not be applicable when the system parameters cannot be accessed completely [2]. Meanwhile, it is indicated in [18] that the solution efficiency of centralized optimization decreases as the system scale increases. Therefore, achieving the optimal scheduling of IES by changing the operational mode from centralized to decentralized is of great necessity. The basic idea of the decentralized optimization method is to decouple the original problem into several sub-problems and solve each sub-problem independently [2]. Various decomposition methods of decentralized optimization have been proposed for the power flow calculation [19,20], optimal operation [21–24], market clearing [25,26], optimal charging for electric vehicles [27], Nash equilibrium [28], and waterfilling problem [29]. The major optimization methods include the alternating direction method of multipliers (ADMM), the analytical target cascading (ATC), the proximal message passing (PMP), the auxiliary problem principle (APP), the optimality condition decomposition (OCD), the consensus + innovation (C+I) and the dual decomposition. Generally speaking, the decomposition methods consist of two categories, namely distributed methods and decentralized methods. The major difference between the distributed method and the decentralized method is whether there is information exchange within the system structure [20].

Generally, different energy systems in the IES are operated by different entities. Therefore, only limited operational information can be exchanged through multi-energy links. To protect the privacy and safety of each sub-system, a distributed scheduling model based on consensus-based ADMM is introduced in this section.

1.4.1   Framework of the distributed solution approach

The general framework of the distributed scheduling, as depicted in Figure 1.7, consists of three primary steps: initialization, self-optimization, and information exchange.

Step 1: The objective of the initialization step is to establish the initial values of state variables for each subsystem and update the coupling relationships between them. This involves setting the basic parameters for each subsystem, including component parameters (such as energy efficiency and the upper/lower boundaries of

Table 1.2   Major features of decomposition methods

Methods	Mathematical formulation	Features
ADMM	Augmented Lagrangian relaxation	✓The model can be decoupled naturally in accordance with the system structure. ✓The model is typically decoupled into two sub-problems. ✓Some oscillations may appear during the iteration process.
ATC	Augmented Lagrangian relaxation	✓The model is decoupled as a hierarchical structure with flexibility. ✓The information is merely exchanged between directly connected upper and lower levels. ✓The communication expense is increased due to the multi-level structure.
PMP	Augmented Lagrangian relaxation	✓The model can be flexibly decoupled into various categories, e.g. the device, terminal and connectivity node [24]. ✓The complexity of system scale and communication structure is increased.
APP	Augmented Lagrangian relaxation	✓The model can be decoupled with flexibility. ✓It requires more algorithm parameters than the other methods. ✓Convergence performance is affected by the algorithm settings.
OCD	Karush–Kuhn–Tucker conditions (KKT conditions)	✓Only the first order optimality constraint of KKT is decoupled and solved in a decentralized manner. ✓The computation expense is lighter than the Lagrangian relaxation-based method. ✓The convergence performance may deteriorate with the increase of couplings among sub-problems.
Approaches	**Mathematical formulation**	**Features**
C+I	KKT conditions	✓The model can be flexibly decoupled. ✓The algorithm may not converge when the model is non-convex.
Dual decomposition	Lagrangian relaxation	✓The reformulated model is easier to form. ✓The computation expense increases rapidly with more control variables and coupling variables.

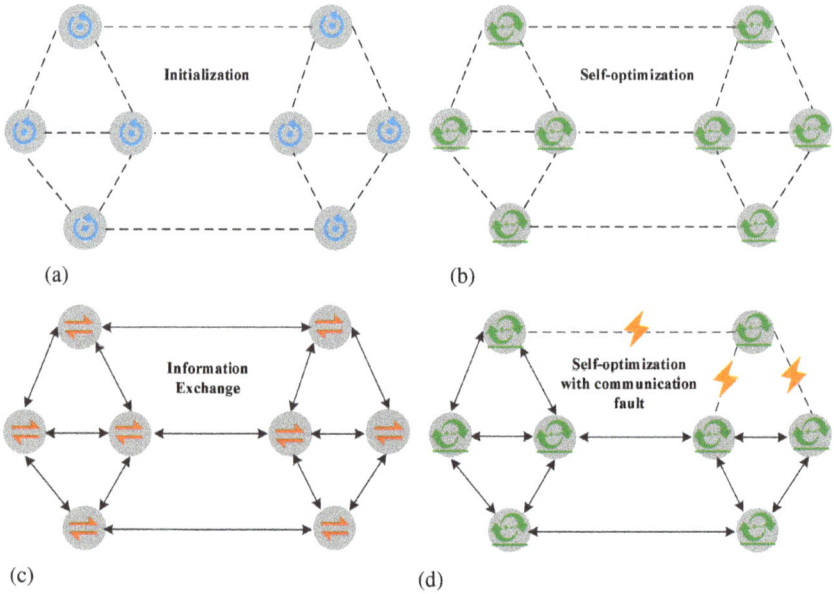

Figure 1.7   Framework of the distributed scheduling method. (a) Initialization for each sub-system. (b) Self-optimization for each sub-system. (c) Information exchange among sub-system. (d) Self-optimization strategy under communication fault circumstance.

energy conversion equipment), system topological structure (such as the topology of the electric system and corresponding branch parameters), load conditions (such as the electric/thermal load of terminal users), and algorithm settings (such as maximum iteration number and convergence errors). Additionally, exchanged information from interconnected subsystems is updated to serve as boundaries for subsequent steps.

Step 2: In the self-optimization step, the main objective is to achieve the optimal operation scheme for each subsystem while maintaining fixed coupling variables in a decentralized manner. The overall optimization problem of the IES is decomposed into several sub-problems based on the structure of the system, including an electric system sub-problem, a natural gas system sub-problem, and an energy hub system sub-problem. Each energy hub system is treated as an independent entity, resulting in the decoupling of the general optimization problem of IES into multiple sub-problems. Furthermore, optimization within each sub-problem is conducted within the respective energy system, leveraging the existing optimization capabilities of the system operator as much as possible.

Step 3: The information exchange step aims to update the key coupling variables between interconnected subsystems to ensure the overall operation and coordination of the IES. Optimal scheduling schemes for each subsystem, including coupling variables and the optimal dispatch of energy conversion equipment, are achieved through distributed optimization. Only limited coupling variables, rather than the overall optimization results, are exchanged and updated between interconnected sub-systems to maintain privacy and confidentiality. If convergence criteria are met, the optimal solution of the distributed optimization is obtained; otherwise, the process returns to the **Step 1** for further iteration.

Furthermore, a self-optimization strategy is proposed to address circumstances involving communication loss. Such situations are common in reality, often caused by communication delays, information loss, or network attacks. When communication faults occur, complete information exchange becomes impossible. In such cases, the connecting subsystem is optimized using the latest information obtained from the faulted subsystems and updated information from the normal subsystems during the period of information loss, as depicted in Figure 1.7(d). Here, normal subsystems and faulted subsystems are optimized separately. Once the information fault is resolved, inter-subsystem communication resumes, and the distributed optimization proceeds according to the standard steps.

1.4.2 The consensus-based ADMM approach

In comparison to centralized optimization methods, distributed optimization involves limited information exchange among each sub-problem, leading to a further simplification of the optimization problem. Consequently, the ADMM-based distributed scheduling method is well-suited for the IES, particularly because different energy systems often belong to distinct entities, making complete information sharing impractical in reality. In the context of the IES, the coupling of multiple energy systems is achieved through the input side of the energy hub system, which also acts as the output side of energy systems. By introducing consensus variables into the model (as depicted in (1.29)), the multi-vector coupling variables are fully decoupled, with the input and output sides being equal to each other, as illustrated in Figure 1.8.

$$\begin{cases} \mathbf{P}_e^{grid} = \mathbf{P}_e^{con}, & \mathbf{P}_e^{con} = \mathbf{P}_e^{hub} \\ \mathbf{P}_g^{grid} = \mathbf{P}_g^{con}, & \mathbf{P}_g^{con} = \mathbf{P}_g^{hub} \end{cases} \tag{1.29}$$

In this case, the augmented Lagrangian function is formulated for each subsystem by adding the consensus constraints into the objective function, as depicted in (1.30)–(1.32), where $\mathbf{x}$ is the vector of coupling variables, $\mathbf{z}$ is the vector of consensus variables, λ is the vector of Lagrangian multipliers. The ADMM-based method is illustrated as **Algorithm 1**.

$$L_e\left(\mathbf{x}_e, \mathbf{z}_e, \lambda_e\right) = \sum_{t=1}^{T} C_{elec,t} P_{elec,t} + \lambda_e\left(\mathbf{x}_e - \mathbf{z}_e\right) + \frac{\rho}{2}\left\|\mathbf{x}_e - \mathbf{z}_e\right\|_2^2 \tag{1.30}$$

$$L_g\left(\mathbf{x}_g, \mathbf{z}_g, \lambda_g\right) = \sum_{t=1}^{T} C_{gas,t} F_{gas,t} + \lambda_g\left(\mathbf{x}_g - \mathbf{z}_g\right) + \frac{\rho}{2}\left\|\mathbf{x}_g - \mathbf{z}_g\right\|_2^2 \tag{1.31}$$

$$L_{hub}\left(\mathbf{x}_{hub}, \mathbf{z}_{hub}, \lambda_{hub}\right) = \lambda_{hub}\left(\mathbf{x}_{hub} - \mathbf{z}_{hub}\right) + \frac{\rho}{2}\left\|\mathbf{x}_{hub} - \mathbf{z}_{hub}\right\|_2^2 \tag{1.32}$$

It is important to highlight that by introducing consensus variables, the original IES model becomes fully decoupled. As a result, the optimization process

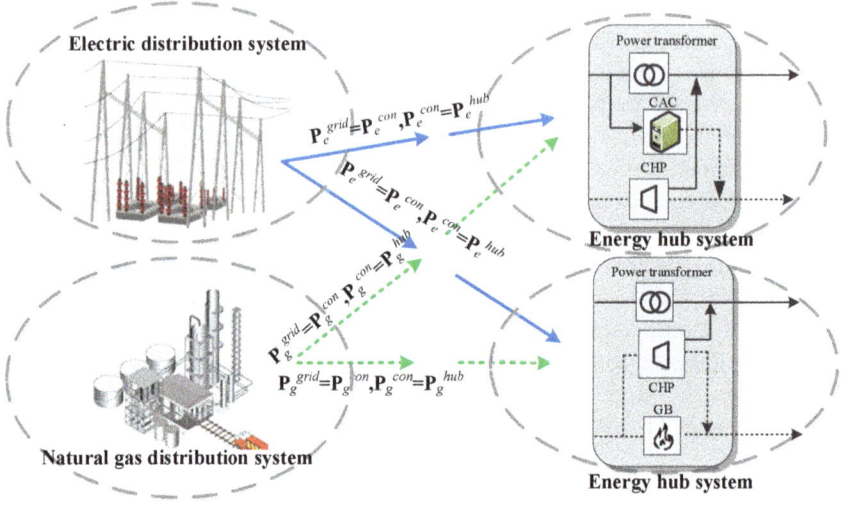

Figure 1.8   IES decoupling by introducing consensus variables

for each subsystem can be carried out independently, without taking into account the influence of multi-energy interactions. This characteristic renders the proposed scheduling method fully distributed. Consequently, the optimal operation of the IES can be decoupled based on the operational entity for each subsystem, making the proposed framework scalable. In each iteration, the coupling variable **x** of the energy sub-system is optimized in parallel with the consensus variable and Lagrangian multiplier fixed. Then, only the latest coupling variable vector **x** is exchanged between the connecting sub-systems. After receiving the latest **x**, the consensus variable vector **z** is updated according to (1.33)–(1.35) and the Lagrangian multipliers vector **λ** is updated according to (1.36)–(1.38). The proof of (1.33)–(1.35) is illustrated in Appendix. It is worth noting that the updating process of **z** and **λ** is conducted within the sub-system so that the proposed scheduling method is decentralized and the information privacy of different entities is guaranteed. Moreover, according to (1.33)–(1.35), the consensus variables share the same value during the iteration although the update process is separately conducted within the connecting sub-system since the coupling variable vector **x** is fully exchanged. The stopping criteria are defined as the primal residual and dual residual being less than the setting tolerance, as shown in (1.39).

$$\mathbf{z}_e^{k+1} = \frac{1}{N_e} \sum_{i=1}^{N_e} \left(\mathbf{x}_e^{k+1} + \frac{1}{\rho}\lambda_e^k \right) = \frac{1}{N_e} \sum_{i=1}^{N_e} \left(\mathbf{x}_e^{k+1} \right) = \bar{\mathbf{x}}_e^{k+1} \tag{1.33}$$

$$\mathbf{z}_g^{k+1} = \frac{1}{N_g} \sum_{i=1}^{N_g} \left(\mathbf{x}_g^{k+1} + \frac{1}{\rho}\lambda_g^k \right) = \frac{1}{N_g} \sum_{i=1}^{N_g} \left(\mathbf{x}_g^{k+1} \right) = \bar{\mathbf{x}}_g^{k+1} \tag{1.34}$$

$$\mathbf{z}_{hub}^{k+1} = \frac{1}{N_{hub}} \sum_{i=1}^{N_{hub}} \left(\mathbf{x}_{hub}^{k+1} + \frac{1}{\rho}\lambda_{hub}^{k} \right) = \frac{1}{N_{hub}} \sum_{i=1}^{N_{hub}} \left(\mathbf{x}_{hub}^{k+1} \right) = \bar{\mathbf{x}}_{hub}^{k+1} \qquad (1.35)$$

$$\lambda_{e}^{k+1} = \lambda_{e}^{k} + \rho \left(\mathbf{x}_{e}^{k+1} - \mathbf{z}_{e}^{k+1} \right) \qquad (1.36)$$

$$\lambda_{g}^{k+1} = \lambda_{g}^{k} + \rho \left(\mathbf{x}_{g}^{k+1} - \mathbf{z}_{g}^{k+1} \right) \qquad (1.37)$$

$$\lambda_{hub}^{k+1} = \lambda_{hub}^{k} + \rho \left(\mathbf{x}_{hub}^{k+1} - \mathbf{z}_{hub}^{k+1} \right) \qquad (1.38)$$

$$\begin{cases} \left\| r^{k+1} \right\|_{2}^{2} = \max \sum_{t=1}^{24} \left\| \mathbf{x}^{k+1} - \mathbf{z}^{k+1} \right\|_{2}^{2} \leqslant \varepsilon^{pri} \\ \\ \left\| s^{k+1} \right\|_{2}^{2} = \max \sum_{t=1}^{24} \rho \left\| \mathbf{z}^{k+1} - \mathbf{z}^{k} \right\|_{2}^{2} \leqslant \varepsilon^{dual} \end{cases} \qquad (1.39)$$

1.4.2.1 Dynamic step size modification of ADMM

The convergence rate of the ADMM is significantly influenced by the step size value. In the original ADMM, the step size is typically fixed throughout the iteration process, which can lead to deteriorating the algorithm performance in later stages. To mitigate this issue, this chapter proposes an improved two-stage dynamic step size modification method.

Stage 1: During each iteration, the change in the values of the primal and dual residuals is calculated, as outlined in (1.40). If the minimum change in both primal and dual residuals exceeds a predefined threshold Δ (e.g., 0.1), the step size remains unchanged, as it is already moving towards reducing both residuals. However, if the current step size risks hindering the algorithm convergence, it is updated according to Stage 2.

$$\Delta \left\| r^{k+1} \right\|_{2}^{2} = \left\| r^{k+1} \right\|_{2}^{2} - \left\| r^{k} \right\|_{2}^{2}, \ \Delta \left\| s^{k+1} \right\|_{2}^{2} = \left\| s^{k+1} \right\|_{2}^{2} - \left\| s^{k} \right\|_{2}^{2} \qquad (1.40)$$

Stage 2: The step size is adjusted based on the current values of the primal and dual residuals, as described in (1.41). If the primal residual significantly exceeds the dual residual, the step size is increased to impose a greater penalty on violations of primal feasibility. Conversely, if the dual residual is notably larger than the primal residual, the step size is decreased to facilitate dual feasibility convergence. This alternating adjustment mechanism helps balance the convergence of primal and dual feasibilities.

$$\rho^{k+1} = \begin{cases} \rho^{k} \left(1 + \lg \frac{\left\| r^{k} \right\|_{\infty}}{\left\| s^{k} \right\|_{\infty}} \right) & if \left\| r^{k} \right\|_{\infty} > 10 \left\| s^{k} \right\|_{\infty} \\ \\ \rho^{k} & otherwise \\ \\ \rho^{k} / \left(1 + \lg \frac{\left\| s^{k} \right\|_{\infty}}{\left\| r^{k} \right\|_{\infty}} \right) & if \left\| s^{k} \right\|_{\infty} > 10 \left\| r^{k} \right\|_{\infty} \end{cases} \qquad (1.41)$$

1.4.2.2   Overall model of distributed scheduling for sub-systems

Based on the proposed distributed scheduling framework and consensus-based ADMM, the overall models for each sub-system are summarized as follows: The overall sub-problem model for electric system is depicted in (1.42), where the control variables $\mathbf{x}_e$ are the coupling electric power of electric system; the state variables $\mathbf{z}_e$ and λ_e are, respectively, the consensus electric power of coupling point and corresponding Lagrangian multipliers.

$$
\begin{cases}
\min\limits_{\substack{\mathbf{x}_e \in \mathbf{R} \\ \mathbf{z}_e \in \mathbf{R},\, \lambda_e \in \mathbf{R}}} & L_e \;=\; \sum\limits_{t-1}^{T} C_{elec,t} P_{elec,t} + \lambda_e \left(\mathbf{x}_e - \mathbf{z}_e\right) + \frac{\rho}{2} \left\| \mathbf{x}_e - \mathbf{z}_e \right\|_2^2 \\
s.t. & (1.33),(1.36)
\end{cases}
\tag{1.42}
$$

The overall sub-problem model for natural gas system is depicted in (1.43), where the control variables $\mathbf{x}_g$ are the coupling natural gas power of natural gas system; the state variables $\mathbf{z}_g$ and λ_g are, respectively, the consensus natural gas power of coupling point and corresponding Lagrangian multipliers.

$$
\begin{cases}
\min\limits_{\substack{\mathbf{x}_g \in \mathbf{R} \\ \mathbf{z}_g \in \mathbf{R},\, \lambda_g \in \mathbf{R}}} & L_g \;=\; \sum\limits_{t-1}^{T} C_{gas,t} F_{gas,t} + \lambda_g \left(\mathbf{x}_g - \mathbf{z}_g\right) + \frac{\rho}{2} \left\| \mathbf{x}_g - \mathbf{z}_g \right\|_2^2 \\
s.t. & (1.34),(1.37)
\end{cases}
\tag{1.43}
$$

The overall sub-problem model for energy hub system is depicted in (1.44), where the control variables $\mathbf{x}_{hub}$ are the coupling input power of energy hub system; the state variables $\mathbf{x}_{hub}$ and λ_{hub} are, respectively, the consensus input power of coupling point and corresponding Lagrangian multipliers.

$$
\begin{cases}
type\ I \begin{cases}
\min\limits_{\substack{\mathbf{x}_{hub} \in \mathbf{R} \\ \mathbf{z}_{hub} \in \mathbf{R},\, \lambda_{hub} \in \mathbf{R}}} & L_{hub} \;=\; \lambda_{hub} \left(\mathbf{x}_{hub} - \mathbf{z}_{hub}\right) + \frac{\rho}{2} \left\| \mathbf{x}_{hub} - \mathbf{z}_{hub} \right\|_2^2 \\
s.t. & (1.35),(1.38)
\end{cases} \\[2em]
type\ II \begin{cases}
\min\limits_{\substack{\mathbf{x}_{hub} \in \mathbf{R} \\ \mathbf{z}_{hub} \in \mathbf{R},\, \lambda_{hub} \in \mathbf{R}}} & L_{hub} \;=\; \lambda_{hub} \left(\mathbf{x}_{hub} - \mathbf{z}_{hub}\right) + \frac{\rho}{2} \left\| \mathbf{x}_{hub} - \mathbf{z}_{hub} \right\|_2^2 \\
s.t. & (1.35),(1.38)
\end{cases}
\end{cases}
\tag{1.44}
$$

The limited exchange of information among subsystems ensures the privacy and safety of various multi-vector links, offering a viable optimal operational solution for the IES involving multiple entities. By incorporating consensus variables among different subsystems, the optimal operation of the IES becomes decoupled based on physical couplings. This allows for distributed optimization, with only the necessary coupling variables being exchanged between sub-systems.

1.5   Conclusion

This chapter first introduces the drivers and concepts of the IES. Then steady-state and dynamic models of the IES as well as the modeling tools are presented in detail. Furthermore, operational methods of the IES, particularly distributed scheduling methods, are presented, considering that the IES may be owned by different entities.

It should be noticed that this chapter mainly gives the models of power, gas, and heating/cooling systems. In practice, other energy vectors, such as hydrogen and steam, may also be a part of the IES. In addition, since this book puts more focus on short-term dynamics, the long-term dynamics of the heating system are not fully considered. Notwithstanding these limitations, this chapter does give an overview of how to model and optimize an IES.

References

[1]   Ramos A, De Jonghe C, Gómez V, *et al.* Realizing the smart grid's potential: Defining local markets for flexibility. *Utilities Policy.* 2016;40:26–35.

[2]   Jin X, Wu Q, and Jia H. Local flexibility markets: Literature review on concepts, models and clearing methods. *Applied Energy.* 2020;261:114387.

[3]   Ma W, Xue X, and Liu G. Techno-economic evaluation for hybrid renewable energy system: Application and merits. *Energy.* 2018;159:385–409.

[4]   Hamilton J, Negnevitsky M, Wang X, *et al.* High penetration renewable generation within Australian isolated and remote power systems. *Energy.* 2019;168:684–692.

[5]   Climate action – 2020 climate and energy package. European Commission. Available from: https://eur-lex.europa.eu/EN/legal-content/summary/2020-climate-and-energy-package.html [cited 2023 Mar 10].

[6]   Climate action – 2030 climate and energy framework. European Commission. Available from: https://www.consilium.europa.eu/en/policies/climate-change/2030-climate-and-energy-framework/ [cited 2023 Mar 10].

[7]   Government TD. The Danish Climate Policy Plan: Towards a low carbon society; 2013. Available from: https://moodle.polymtl.ca/pluginfile.php/413972/mod_page/content/70/Denmark_Climate_policy_plan_2013.pdf.

[8]   Switching to the electricity of the future. The Press and Information Office of the Federal Government; 2011. Available from: http://archiv.bundesregierung.de/ContentArchiv/EN/Archiv17/Artikel/_2011/06/2011-06-09-regierungserklaerung_en.html. [updated 2021 Dec 19; cited 2023 Mar 10].

[9]   Zhou Y, Wu J, and Long C. Evaluation of peer-to-peer energy sharing mechanisms based on a multiagent simulation framework. *Applied Energy.* 2018;222:993–1022.

[10] Eid C, Codani P, Perez Y, *et al.* Managing electric flexibility from distributed energy resources: A review of incentives for market design. *Renewable and Sustainable Energy Reviews.* 2016;64:237–247.

[11] Zhang C, Wu J, Zhou Y, *et al.* Peer-to-Peer energy trading in a Microgrid. *Applied Energy.* 2018;220:1–12.

[12] Long C, Wu J, Zhou Y, *et al.* Peer-to-peer energy sharing through a two-stage aggregated battery control in a community Microgrid. *Applied Energy.* 2018;226:261–276.

[13] Khorasany M, Mishra Y, and Ledwich G. Market framework for local energy trading: A review of potential designs and market clearing approaches. *IET Generation, Transmission & Distribution.* 2018;12(22):5899–5908.

[14] Jin X, Jiang T, Mu Y, *et al.* Scheduling distributed energy resources and smart buildings of a microgrid via multi-time scale and model predictive control method. *IET Renewable Power Generation.* 2018;13(6):816–833.

[15] Jiang T, Li Z, Jin X, *et al.* Flexible operation of active distribution network using integrated smart buildings with heating, ventilation and air-conditioning systems. *Applied Energy.* 2018;226:181–196.

[16] Agency IE. World Energy Outlook 2017; 2017. Available from: https://www.oecd-ilibrary.org/content/publication/weo-2017-en.

[17] Wu J, Yan J, Jia H, *et al.* Integrated energy systems. *Applied Energy.* 2016;167:155–157.

[18] Liu G, Jiang T, Ollis TB, *et al.* Distributed energy management for community microgrids considering network operational constraints and building thermal dynamics. *Applied Energy.* 2019;239:83–95.

[19] Zhang L, Chen S, Yan Z, *et al.* Distributed multi-area optimal power flow algorithm based on blockchain consensus mechanism. *Proceedings of the CSEE.* 2020;40(20):6433–6442.

[20] Molzahn DK, Dörfler F, Sandberg H, *et al.* A survey of distributed optimization and control algorithms for electric power systems. *IEEE Transactions on Smart Grid.* 2017;8(6):2941–2962.

[21] Lin W, Jin X, Jia H, *et al.* Decentralized optimal scheduling for integrated community energy system via consensus-based alternating direction method of multipliers. *Applied Energy.* 2021;302:117448.

[22] Qu K, Huang L, Yu T, *et al.* Decentralized dispatch of multi-area integrated energy systems with carbon trading. *Proceedings of the CSEE.* 2018;38(3):697–707.

[23] Hug G, Kar S, and Wu C. Consensus + innovations approach for distributed multiagent coordination in a microgrid. *IEEE Transactions on Smart Grid.* 2015;6(4):1893–1903.

[24] Peijie L, Yong L, Xiaoqing B, *et al.* Decentralized optimization for dynamic economic dispatch based on alternating direction method of multipliers. *Proceedings of the CSEE.* 2015;35(10):2428–2435.

[25] Deng R, Yang Z, Hou F, *et al.* Distributed real-time demand response in multiseller–multibuyer smart distribution grid. *IEEE Transactions on Power Systems.* 2014;30(5):2364–2374.

[26] Samadi P, Mohsenian-Rad AH, Schober R, *et al.* Optimal real-time pricing algorithm based on utility maximization for smart grid. In: *2010 First IEEE international conference on smart grid communications.* IEEE; 2010. p. 415–420.

[27]   Carli R, and Dotoli M. A distributed control algorithm for optimal charging of electric vehicle fleets with congestion management. *IFAC-PapersOnLine*. 2018;51(9):373–378.

[28]   Parise F, Grammatico S, Gentile B, *et al.* Distributed convergence to Nash equilibria in network and average aggregative games. *Automatica*. 2020;117: 108959.

[29]   Carli R, and Dotoli M. A distributed control algorithm for waterfilling of networked control systems via consensus. *IEEE Control Systems Letters*. 2017;1(2):334–339.

Chapter 2
Energy-based modeling and dissipativity analysis of district heating systems

Juan E. Machado[1], Michele Cucuzzella[2] and
Jacquelien M. A. Scherpen[2]

The decarbonization of the heating sector has the potential for considerably reducing the amount of anthropogenic emissions contributing to global warming. Modern district heating systems, which can incorporate various types of energy sources and storage devices, are considered key enablers for transitioning toward a more sustainable heating sector. However, in contrast to conventional system setups based on fossil fuels, the robust and efficient incorporation of heat from distributed and renewable sources requires an increased coordination among heat production, storage, and consumption devices, which is not offered by classic district heating system control standards. In this chapter, we contribute toward the advancement of modern district heating system control frameworks via physics and graph-theoretic based district heating systems modeling and through the analysis (and synthesis) of dissipativity-based qualitative properties (controllers) that have been crucial for controlling complex and large-scale networked systems in general by virtue of the offered scalability, robustness, and stability guarantees.

2.1 Introduction

Today, it is widely accepted that the decarbonization of the energy sector is an important measure for reducing emissions [1] and supporting the endeavors aimed at preventing the catastrophic consequences of global warming [2]. The decarbonization of the energy sector entails fundamental transformative changes in the way energy is produced, transported, and consumed [3]. Energy production wise, in the last couple of decades, some progress has been made toward the integration of ever more low carbon and renewable energy sources into the energy system, such as solar and wind energy [4]. However, much more efforts are still needed, considering that world-wide fossil-based energy generation is still predominant [4].

[1]Chair of Control Systems and Network Control Technology, Faculty 3 Mechanical Engineering, Electrical and Energy Systems, Brandenburg University of Technology Cottbus–Senftenberg, Germany
[2]Jan C. Willems Center for Systems and Control, ENTEG, Faculty of Science and Engineering, University of Groningen, The Netherlands

In the heating sector the landscape is particularly adverse, since about two-thirds of the total heat production is based on fossil fuels, such as coal and natural gas [5]. Critically, in residential and commercial contexts, where the energy use for heating can be significantly larger than for other end uses, the adoption of solutions for local sustainable heat production may be technically or financially challenging [3]. Nonetheless, a number of countries, particularly in Europe, have managed to further integrate renewable energy into the heating sector by implementing district heating systems [6].

District heating systems are networks of underground pipes that distribute, within a set of neighborhoods or a whole city, the heat generated by one or a few power plants at a distance [7,8]. One of the characteristics that make these systems suitable candidates for incorporating significant amounts of (unpredictable and intermittent) renewable energy is their ability to efficiently store substantial amounts of heat which, at the same time, increases the operational flexibility, allowing for a more robust balance between heat supply and demand [9,10].

From the design and operational viewpoints, the massive incorporation of more diversified, distributed, and sustainable heat sources will lead to more complex district heating configurations that will require a greater coordination among production, storage, and consumer entities for robustly and efficiently achieving sophisticated control goals, such as demand-side flexibility, peak heat demand shaving and fair heat distribution among consumers [7,9]. These advanced operational attributes cannot be addressed via classic district heating systems control strategies, as they allow for little to none coordination among heat production and consumption units [9]. Then, it is necessary to develop new control systems that make use of pervasive metering and communication networks to flexibly and robustly manage the available energy sources to guarantee the supply of quality heat to consumers [9].

However, to develop such types of control systems, it will be necessary to deal with a thermo-hydraulic system that is highly nonlinear, spatially distributed, it has different time scales and that is subject to disturbances and uncertainties introduced by ever-changing weather conditions and heat demand patterns. Therefore, this chapter contributes toward the advancement of modern district heating control frameworks by developing a dynamic model describing the hydraulic and thermal behavior for system configurations containing multiple heat sources, storage units, and consumers that are interconnected through (potentially) meshed pipe networks. Furthermore, we present conditions under which the proposed model exhibits a number of dissipativity-based qualitative properties [11]. Such a system-theoretic analysis represents in general a stepping stone in the design of controllers of complex and large-scale networked systems by virtue of their scalability, robustness, and stability guarantees (see, e.g., [12–15] for recent applications). For the proposed district heating model, in particular, we present simple and physically interpretable dissipativity-based controllers that provably achieve system stabilization.

In view of the scope of this book, we would like to note that the district heating systems' modeling and the model-based dissipativity properties presented in this chapter can be seamlessly exploited for the modular modeling and control of general multi-energy microgrids. Indeed, the generality of the modeling procedure, founded on graph theory, makes it easy to represent arbitrarily sized district heating systems of complex topology, which aligns well with the traits of general microgrids models [16]. Moreover, by letting heat exchangers be at the boundary between district heating and other energy carriers, it is possible to directly link our model with those of complex electro-thermal or gas-based devices for the production or consumption of heat, effectively linking various energy systems domains in a modular fashion (see, e.g., [17] for the treatment of electro-thermal microgrids). Modularity moreover extends also to the identified dissipativity—and in particular passivity—properties since, as it is well known in the control systems community, complex passive systems can be formed from smaller passive subsystems under suitable interconnection laws. Therefore, the identification of analogous properties for the energy carriers interacting with district heating systems opens the possibility for the design of scalable and decentralized controllers for arbitrarily sized and complex, interconnected multi-energy microgrids; we refer the reader to [18] and [19] for the dissipativity-based modeling of electric microgrids and gas networks, respectively.

2.1.1   Notation and list of abbreviations

The following notation is adopted in the chapter: $\mathbb{R}$ is the set of real numbers; $M \in \mathbb{R}^{m \times n}$ is an m-by-n real matrix; $0_{m \times n}$ ($1_{m \times n}$) is a matrix of size m-by-n with all entries equal to zero (one); for any $x = (x_1, x_2, \ldots, x_n) \in \mathbb{R}^n$, sign($x$) returns, entry-wise, the sign of the components of x, respectively; for any ordered set of indices $\mathscr{I} = \{i_1, i_2, \ldots, i_n\} \subset \mathbb{N}$, $\mathrm{col}(x_i)_{i \in \mathscr{I}} = (x_{i_1}, x_{i_2}, \ldots x_{i_n})$; for $x \in \mathbb{R}^n$, diag(x) is a diagonal matrix with the elements of x on its main diagonal; the same operator is used for representing block diagonal matrices; if $\mathscr{M}$ and $\mathscr{N}$ are two sets of ordered indices, with cardinalities m and n, respectively, then $A \in \mathbb{R}^{m \times n}$, defined as $[A_{i,j}]_{(i,j) \in \hat{\mathscr{M}} \times \hat{\mathscr{N}}}$ is a block matrix formed with the rows and columns of A associated to the indices $\hat{\mathscr{M}} \subset \mathscr{M}$ and $\hat{\mathscr{N}} \subset \mathscr{N}$, respectively; for any function $(t, x) \mapsto u(t, x)$, $\partial_t u(t, x)$, and $\partial_x u(t, x)$, respectively, denote the partial derivatives of u with respect to the scalars t and x; whenever it is clear from the context we will omit writing functions arguments.

Additionally, we use the following abbreviations and symbols:

- DHS: District Heating System
- HX: Heat Exchanger
- $\mathscr{G}$, $\mathscr{N}$, $\mathscr{E}$: DHS graph, nodes, and edges
- $\mathscr{B}_0$: Arbitrary incidence matrix of $\mathscr{G}$
- $\mathscr{B}$: Flow-rate dependent incidence matrix of $\mathscr{G}$, where the orientation of the edges matches the streams' directions
- n_{pr} (n_{c}): Number of heat producers (consumers) in the DHS
- n_{ST}: Number of storage tanks in the DHS
- $P_{\mathrm{pr},i}$: Rate of heat transfer by ith producer (in Watts)
- $P_{\mathrm{c},i}$: Rate of heat transfer by ith consumer (in Watts)
- $q_{\mathrm{E},i}$: Volumetric flow rate through edge $i \in \mathscr{E}$ (in cubic meters per second)
- $p_{\mathrm{N},i}$: Pressure of node $i \in \mathscr{N} \cup \mathscr{E}$ (in Pascal)
- $V_{\mathrm{N},i}$ ($V_{\mathrm{E},j}$): Volume of $i \in \mathscr{N}$ ($j \in \mathscr{E}$) (in cubic meters)
- $T_{\mathrm{N},i}$ ($T_{\mathrm{E},j}$): Temperature of $i \in \mathscr{N}$ ($j \in \mathscr{E}$) (in Kelvin)

- $f_{E,i}(q_{E,i})$: Pressure drop due to viscous friction across $i \in \mathscr{E}$ (in Pascal)
- $w_{E,i}$: Pressure difference across any pump $i \in \mathscr{E}$ (in Pascal)

2.2 System setup

Consider a DHS with $n_{pr} \geq 1$ heat producers, $n_c \geq 1$ consumers and $n_{ST} \geq 0$ storage tanks. These elements are interconnected through a common, possibly meshed distribution network as shown in Figure 2.1. Water is the working fluid, and we assume it to be always and everywhere in liquid state. The distribution network is considered to be conformed by two independent systems of pipelines, namely the supply layer and the return layer, and they, respectively, transport hot and cold water: hot water flows from producers to consumers via the supply layer and cold water flows in a converse manner through the return layer. Without loss of generality, it is assumed that the distribution network is symmetric, i.e., the supply and return layers are topologically equivalent. We also consider the practical assumption that the physical interconnection between the two distribution layers is done exclusively via producers, consumers, or storage devices. The overall system is assumed to be leak free without any loss of the fluid mass [22].

Based on [23], next we describe with more detail the composition of producers, consumers, and storage tanks:

- **Producers (and consumers):** Any producer (or consumer) is represented as a connection in series of a pump, a heat exchanger, and a valve. The pump is assumed to be of variable speed type and is considered to be the main hydraulic actuator. Generally, a heat exchanger has a primary side (stream of hot water to be cooled down) and a secondary side (stream of cold water to be heated up) with heat transfer occurring between primary and secondary sides via conduction and without mixing the streams of water. The amount of heat transferred between the sides of a heat exchanger depends on the temperature of the streams and their volumetric flow rate. However, following [24,25], here we will only consider the dynamics of the secondary side for producers and of the primary side for consumers, which are the sides that are hydraulically coupled with the

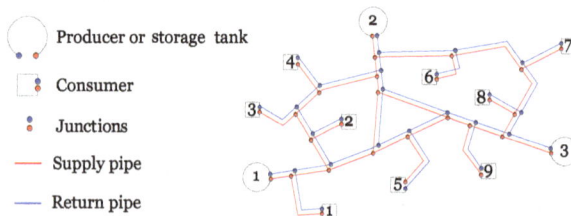

Figure 2.1   Sketch of a DHS in which a meshed distribution pipe network interconnects three heat producers and nine consumers (c.f., [20]). Figure courtesy of [21].

distribution network of the DHS. Moreover, we will treat the rate of heat transfer through the conducting wall—denoted by $P_{\mathrm{pr},i}$ for producers and by $P_{\mathrm{pr},i}$ for consumers—respectively, as a control input and a disturbance. For a more detailed model of producer or consumer heat exchangers, the reader is referred to [26] and the references therein.

- **Storage tanks:** Any tank is assumed to store a mixture of hot and cold water, hot layer on top and cold one at the bottom, and separated by a thermocline, through which no heat or mass exchange can occur. We consider that any tank has either two or four valves. In the two valve scenario, one valve is at the top and the other one at the bottom, and they serve as inlets or outlets of water. Analogously, in the four valve scenario, two valves are at the top and two at the bottom. In the four valve scenario, we assume that one inlet–outlet pair is reserved for directly connecting a producer or a consumer and that the remaining inlet–outlet pair is connected to the DHS distribution network, as shown in Figure 2.2.

The following are standing assumptions in this chapter:

Assumption 1.

(i) *The density $\rho > 0$ and specific heat $c_{\mathrm{s.h.}} > 0$ of water are spatially uniform and constant in time.*
(ii) *All pipes are cylindrical.*
(iii) *The flow through any edge $i \in \mathcal{E}$ is (spatially) one-dimensional.*
(iv) *Gravitational forces are irrelevant.*
(v) *The pressure of any simple junction or storage tank layer is spatially uniform, and for each tank the pressure of its layers is equal.*
(vi) *All devices (pipes, heat exchangers, valves, pumps, storage tanks, junctions) are completely filled with water all the time.*
(vii) *Any valve or pump is in series either with a pipe segment or a heat exchanger.*

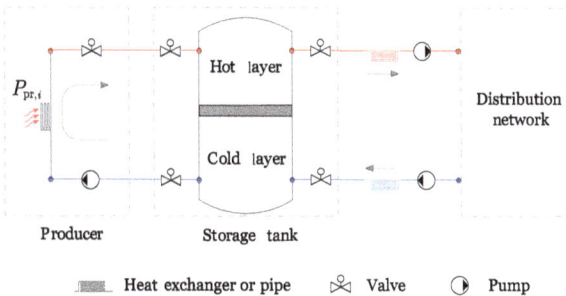

Figure 2.2   *Sketch of a four valve storage tank with a producer adjacent to it. We illustrate a potential placement for pumps, but their actual placement is dependent on the desired control goals and operation modes of the overall DHS. Figure courtesy of [21].*

Assumptions 1(i)–(iii) are fairly common in DHS modeling (see, e.g., [23,27]). Assumption 1(iv) is taken for simplification purposes. Via Assumption 1(v) we discard inertial and viscous forces in the equations of balance of momentum at each node and thus simplify the modeling procedure. Assumption 1(vi) is also common and, in particular, it implies that for each tank the sum of the volume of the hot and cold layer is constant (see, e.g., [23,28]). Assumption 1(vii) is introduced mainly for technical reasons, but it is practically sensible.

2.2.1 DHS as a graph

Following [20,24,27,29], we model the overall DHS as a connected graph $\mathcal{G} = (\mathcal{N}, \mathcal{E})$ with no self-loops. All pipes, heat exchangers, pumps, and valves conform the set of edges $\mathcal{E}$. All physical junctions as well as the layers (hot and cold) of all storage tanks conform the set of nodes $\mathcal{N}$. We find it convenient to split the sets of nodes and edges as follows:

$$\mathcal{N} = \mathcal{N}_{\mathrm{sh}} \cup \mathcal{N}_{\mathrm{sc}} \cup \mathcal{N}_{\mathrm{sj}}, \tag{2.1}$$

$$\mathcal{E} = \mathcal{P} \cup \mathcal{H}_{\mathrm{pr}} \cup \mathcal{H}_{\mathrm{c}} \cup \mathcal{E}_{\mathrm{valves}} \cup \mathcal{E}_{\mathrm{pumps}}, \tag{2.2}$$

where $\mathcal{P}$ is the set of pipes, $\mathcal{H}_{\mathrm{pr}}$ ($\mathcal{H}_{\mathrm{c}}$) is the set of producers' (consumers') heat exchangers, $\mathcal{E}_{\mathrm{valves}}$ is the set of valves, and $\mathcal{E}_{\mathrm{pumps}}$ the set of pumps. Also, $\mathcal{N}_{\mathrm{sh}}$ ($\mathcal{N}_{\mathrm{sc}}$) is the set of storage tanks' hot (cold) layers and $\mathcal{N}_{\mathrm{sj}}$ is the set of simple junctions. The cardinalities of $\mathcal{N}$ and $\mathcal{E}$ are denoted by n_{N} and n_{E}, respectively. The variables $q_{\mathrm{E},i}$, $V_{\mathrm{E},i}$, $T_{\mathrm{E},i}$, and $p_{\mathrm{E},i}$ denote the flow rate, volume, temperature, and pressure of the stream through $i \in \mathcal{E}$. Analogous descriptions follow for the variables $V_{\mathrm{N},k}$, $T_{\mathrm{N},k}$, and $p_{\mathrm{N},k}$, $k \in \mathcal{N}$.

For establishing a reference direction for the streams of water through the DHS, we fix an arbitrary orientation for every edge of $\mathcal{G}$. Then, it is possible to define the following sets, which are instrumental for later writing conservation laws for nodes and edges (see [27,29]):

- For any $i \in \mathcal{E}$, we denote by $\mathcal{N}_i^-$ ($\mathcal{N}_i^+$) the tail (head) node of i.
- For any $k \in \mathcal{N}$, we denote by $\mathcal{E}_k^- \subset \mathcal{E}$ ($\mathcal{E}_k^+ \subset \mathcal{E}$) the set of edges $i \in \mathcal{E}$ whose tail (head) node is k.

Considering the assigned orientation for the edges of $\mathcal{G}$ and the definition of the above sets, we can define a constant incidence matrix $\mathcal{B}_0$ as follows:

$$(\mathcal{B}_0)_{i,j} = \begin{cases} 1, & \text{if } i = \mathcal{N}_j^+ \in \mathcal{N}, j \in \mathcal{E}, \\ -1, & \text{if } i = \mathcal{N}_j^- \in \mathcal{N}, j \in \mathcal{E}, \\ 0, & \text{otherwise.} \end{cases} \tag{2.3}$$

This matrix will be useful for writing vector forms of some of the derived models.

2.3 Hydraulic dynamics

In this section we present a model to describe the dynamic behavior of the hydraulic variables of the DHS. First, by invoking first principles, a hydraulic model based on a differential algebraic equation (DAE) is defined. Subsequently, and following the ideas in [24] and [20], we proceed to identify a set of independent flows from which the entire hydraulic state of the DHS can be determined from an ODE.

2.3.1 Edges

Let $i \in \mathscr{P} \cup \mathscr{H}_{pr} \cup \mathscr{H}_c$ be an arbitrary pipe or heat exchanger. Since water has been assumed to be incompressible and its density ρ, constant, the differential forms of the balance of mass and momentum can be written as follows (see [29]):

$$0 = \partial_x v_{E,i}(t, x) \tag{2.4a}$$

$$0 = \rho \partial_t v_{E,i}(t, x) + \partial_x p_{E,i}(t, x) + \rho \frac{\lambda_{E,i}}{2D_{E,i}} |v_{E,i}(t, x)| v_{E,i}(t, x), \tag{2.4b}$$

where $v_{E,i}(t, x)$ and $p_{E,i}(t, x)$ are the velocity and the pressure of the stream along the axis of the pipe ($0 \le x \le \ell_{E,i}$), respectively. The parameters $\lambda_{E,i}$ and $D_{E,i}$ are the friction factor and the pipe's diameter, respectively; both parameters are assumed to be constant.

The first subequation in (2.4) implies that the velocity of the stream of water through the pipe depends only on time. Then, we can further simplify the second subequation by integrating it with respect to the spatial variable x. If we work with the flow rate $q_{E,i} = A_{E,i}v_{E,i}$ instead of the velocity and define the boundary conditions $p_{E,i}^{in} := p_{E,i}(t, 0)$ and $p_{E,i}^{out} := p_{E,i}(t, \ell_{E,i})$, then the integration of (2.4) with respect to x results in the following equation (see [24,30]):

$$p_{E,i}^{in} - p_{E,i}^{out} = J_{E,i}\dot{q}_{E,i} + f_{E,i}(q_{E,i}), \tag{2.5}$$

where

$$J_{E,i} = \rho \ell_{E,i}/A_{E,i}, \quad f_{E,i}(q_{E,i}) = K_{E,i}|q_{E,i}|q_{E,i}, \quad K_{E,i} = \lambda_{E,i}\rho \ell_{E,i}/2D_{E,i}A_{E,i}^2. \tag{2.6}$$

For any valve $i \in \mathscr{E}_{valves}$, we also use (2.5) to model the pressure drop through it. Similar to [24], we neglect the inertial forces within the valve by taking $J_{E,i} = 0$. Also $K_{E,i} > 0$ is in this case considered to be an unknown constant whose value depends on generic valve parameters, such as flow capacity [20].

For any pump $i \in \mathscr{E}_{pumps}$, we model the pressure difference between inlet and outlet by $p_{E,i}^{in} - p_{E,i}^{out} = -w_{E,i}$, where $w_{E,i}$ is the pressure difference produced by the pump. Note that this model is also analogous to (2.5), but with $J_{E,i} = K_{E,i} = 0$ and adding a (negative) pump pressure difference in the right-hand side. From the above developments, it is possible (see [24]) to write the momentum balance equation for all edges in a single vector equation as $p_E^{in} - p_E^{out} = \mathrm{diag}(J_E)\dot{q}_E + f_E(q_E) - w_E$. Now, let us bring the following constraints

$$p_{E,i}^{in} = p_{N,k}, \; \forall i \in \mathscr{E}, \; k \in \mathscr{N}_i^- \quad \text{and} \quad p_{E,j}^{out} = p_{N,k}, \; \forall i \in \mathscr{E}, \; k \in \mathscr{N}_i^+,$$

which for any edge $i \in \mathscr{E}$ assign the pressure at its inlet and outlet with the corresponding tail or head node in $\mathscr{N}$, thus ensuring that the pressure profile throughout the DHS is continuous in space [29] (see also [27,31]). Then, considering the incidence matrix defined in (2.3), it is possible to write that $p_{\mathrm{E}}^{\mathrm{in}} - p_{\mathrm{E}}^{\mathrm{out}} = -\mathscr{B}_0^\top p_{\mathrm{N}}$ (see [31]), which leads to

$$-\mathscr{B}_0^\top p_{\mathrm{N}} = \mathrm{diag}(J_{\mathrm{E}})\dot{q}_{\mathrm{E}} + f_{\mathrm{E}}(q_{\mathrm{E}}) - w_{\mathrm{E}}. \tag{2.7}$$

In the sequel this equation will be referred back to produce an ODE for a selected set of flows $q_{\mathrm{E},i}, i \in \mathscr{C} \subset \mathscr{E}$.

2.3.2   Nodes

Consider the standing assumption that the pressure is uniform in each tank (see Assumption 1(v)). Then, to describe the dynamics of any node $k \in \mathscr{N}$, it is enough to invoke the mass balance equation (see [27,29,31]):

$$\rho \dot{V}_{\mathrm{N},k} = \rho \sum_{i \in \mathscr{E}_k^+} q_{\mathrm{E},i} - \rho \sum_{i \in \mathscr{E}_k^-} q_{\mathrm{E},i}, \quad k \in \mathscr{N}. \tag{2.8}$$

According to Assumption 1(vi), the left-hand side of (2.8) is zero for all simple junctions. Then, the collection of (2.8) for all $k \in \mathscr{N}$ can be written as

$$\mathrm{diag}(\eta_k)_{k \in \mathscr{N}} \dot{V}_{\mathrm{N}} = \mathscr{B}_0 q_{\mathrm{E}}, \tag{2.9}$$

where $\eta_k = 0$ if $k \in \mathscr{N}_{\mathrm{sj}}$ is a simple junction and $\eta_k = 1$ if $k \in \mathscr{N}_{\mathrm{sh}} \cup \mathscr{N}_{\mathrm{sc}}$ is any layer of a storage tank.

2.3.3   Overall hydraulic model as a DAE

Equations (2.7) and (2.9) compose the hydraulic model of the overall, interconnected DHS. This model is a DAE in view of the presence of the algebraic variables P_{N} and the fact that the coefficient matrix in the left-hand side of (2.9) is singular. Next we explain the steps that will lead to a hydraulic model based on an ODE.

2.3.4   Overall hydraulic model as an ODE

In order to find an ODE that is equivalent to the obtained DAE model (2.7) and (2.9), we proceed to reduce the DHS graph $\mathscr{G}$ by viewing each tank, as a whole, as an individual node. The benefit of this reduction will be the possibility to write the mass balance at any tank as an algebraic constraint only on the flows, without requiring any additional state for the volume, which will be important for identifying a set of independent flows.

Let us define the graph $G = (N, \mathscr{E})$, where $N = \mathscr{N}_{\mathrm{ST}} \cup \mathscr{N}_{\mathrm{sj}}$, $\mathscr{N}_{\mathrm{ST}} := \mathscr{N}_{\mathrm{sh}} = \mathscr{N}_{\mathrm{sc}}$. Note that G has the same edges as the original DHS graph $\mathscr{G}$, but the nodes representing hot and cold layers of storage tanks have been merged into a single set, $\mathscr{N}_{\mathrm{ST}}$. Then, G is a reduced version of $\mathscr{G}$. We let $|N| = |\mathscr{N}| - n_{\mathrm{ST}} =: n_{\mathrm{n}}$ and assume that the orientation of the edges is preserved in the reduced graph. With G we associate an incidence matrix B_0 which is analogous to (2.3). In addition, since $\mathscr{G}$ is connected (by assumption), then G is also connected.

Considering the steps leading to the momentum and balance equations (2.7) and (2.9), respectively, it is possible to obtain equations analogous to these for the reduced graph G and obtain:

$$-B_0^\top p_n = \text{diag}(J_E)\dot{q}_E + f_E(q_E) - w_E, \tag{2.10a}$$

$$0 = B_0 q_E, \tag{2.10b}$$

where all flow-related variables and parameters are as before, yet we use the incidence matrix B_0 and the node pressure vector p_n associated with the reduced graph. Although (2.10) is still a DAE, this form is amenable for algebraic manipulations that will lead to an equivalent ODE, as done in [24] for DHSs without storage tanks.

Before proceeding, further notions from graph theory are recalled. Since G is connected, it has a (possibly non-unique) spanning tree (see [24,32]), which is a connected subgraph of G formed with all its nodes but with a subset of edges $\mathcal{E}' \subset \mathcal{E}$ such that no loops exist. Let us denote by $S = (N, \mathcal{E}')$ any spanning tree of G. Any $i \in \mathcal{E} \setminus \mathcal{E}'$ is referred to as a *chord* of S and the set of all chords is denoted by $\mathcal{C} = \mathcal{E} \setminus \mathcal{E}'$. Let $n_f = n_E - (n_n - 1)$ denote the cardinality of $\mathcal{C}$. A fundamental loop, denoted here by $\mathcal{L}_i$, is the sequence of edges associated with the loop that is formed when a chord is added to the spanning tree S. It is assumed that each $\mathcal{L}_i$ has an orientation matching that of the chord which generates it. Then, the fundamental loop matrix $F \in \mathbb{R}^{n_f \times n_E}$ can be defined by components as (see [24,33])

$$(F)_{i,h} = \begin{cases} 1, & \text{if } h \in \mathcal{L}_i \text{ and orientations agree,} \\ -1, & \text{if } h \in \mathcal{L}_i \text{ and orientations disagree,} \\ 0, & \text{if } h \notin \mathcal{L}_i. \end{cases} \tag{2.11}$$

Since G is connected, then F is a full row rank matrix ([24,33]).

In the considered DHS setup, based on the analysis in [20,21,24], it can be shown that a suitable selection of the set of chords is the following:

$$\mathcal{C} = \mathcal{H}_c \cup \hat{\mathcal{H}}_{pr} \cup \mathcal{P}_d \cup \hat{\mathcal{P}}_{ST} \subset \mathcal{E}, \tag{2.12}$$

where $\mathcal{H}_c$ are the heat exchangers associated with consumers, as defined in (2.1), and $\mathcal{P}_d$ are the minimal set of pipes either in the supply or return distribution networks which, if removed, would render such networks loop-less, yet preserving connectivity. The definition of $\hat{\mathcal{H}}_{pr} \subset \mathcal{H}_{pr}$ and $\hat{\mathcal{P}}_{ST} \subset \mathcal{P}$ is more involved and depends on the configuration of the DHS. On the one hand, to define $\hat{\mathcal{H}}_{pr}$ let us denote by $\mathcal{H}_{pr}^{ST} \subset \mathcal{H}_{pr}$ the heat exchangers of all the producers which are directly connected to storage tanks and not to the DHS's distribution network (see Figure 2.2), and, additionally, let us denote by $\mathcal{H}_{pr}^{DN} = \mathcal{H}_{pr} \setminus \mathcal{H}_{pr}^{ST}$ the heat exchangers of the remaining producers, i.e., those which are directly connected to the distribution network. Then, $\mathcal{H}_{pr} = \mathcal{H}_{pr}^{ST} \cup \hat{\mathcal{H}}_{pr}^{DN}$, where $\hat{\mathcal{H}}_{pr}^{DN}$ is formed from $\mathcal{H}_{pr}^{DN}$ after removal of an arbitrary element, which would instead belong to the spanning tree S. On the other hand, let $\mathcal{P}_{ST} \subset \mathcal{P}$ be the set of pipes connecting (either) the hot (or the cold) layers of storage tanks with the hot (or the cold) layer of the distribution network. Then, $\hat{\mathcal{P}}_{ST} = \mathcal{P}_{ST}$, if $\hat{\mathcal{H}}_{pr}^{DN}$ is not empty, and, alternatively, $\hat{\mathcal{P}}_{ST}$ is formed from $\mathcal{P}_{ST}$ after removal of an arbitrary element, if $\hat{\mathcal{H}}_{pr}^{DN}$ is empty.

Note that with the above considerations, the spanning tree S that results from the removal of the chords $\mathscr{C}$ in (2.12) from the DHS's graph G, is both loop-less and connected for any sensible configuration of the DHS. Without loss of generality, we assume henceforth that the elements in $\mathscr{P}_{\text{pr}}$ ($\mathscr{P}_{\text{c}}$) are oriented toward the supply (return) layer of the DHS.

Now that a set of chords has been identified, it is possible to compute the associated fundamental loop matrix according to (2.11) and invoke Kirchhoff's voltage and current laws [24,33], which for general hydraulic networks can, respectively, be written as follows [24]:

$$FB_0^\top p_{\text{n}} = 0, \tag{2.13a}$$

$$q_{\text{E}} = F^\top q_{\text{f}}, \tag{2.13b}$$

where the vector $q_{\text{f}} = \text{col}(q_{\text{E},i})_{i \in \mathscr{C}}$, which groups the chords flows, is an independent variable. Note that all the DHS flows can be written as linear combinations of the components of q_{f}. By combining (2.13) with (2.7) it is possible to get the following ODE for q_{f}:

$$F\text{diag}(J_{\text{E}})F^\top \dot{q}_{\text{f}} = -Ff_{\text{E}}(F^\top q_{\text{f}}) + Fw_{\text{E}}, \tag{2.14a}$$

or, equivalently,

$$\mathscr{J}_f \dot{q}_{\text{f}} = -f_{\text{f}}(q_{\text{f}}) + Fw_{\text{E}}, \quad \text{where} \tag{2.15a}$$

$$\mathscr{J}_f = F\text{diag}(J_{\text{E}})F^\top \tag{2.15b}$$

$$f_{\text{f}}(q_{\text{f}}) = F f_{\text{E}}(q_{\text{E}})|_{q_{\text{E}}=F^\top q_{\text{f}}}. \tag{2.15c}$$

In the sequel we will elaborate on the properties of $\mathscr{J}_f$ and f_{f}. For now we proceed to give a more meaningful representation for the term Fw_{E}, whose ith component represents the sum of the pressure differences produced by the pumps belonging to the ith fundamental loop.

Assume there is an independently controlled pump adjacent (or in series) to each chord $i \in \mathscr{C}$. Let $\mathscr{W}_{\text{f}} \subset \mathscr{E}_{\text{pumps}}$ and $w_{\text{f},i}$ denote the set of these pumps and the pressure difference produced by any of them, respectively. If the orientation of each pump in $\mathscr{W}_{\text{f}}$ matches that of its adjacent chord, then $F_{i,j} = 1$ for all $i \in \mathscr{C}$ and for all $j \in \mathscr{W}_{\text{f}} \cap \mathscr{L}_i$, where we recall that $\mathscr{L}_i$ is the fundamental loop generated by the chord i. Consequently, the product Fw_{E} can be equivalently written as $Fw_{\text{E}} = w_{\text{f}} + B_{\text{b}}w_{\text{b}}$, where the ith component of $B_{\text{b}}w_{\text{b}}$ is the direct sum of the pressure differences produced by any pump $k \in \mathscr{E}_{\text{pumps}} \setminus \mathscr{W}_{\text{f}}$ over the ith fundamental loop $\mathscr{L}_i$. For completeness we note that $w_{\text{b}} = \text{col}(w_{\text{E},i})_{i \in \mathscr{E}_{\text{pumps}} \setminus \mathscr{W}_{\text{f}}}$ and that $B_{\text{b}} = [F_{i,j}]_{(i,j) \in \mathscr{C} \times \mathscr{E}_{\text{pumps}} \setminus \mathscr{W}_{\text{f}}}$. Therefore, the DHS flow dynamics (2.15) can be equivalently represented by

$$\mathscr{J}_f \dot{q}_{\text{f}} = -f_{\text{f}}(q_{\text{f}}) + w_{\text{f}} + B_{\text{b}}w_{\text{b}}. \tag{2.16}$$

Through this equation we can describe the evolution of the flow through all the DHS edges. However, for a complete hydraulic DHS model, we still need to incorporate a model describing the evolution of the layers of the storage tanks. This is done next.

Consider once again the full DHS graph $\mathscr{G}$ and the assumed ordering of its nodes (see (2.1)). In view of (2.13), the mass balance equations for all nodes in (2.9) can be equivalently written as $\mathrm{diag}(\eta_k)_{k \in \mathscr{N}} \dot{V}_N = \mathscr{B}_0 F^\top q_f$. An implication of this equation is that the volume dynamics of the hot layers of the tanks can be written as $\dot{V}_{sh} = (W' \mathscr{B}_0 F^\top) q_f$, where $V_{sh} = \mathrm{col}(V_{N,k})_{k \in \mathscr{N}_{sh}}$ and $W' = \left[I_{n_{ST}}, \quad 0_{n_{ST} \times (n_N - n_{ST})} \right]$. Since the total volume of each tank remains constant all the time due to (2.10), then it is possible to write the volume dynamics of the cold layers of the tanks simply as $\dot{V}_{sc} = -(W' \mathscr{B}_0 F^\top) q_f$, where $V_{sc} = \mathrm{col}(V_{N,k})_{k \in \mathscr{N}_{sc}}$. However, V_{sc} is not an independent variable as it can be written as $V_{sc} = V_{ST} - V_{sh}$, where $V_{ST} \in \mathbb{R}^{n_{ST}}_{>0}$ is a constant vector with the volume capacities of the tanks.

Summarizing, the full hydraulic dynamics of the DHS are defined by the following ODE:

$$\mathscr{J}_f \dot{q}_f = -f_f(q_f) + w_f + B_b w_b, \tag{2.17a}$$

$$\dot{V}_{sh} = (W' \mathscr{B}_0 F^\top) q_f. \tag{2.17b}$$

The vector of all edges flows q_E can be computed via (2.13) and the evolution of the tanks cold layers volumes is determined by $V_{sc} = V_{ST} - V_{sh}$.

Before concluding this subsection, we find it convenient to introduce two properties about the q_f-dynamics that will be relevant to later analyze its dissipativity properties. The first property refers to the positive definiteness of $\mathscr{J}_f$. This condition, which can be established following the developments in [24], will be helpful to define a positive-definite storage function based on the kinetic energy of the streams through the pipes. The second property refers to the monotonicity of the function f_f. In general, a mapping $\mathscr{F} : \mathbb{R}^n \to \mathbb{R}^n$ is said to be monotone if $(u - v)^\top (\mathscr{F}(u) - \mathscr{F}(v)) \geq 0$, $\forall u, v \in \mathbb{R}^n$ [34, Section 4]. In the situation that $\mathscr{F}$ is differentiable, then a necessary and sufficient condition for monotonicity is given by $\nabla \mathscr{F}(x) + \nabla \mathscr{F}(x)^\top \geq 0$, $\forall x \in \mathbb{R}^n$ [34, p.12]. In our case, f_f is differentiable, and the symmetric part of its Jacobian can be shown (using the developments in [24]) to be positive semi-definite for any real vector q_f. Then, f_f is monotone, i.e., the following holds:

$$(q_f - \bar{q}_f)^\top (f_f(q_f) - f_f(\bar{q}_f)) \geq 0, \quad \forall q_f, \bar{q}_f \in \mathbb{R}^{n_f}. \tag{2.18}$$

2.3.5   *Further notes on the DHS hydraulic dynamics*

Consider the following:

(i)   Concerning the set of pumps $\mathscr{W}_f$, we note that it is common that independently controlled pumps are placed at heat producers [20], consumers [24,35], and storage units [10]. Although some consumers or storage units might alternatively have control valves [20,35], in this chapter we restrict ourselves to the multi-pump case only. For a more detailed DHS hydraulic modeling which considers both pump dynamics as well as actuating valves, the reader is referred to our recent work [15].

(ii)   Even though the total mass in each tank remains constant all the time, this does not imply that the volume of the hot or cold layer are always simultaneously positive or constant. Then, we assume henceforth that $V_{\mathrm{sh},j} > 0$ and $V_{\mathrm{sh},k} > 0$ for any $j \in \mathcal{N}_{\mathrm{sh}}$ and any $k \in \mathcal{N}_{\mathrm{sc}}$.

2.4   Thermal dynamics

In this section we present a model that describes the dynamic behavior of temperatures of the overall DHS.

2.4.1   Edges

Let $i \in \mathscr{P}$ be an arbitrary pipe. For simplicity of exposition, let us assume that the velocity $v_{\mathrm{E},i}$ of the water stream through i is non-negative all the time. Then, following [29,36] (see also [27]) it is possible to write the differential form of the one-dimensional energy balance at an infinitesimally small mass point in i as

$$\rho A_{\mathrm{E},i} c_{\mathrm{s.h.}} \frac{\partial T_{\mathrm{E},i}(t,x)}{\partial t} + \rho A_{\mathrm{E},i} c_{\mathrm{s.h.}} v_{\mathrm{E},i}(t) \frac{\partial T_{\mathrm{E},i}(t,x)}{\partial x} = \rho \frac{\lambda_{\mathrm{E},i} A_{\mathrm{E},i}}{2 D_{\mathrm{E},i}} |v_{\mathrm{E},i}(t)| v_{\mathrm{E},i}^2(t)$$

$$+ \frac{1}{R_{\mathrm{env},i}} (T_{\mathrm{env},i} - T_{\mathrm{E},i}(t,x)),$$

$$(2.19)$$

where $T_{\mathrm{E},i}(t,x)$ is cross-section averaged temperature of the stream along the axis of the pipe $(0 \leq x \leq \ell_{\mathrm{E},i})$ and we recall that $p_{\mathrm{E},i}(t,x)$ is the stream's pressure. Note that in (2.19) we have already incorporated the knowledge from (2.4) that $v_{\mathrm{E},i}$ is independent of x. Combined, the two terms in the left-hand side of (2.19) represent the material derivative of the internal energy per unit length of the stream. Note that we have assumed that the internal energy of water depends linearly with respect to the temperature, as done, e.g., in [25,37,38]; the proportionality factor depends on the specific heat $c_{\mathrm{s.h.}}$ of water, which here is assumed to be uniform and constant.

In the right-hand side of (2.19), the first term is the dissipation per unit length produced by frictional forces, whereas the second term is the heat transfer per unit length between the stream of water and the environment; $R_{\mathrm{env},i}$ is the total thermal resistance per unit length, which groups both convective and conductive terms. For more details on how $R_{\mathrm{env},i}$ is defined for insulated and encased heating pipes we refer to [39]; for simplicity here we assume that both $R_{\mathrm{env},i}$ and environment's temperature $T_{\mathrm{env},i}$ are constant and uniform. Note that heat conduction along the axis of the pipe is not considered in (2.19): in [39] it is suggested that it is a comparatively small term. Even though the heat dissipation term might also be small compared to the other terms, we will keep it in (2.19) since it will be relevant for establishing that the overall thermo-hydraulic model is cyclo-dissipative (see Section 2.5 for more details).

Under certain assumptions it is possible to explicitly solve (2.19) for $T_{\mathrm{E},i}$ (see, e.g., [25,36,37]). Instead, here we proceed following the ideas of [27,29] and, based on the analysis in [29], write a spatially discretized approximation of (2.19). As in [29], we proceed via the finite volume method. Assume that the pipe is divided into $\Omega_{\mathrm{E},i}$ segments of equal length $\ell_{\mathrm{E},i}/\Omega_{\mathrm{E},i}$. Let $T_{\mathrm{E},i}^{\alpha}(t)$ denote the average of $T_{\mathrm{E},i}(t,x)$

over the length of the αth pipe segment. Then, $T_{\mathrm{E},i}^{\alpha}(t)$ evolves according to the following ODE:

$$\rho c_{\mathrm{s.h.}} A_{\mathrm{E},i} \frac{\ell_{\mathrm{E},i}}{\Omega_{\mathrm{E},i}} \dot{T}_{\mathrm{E},i}^{(\alpha)} = \rho c_{\mathrm{s.h.}} q_{\mathrm{E},i} (T_{\mathrm{E},i}^{(\alpha-1)} - T_{\mathrm{E},i}^{(\alpha)}) + \frac{K_{\mathrm{E},i}}{\Omega_{\mathrm{E},i}} |q_{\mathrm{E},i}| q_{\mathrm{E},i}^2,$$

$$+ \frac{\ell_{\mathrm{E},i}}{R_{\mathrm{env},i} \Omega_{\mathrm{E},i}} (T_{\mathrm{env},i} - T_{\mathrm{E},i}^{(\alpha)}), \quad \alpha = 1, 2, ..., \Omega_{\mathrm{E},i}, \qquad (2.20)$$

where we have written the stream's velocity $v_{\mathrm{E},i}$ in terms of the flow rate $q_{\mathrm{E},i}$ (via $q_{\mathrm{E},i} = A_{\mathrm{E},i} v_{\mathrm{E},i}$) and introduced the friction-related coefficient $K_{\mathrm{E},i}$, which is defined in (2.6). By associating the boundary condition $T_{\mathrm{E},i}(t, 0) = T_{\mathrm{E},i}^{\mathrm{in}}$ to (2.19), we can also specify $T_{\mathrm{E},i}^{(0)} = T_{\mathrm{E},i}^{\mathrm{in}}$.

Note that the larger the number $\Omega_{\mathrm{E},i}$ of segments into which the pipe is subdivided, the more accurate the visualization of the temperature profile within the pipe will be. For simplicity and without sacrificing generality of the results in Section 2.5, we will assume that $\Omega_{\mathrm{E},i} = 1$. This means that $T_{\mathrm{E},i}^{(1)}$ represents the average temperature of the pipe over its full length. Due to the considered (upwind) finite volume discretization scheme, we will have that $T_{\mathrm{E},i}^1$ is also the temperature at the outlet of the pipe. Then, the overall simplified pipe model can be written as

$$\rho c_{\mathrm{s.h.}} A_{\mathrm{E},i} \ell_{\mathrm{E},i} \dot{T}_{\mathrm{E},i} = \rho c_{\mathrm{s.h.}} q_{\mathrm{E},i} (T_{\mathrm{E},i}^{\mathrm{in}} - T_{\mathrm{E},i}) + K_{\mathrm{E},i} |q_{\mathrm{E},i}| q_{\mathrm{E},i}^2 + \frac{\ell_{\mathrm{E},i}}{R_{\mathrm{env},i}} (T_{\mathrm{env},i} - T_{\mathrm{E},i}),$$

$$(2.21)$$

where, abusing the notation, we have dropped the use of the superindex $(\alpha) = 1$ in $T_{\mathrm{E},i}$.

Having established the model to describe the temperature dynamics of any pipe of the DHS, we continue with the remaining elements of the set of edges $\mathscr{E}$. Let $i \in \mathscr{H}_{\mathrm{pr}} \cup \mathscr{H}_{\mathrm{c}}$ be an arbitrary heat exchanger. We recall that for any heat exchanger we consider only the side which is hydraulically coupled with the distribution network of the DHS. Then, we can treat any heat exchanger as a pipe and, using a procedure analogous to the one used to obtain (2.21), write the following simplified dynamics for the temperature of the stream (averaged over the heat exchanger's length):

$$\rho c_{\mathrm{s.h.}} A_{\mathrm{E},i} \ell_{\mathrm{E},i} \dot{T}_{\mathrm{E},i} = \rho c_{\mathrm{s.h.}} q_{\mathrm{E},i} (T_{\mathrm{E},i}^{\mathrm{in}} - T_{\mathrm{E},i}) + K_{\mathrm{E},i} |q_{\mathrm{E},i}| q_{\mathrm{E},i}^2 + P_{\mathrm{pr},i} - P_{\mathrm{c},i}, \quad (2.22)$$

where we recall that $P_{\mathrm{pr},i}$ ($P_{\mathrm{c},i}$) is the heat transfer rate of a given producer (consumer). Note that if $i \in \mathscr{E}$ is a producer's heat exchanger, then $P_{\mathrm{c},i} = 0$. Analogously, if $i \in \mathscr{E}$ is a consumer's heat exchanger, then $P_{\mathrm{pr},i} = 0$. Note that we adopt the convention that the heat transfer rate is positive when flowing from primary to secondary side.

It remains to define the temperature dynamics of valves and pumps. For any valve $i \in \mathscr{E}_{\mathrm{valves}}$ we use a model analogous to (2.22), but with $P_{\mathrm{p},i} = P_{\mathrm{c},i} = 0$. Similarly, the temperature dynamics of any pump $i \in \mathscr{E}_{\mathrm{pumps}}$ is assumed to be described by (2.22), but with $K_{\mathrm{E},i} = P_{\mathrm{p},i} = P_{\mathrm{c},i} = 0$.

Based on the above, it is possible to write a general form for the temperature dynamics of any edge $i \in \mathcal{E}$ as follows:

$$\rho c_{\text{s.h.}} A_{\text{E},i} \ell_{\text{E},i} \dot{T}_{\text{E},i} = \rho c_{\text{s.h.}} q_{\text{E},i} (T_{\text{E},i}^{\text{in}} - T_{\text{E},i}) + \frac{K_{\text{E},i}}{A_{\text{E},i}} |q_{\text{E},i}| q_{\text{E},i}^2 + \frac{\ell_{\text{E},i}}{R_{\text{env},i}} (T_{\text{env}} - T_{\text{E},i}),$$
$$+ P_{\text{pr},i} - P_{\text{c},i}, \; i \in \mathcal{E}, \tag{2.23}$$

where the set of parameters is defined according to the type of edge i and in accordance to the developments in this subsection so far. Note that to reflect the constancy and uniformity of the environment's temperature we have dropped the subindexing for T_{env} in (2.23).

2.4.2   Nodes

Let $k \in \mathcal{N}$ be an arbitrary node of the DHS. For simplicity, we assume that its temperature $T_{\text{N},k}$ is spatially uniform. For ease of exposition we additionally assume that $q_{\text{E},i} \geq 0$ for all $i \in \mathcal{E}$. Then, the energy balance at k can be written as

$$\frac{\mathrm{d}}{\mathrm{d}t} (\rho c_{\text{s.h.}} V_{\text{N},k} T_{\text{N},k}) = \sum_{j \in \mathcal{E}_k^+} \rho c_{\text{s.h.}} q_{\text{E},j} T_{\text{E},j} - \sum_{j \in \mathcal{E}_k^-} \rho c_{\text{s.h.}} q_{\text{E},j} T_{\text{E},j}^{\text{in}}, \tag{2.24}$$

where we recall that $\mathcal{E}_k^+ \subset \mathcal{E}$ ($\mathcal{E}_k^- \subset \mathcal{E}$) is the set of edges whose head (tail) is k. The term in the left-hand side of (2.24) represents the rate of change of the internal energy stored at node k: similar to the modeling of the edges, the internal energy has been assumed to depend linearly on the temperature, with $c_{\text{s.h.}}$ representing the specific heat. In the right-hand side we have the balance between the (convective) heat flows entering and leaving k. We note that for writing the first term in the right-hand side of (2.24) we have used the fact that the temperature at the outlet of any edge is the same as the spatially averaged temperature over the length of the edge. Additionally, note that heat transfer over any tank surface and thermocline is neglected. Also, if k represents the layer of a storage tank, it is more precise to say that $T_{\text{N},k}$ is the temperature of the water inside the tank averaged over a varying control volume representing the layer of the tank; nonetheless, we would still write the dynamics as in (2.24).

Based on the nodal constraints described in [27], we impose that the temperature at the inlet of any edge should be equal to the temperature of the tail node. Then, the following should hold all the time:

$$T_{\text{E},j}^{\text{in}} = T_{\text{N},k}, \; j \in \mathcal{E}_k^-, \; k \in \mathcal{N}. \tag{2.25}$$

This equation couples the temperature dynamics in (2.24) of the nodes with that of the edges in (2.23). Using (2.25) we can write a simpler, equivalent form of (2.24). Indeed, if we now recall that the mass balance equation at node $k \in \mathcal{N}$ can be written as $\rho \dot{V}_{\text{N},k} = \sum_{j \in \mathcal{E}_k^+} \rho q_{\text{E},j} - \sum_{j \in \mathcal{E}_k^-} \rho q_{\text{E},j}$, then (2.24) is equivalent to

$$\rho c_{\text{s.h.}} V_{\text{N},k} \dot{T}_{\text{N},k} = \sum_{j \in \mathcal{E}_k^+} \rho c_{\text{s.h.}} q_{\text{E},j} T_{\text{E},j} - \left(\sum_{j \in \mathcal{E}_k^+} \rho c_{\text{s.h.}} q_{\text{E},j} \right) T_{\text{N},k}, \; k \in \mathcal{N}. \tag{2.26}$$

Remark 1.

(i) *Throughout this section we have assumed that $q_{E,i} \geq 0$ for all time and for all edges $i \in \mathcal{E}$. If for a given $i \in \mathcal{E}$, $q_{E,i}(t) \leq 0$ for certain t, then (2.23), (2.25), and (2.26) remain valid if we substitute $q_{E,i}$ by its absolute value and re-define the sets $\mathcal{E}_k^-$ and $\mathcal{E}_k^+$ as follows [27,29], (c.f., [31])*

$$\hat{\mathcal{E}}_k^- = \{i \in \mathcal{E} : (k \text{ is the tail of } i \text{ and } q_{E,i} \geq 0)$$
$$or \ (k \text{ is the head of } i \text{ and } q_{E,i} < 0)\}, \tag{2.27a}$$

$$\hat{\mathcal{E}}_k^+ = \{i \in \mathcal{E} : (k \text{ is the tail of } i \text{ and } q_{E,i} < 0)$$
$$or \ (k \text{ is the head of } i \text{ and } q_{E,i} \geq 0)\}. \tag{2.27b}$$

In the sequel we say that node k is the source of the stream through any $i \in \hat{\mathcal{E}}_k^-$ and that k is the target of the stream through any $i \in \hat{\mathcal{E}}_k^+$.

(ii) *For simplicity we henceforth let $\rho = c_{s.h} = 1$.*

2.4.3 Thermal model in vector form

We proceed to write the system conformed by (2.23), (2.25), and (2.26) in vector form. Then, for the DHS graph $\mathcal{G}$ we introduce a flow-dependent incidence matrix $\mathcal{B}$ capturing the actual direction of the water stream through any edge $i \in \mathcal{E}$, which is defined as follows:

$$(\mathcal{B})_{i,j} = \begin{cases} 1, & \text{if } j \in \hat{\mathcal{E}}_i^+ \subset \mathcal{E}, \ i \in \mathcal{N}, \\ -1, & \text{if } j \in \hat{\mathcal{E}}_i^- \subset \mathcal{E}, \ i \in \mathcal{N}, \\ 0, & \text{otherwise}, \end{cases} \tag{2.28}$$

where we have used the sets introduced in (2.27). More precisely, we will work with the positive and negative parts of $\mathcal{B}$, which are, respectively, denoted by $\mathcal{B}^+$ and $\mathcal{B}^-$. Then, it is possible to claim that the system (2.23), (2.25) and (2.26) describing the temperature dynamics of the DHS (edges and nodes) can be written in vector form as follows:

$$\text{diag}(V_E, V_N) \begin{bmatrix} \dot{T}_E \\ \dot{T}_N \end{bmatrix} = \mathcal{A}(q_E) \begin{bmatrix} T_E \\ T_N \end{bmatrix} + B_E \text{diag}(q_E) f_E(q_E) + B_{pr} P_{pr} - B_c P_c + d_{env}, \tag{2.29}$$

where:

$$\mathcal{A}(q_E) = \begin{bmatrix} -\text{diag}(|q_E|) & \text{diag}(|q_E|)(\mathcal{B}^-)^\top \\ \mathcal{B}^+ \text{diag}(|q_E|) & -\text{diag}(\mathcal{B}^+|q_E|) \end{bmatrix} - \begin{bmatrix} \text{diag}(\frac{\ell_{E,i}}{R_{env,i}}) & 0_{n_E \times n_N} \\ 0_{n_N \times n_E} & 0_{n_n \times n_N} \end{bmatrix}; \tag{2.30}$$

$(B_E)_{i,j} = 1$ if $i = j \in \mathcal{E}$ and $(B_E)_{i,j} = 0$ otherwise; f_E is defined by components in (2.6); $(B_{pr})_{i,j} = 1$ if $i = j \in \mathcal{H}_{pr}$ and $(B_{pr})_{i,j} = 0$ otherwise; $P_{pr} = \text{col}(P_{pr,i})_{i \in \mathcal{H}_{pr}}$; $(B_c)_{i,j} = 1$ if $i = j \in \mathcal{H}_c$ and $(B_c)_{i,j} = 0$ otherwise; $P_c = \text{col}(P_{c,i})_{i \in \mathcal{H}_c}$; and $d_{env} = \frac{T_{env}\ell_{E,i}}{R_{env,i}}$ if $i \in \mathcal{P}$ and $d_{env,i} = 0$ otherwise.

2.4.4   *Further notes on the DHS thermal dynamics*

The following remarks are in order:

(i)   The model (2.29) is a nonlinear ODE since $\mathscr{A}$ depends on the flow state variable q_f (recall from (2.13) that $q_E = F^\top q_f$) and due to the fact that the components of V_N associated to storage tanks are time-varying.

(ii)   It is possible to assume that the thermal inertia of valves, pumps, and junctions are negligible when compared to the thermal inertia of pipes and storage tanks. If we were to assign $V_{E,i} = 0$ for any $i \in \mathscr{E}_{valves} \cup \mathscr{E}_{pumps}$ associated with a valve or pump, and $V_{N,j} = 0$ for any $j \in \mathscr{N}_{sj}$ associated with a simple junction, then the thermal inertia of these devices would be neglected from (2.29). However, the model would become a semi-explicit DAE. The algebraic variables of this DAE would correspond to the (outlet) temperatures of valves, pumps, and junctions. Such algebraic variables can be explicitly solved in terms of the remaining dynamic (state) variables provided that a rank condition is met. Due to space constraints and for simplicity, we proceed with the ODE formulation of the thermal dynamics given in (2.29). More details about the DAE formulation and its subsequent transformation into an equivalent ODE are discussed in [21] for a slight variant of (2.29) which neglects heat dissipation and heat conduction through pipe walls.

2.5   Dissipativity properties

In this section we investigate dissipativity properties exhibited by the developed thermo-hydraulic DHS model. For ease of reference, let us write the overall DHS model in vector form as follows (see (2.17) and (2.29)):

$$\mathscr{J}_f \dot{q}_f = -f_f(q_f) + w_f + B_b w_b, \tag{2.31a}$$

$$\dot{V}_{sh} = (W' \mathscr{B}_0 F^\top) q_f, \tag{2.31b}$$

$$\dot{V}_{sc} = -(W' \mathscr{B}_0 F^\top) q_f, \tag{2.31c}$$

$$\operatorname{diag}(V_E, V_N) \begin{bmatrix} \dot{T}_E \\ \dot{T}_N \end{bmatrix} = \mathscr{A}(q_E) \begin{bmatrix} T_E \\ T_N \end{bmatrix} + B_E \operatorname{diag}(q_E) f_E(q_E) + B_{pr} P_{pr} - B_c P_c + d_{env}, \tag{2.31d}$$

where all the defining parameters, variables, and functions have already been identified in the previous two sections. To improve clarity in subsequent computations, note that we have also brought into (2.31) the hydraulic dynamics of the cold layers' volumes V_{sc} (even though V_{sc} is not an independent vector variable).

For making this section self-contained, let us recall from [11] some notions about dissipative systems. Consider a general control system of the form $\dot{x} = f(x, u)$ with state $x \in \mathbb{R}^n$, input $u \in \mathbb{R}^m$ and output $y = h(x, u) \in \mathbb{R}^m$. Then:

(i)   The system is dissipative with respect to a supply rate $w(u, y)$ if there exists a non-negative, scalar storage function $S : \mathbb{R}^n \to \mathbb{R}_{\geq 0}$ such that

$$\dot{S}(x) = \nabla^\top S(x) f(x, u) \leq w(u, y). \tag{2.32}$$

(ii)   The system is passive if the supply rate is of the form $w(u, y) = u^\top y$.
(iii) The system is shifted (or equilibrium-independent) passive if the supply rate
       is of the form $w(u, y) = (u - \bar{u})^\top (y - \bar{y})$, where $(\bar{u}, \bar{y})$ is any equilibrium pair
       of the control system.
(iv)   If any of the above properties hold with a storage function which is bounded
       from below, but not necessarily non-negative, then we precede the property
       with the identifier "cyclo", e.g., cyclo-passivity. Moreover, if (2.32) holds
       with an equality, then the system is said to be lossless or conservative.

Remark 2 (see, e.g., [40]). *If we give to the storage function S the interpretation
of being the "energy" stored in a dynamic control system, then the inequality (2.32)
states that the rate of change of S cannot be greater than the rate at which the system
exchanges "energy" with the environment. Then, dissipativity can be understood as
a system-theoretic generalization of the conservation of energy postulate. Note in
particular that passive systems are intrinsically stable at the origin in open-loop
$(u = 0)$.*

We are then in position to state the following.

Proposition 1. *Consider the system (2.31). Then the following claims are provable:*

(i)   *Let us extend the dynamics (2.31) by introducing the entropy s_{env} of the
       environment, with dynamics*

$$\dot{s} = \sum_{i \in \mathscr{E}} \frac{\ell_{E,i}}{R_{env,i} T_{env}} (T_{E,i} - T_{env}). \tag{2.33}$$

*Then the overall thermo-hydraulic DHS model (2.31), extended with (2.33),
is cyclo-dissipative (in fact cyclo-lossless) with storage function*

$$\mathbb{H} = \frac{1}{2} q_f^\top \mathscr{J}_f q_f + V_E^\top T_E + V_N^\top T_N + T_{env} s. \tag{2.34}$$

and with respect to the supply rate $q_f^\top w_f + 1_{n_{pr}}^\top P_{pr} - 1_{n_c}^\top P_c$.

(ii)   *For the q_f-dynamics, assume that w_b is a constant vector. Let $(\bar{q}_f, \bar{w}_f)$ denote
       any equilibrium pair of this system, provided it exists. Then, the q_f-dynamics
       are shifted passive with storage function*

$$\mathbb{K} = \frac{1}{2} (q_f - \bar{q}_f)^\top \mathscr{J}_f (q_f - \bar{q}_f) \tag{2.35}$$

and with respect to the supply rate $(w_f - \bar{w}_f)^\top (q_f - \bar{q}_f)$.

(iii) *Assume that $q_f = \bar{q}_f$ all the time, where $(\bar{q}_f, \bar{w}_f)$ is any feasible equilibrium
       pair of the q_f-dynamics, provided it exists. Then, the (T_E, T_N)-dynamics are
       shifted passive with storage function*

$$\mathbb{E} = \frac{1}{2} (T_E - \bar{T}_E)^\top V_E (T_E - \bar{T}_E) + \frac{1}{2} (T_N - \bar{T}_N)^\top V_N (T_N - \bar{T}_N) \tag{2.36}$$

and with respect to the supply rate $(P_{pr} - \bar{P}_{pr})^\top (T_{pr} - \bar{T}_{pr}) - (P_c - \bar{P}_c)^\top (T_c - \bar{T}_c)$.

Proof sketch: For establishing the first claim, consider the time derivative of $\mathbb{H}$ along solutions of (2.31):

$$\dot{\mathbb{H}} = q_f^\top \mathscr{J}_f \dot{q}_f + V_E^\top \dot{T}_E + V_N^\top \dot{T}_N + \dot{V}_N^\top T_N + T_{env}\dot{s}$$

$$= q_f^\top \left(-f_f(q_f) + w_f + B_b w_b\right) + 1_{n_E+n_N}^\top \mathrm{diag}(V_E, V_N)\begin{bmatrix} \dot{T}_E \\ \dot{T}_N \end{bmatrix} + \dot{V}_N^\top T_N$$

$$+ \sum_{i\in\mathscr{E}} \frac{\ell_{E,i}}{R_{env,i}}(T_{E,i} - T_{env}). \tag{2.37}$$

By direct computations, it can be shown that

$$1_{n_E+n_N}^\top \mathrm{diag}(V_E, V_N)\begin{bmatrix} \dot{T}_E \\ \dot{T}_N \end{bmatrix} = 1_{n_E+n_N}^\top \mathscr{A}(q_E)\begin{bmatrix} T_E \\ T_N \end{bmatrix} + 1_{n_E+n_N}^\top B_E \mathrm{diag}(q_E)f_E(q_E)$$

$$+ 1_{n_E+n_N}^\top B_{pr}P_{pr} - 1_{n_E+n_N}^\top B_c P_c + 1_{n_E+n_N}^\top d_{env}$$

$$= -\dot{V}_N^\top T_N + q_f^\top f_f(q_f) + 1_{n_{pr}}^\top P_{pr} - 1_{n_c}^\top P_c$$

$$- \sum_{i\in\mathscr{E}} \frac{\ell_{E,i}}{R_{env,i}}(T_{E,i} - T_{env}), \tag{2.38}$$

where we have used

- $1_{n_E+n_N}^\top \mathscr{A}(q_E)\begin{bmatrix} T_E \\ T_N \end{bmatrix} = -\dot{V}_N^\top T_N - \sum_{i\in\mathscr{E}} \frac{\ell_{E,i}}{R_{env,i}} T_{E,i}$ (see the developments in the proof of [21, Lemma 2]),
- $1_{n_E+n_N}^\top B_E \mathrm{diag}(q_E)f_E(q_E) = q_E^\top f_E(q_E) = q_f^\top f_f(q_f)$ (see (2.15)),
- $1_{n_E+n_N}^\top d_{env} = T_{env}\sum_{i\in\mathscr{P}} \frac{\ell_{E,i}}{R_{env,i}}$.

Then, (2.37) is simplified to

$$\dot{\mathbb{H}} = q_f^\top w_f + 1_{n_{pr}}^\top P_{pr} - 1_{n_c}^\top P_c. \tag{2.39}$$

Consequently, the thermo-hydraulic DHS model (2.31), extended with (2.33), is cyclo-dissipative with storage function $\mathbb{H}$ and supply rate $q_f^\top w_f + 1_{n_{pr}}^\top P_{pr} - 1_{n_c}^\top P_c$.

For establishing the second claim, we can proceed similarly. However, before computing the time derivative of $\mathbb{K}$, let us write an equivalent representation of the q_f-dynamics. Assume that $(\bar{q}_f, \bar{w}_f)$ is a feasible equilibrium pair for the q_f-dynamics. Then, the following holds:

$$\mathscr{J}_f \dot{q}_f = -(f_f(q_f) - f_f(\bar{q}_f)) + w_f - \bar{w}_f.$$

Now, by computing $\mathbb{K}$ along solutions of this equivalent representation of the q_f-dynamics, we obtain the following:

$$\dot{\mathbb{K}} = -(q_f - \bar{q}_f)^\top (f_f(q_f) - f_f(\bar{q}_f)) + (q_f - \bar{q}_f)^\top (w_f - \bar{w}_f).$$

Using the fact that f_f is a monotone function and hence it satisfies (2.18), we directly get that

$$\dot{\mathbb{K}} \le (q_f - \bar{q}_f)^\top (w_f - \bar{w}_f).$$

Thus, the q_f-dynamics are shifted passive with storage function $\mathbb{K}$ and with respect to the supply rate $(q_f - \bar{q}_f)^\top (w_f - \bar{w}_f)$.

For establishing the third claim, assume that q_f is fixed to a feasible equilibrium value $\bar{q}_f$. Then, the coefficient matrices $\mathrm{diag}(V_E, V_N)$ and $\mathscr{A}(q_E)$, as well as the disturbance term $B_E \mathrm{diag}(q_E) f_E(q_E)$ of the (T_E, T_N)-dynamics in (2.31) become constant. Under these conditions, assume that $(\bar{T}_E, \bar{T}_N, \bar{P}_{pr}, \bar{P}_c)$ is a feasible equilibrium tuple of the (T_E, T_N)-dynamics. Then, the following holds:

$$\mathrm{diag}(V_E, V_N) \begin{bmatrix} \dot{T}_E \\ \dot{T}_N \end{bmatrix} = \mathscr{A}(\bar{q}_E) \begin{bmatrix} T_E - \bar{T}_E \\ T_N - \bar{T}_N \end{bmatrix} + B_{pr}(P_{pr} - \bar{P}_{pr}) - B_c(P_c - \bar{P}_c).$$

(2.40)

Considering this equivalent representation of the thermal dynamics and the proposed storage function $\mathbb{E}$ in (2.36), to establish the Proposition's third claim, it is sufficient to show that the symmetric part of $\mathscr{A}(\bar{q}_E)$ is negative semi-definite. In [21] and [41], it is shown that the q_E-dependent portion of $\mathscr{A}$ is negative semi-definite when q_E is fixed to an equilibrium value. Considering the definition of $\mathscr{A}$ in our case (see (2.30)), it is straightforward to verify that the time-derivative of $\mathbb{E}$ along solutions of (2.40) satisfies the following:

$$\dot{\mathbb{E}} \leq -(T_{\mathscr{P}} - \bar{T}_{\mathscr{P}})^\top \mathrm{diag}(\frac{\ell_{E,i}}{R_{env,i}})_{i \in \mathscr{P}} (T_{\mathscr{P}} - \bar{T}_{\mathscr{P}}) + (P_{pr} - \bar{P}_{pr})^\top (T_{pr} - \bar{T}_{pr})$$
$$- (P_c - \bar{P}_c)^\top (T_c - \bar{T}_c),$$

where $T_{\mathscr{P}} = \mathrm{col}(T_{E,i})_{i \in \mathscr{P}}$. From the above inequality the shifted passivity claim of the thermal dynamics is established.   ∎

2.5.1 Further notes on the DHS dissipativity properties

The following remarks are in order:

(i)   The idea for establishing the cyclo-dissipativity property of the thermal dynamics (2.31) comes from the developments in [42] for representing general cyclo-passive systems as cyclo-lossless systems with the addition of an extra variable s, defined analogously to (2.33) and which can be interpreted as the entropy of the environment. In [42] such cyclo-lossless systems have the structural properties of irreversible thermodynamic systems, which is the class of systems to which the developed thermo-hydraulic DHS model (2.31) belongs to. Moreover, the storage function $\mathbb{H}$ in (2.34) corresponds to the total energy stored in the DHS, i.e., the sum of total kinetic energy, total internal energy of the DHS and total internal energy of the environment.

(ii) The storage function $\mathbb{K}$ of the q_f-dynamics is the total kinetic energy of the streams through the DHS's pipes and heat exchangers, shifted with respect to an equilibrium value; the passive output is the vector of independent flows q_f (recall (2.13)).

(iii) The storage function $\mathbb{E}$ in (2.36) for the DHS thermal dynamics is the total system's ectropy, shifted with respect to any feasible equilibrium value. The

concept of ectropy was introduced in [43] for general classes of thermodynamic systems and is defined as a quadratic function of the total energy
stored within the system; see [44] for an application of this concept for control synthesis in networks of heat exchangers, and [29, Theorem 3.1] for a
(port-)Hamiltonian formulation of the thermal dynamics of a DHS using a
function quadratic on the system internal energy as a Hamiltonian. In our case,
the corresponding (shifted) passive output is a vector stacking the (shifted)
internal energies of the producers' and consumers' heat exchangers.

2.6 Dissipativity-based stabilization

The identification of dissipativity properties sets in general the basis for dissipativity-
based control design and subsequent closed-loop stability analysis. In this section we
illustrate how the dissipativity properties of the overall thermo-hydraulic DHS model
are relevant for designing controllers of simple structure that can asymptotically
stabilize feasible equilibrium points.

Proposition 2. *Consider the overall thermo-hydraulic DHS model in* (2.31). *Assume
that* w_b *and* P_c *are constant disturbance vectors. Let* $(\bar{q}_f, \bar{V}_{sh}, \bar{T}_E, \bar{T}_N, \bar{w}_f, \bar{P}_{pr})$ *denote
a feasible equilibrium tuple, provided it exists. Then, the following claims hold true:*

(i)   *Consider the* (q_f, V_{sh})-*dynamics. Assume that the constant coefficient matrix*
      $W'\mathbb{B}_0F^\top$ *has a full row rank. Then the* (q_f, V_{sh})-*dynamics, in closed-loop with
      the control law*

$$w_f = \bar{w}_f - k_f(q_f - \bar{q}_f) - (W'\mathbb{B}_0F^\top)^\top(V_{sh} - \bar{V}_{sh}), \tag{2.41}$$

are asymptotically stable at the equilibrium point $(\bar{q}_f, \bar{V}_{sh})$. *For simplicity we
take the control parameter* k_f *as a positive scalar. However, the claim holds if*
k_f *is any symmetric positive-definite matrix.*

(ii) *Assume that* $(q_f, w_f) = (\bar{q}_f, \bar{w}_f)$, $\forall t \geq 0$, *where* $(\bar{q}_f, \bar{w}_f)$ *is a feasible equi-
      librium pair of the* q_f-*dynamics, provided it exists. Assume moreover that
      consumers are not connected in parallel and that:*

$$|\bar{q}_{E,i}| > 0, \forall i \in \mathscr{E}, \text{ and } \left| \sum_{j\in\hat{\mathscr{E}}_k^+} \bar{q}_{E,j} \right| > 0, \forall k \in \mathscr{N}. \tag{2.42}$$

(Recall that $q_E = F^\top q_f$.) *Then, the* (T_E, T_N)-*dynamics, in closed-loop with
the control law*

$$P_{pr} = \bar{P}_{pr} - k_{pr}(T_{pr} - \bar{T}_{pr}), \tag{2.43}$$

are asymptotically stable at the equilibrium tuple $(\bar{T}_E, \bar{T}_N)$. *For simplicity we
take the control parameter* k_{pr} *as a positive scalar. However, the claim holds
if* k_{pr} *is any symmetric positive-definite matrix.*

Proof sketch: For establishing the first claim, let us first write the (q_f, V_{sh})-dynamics in open-loop as follows:

$$\mathscr{J}_f \dot{q}_f = -f_f(q_f) w_f + B_b w_b, \tag{2.44}$$

$$\dot{V}_{sh} = (W' \mathscr{B}_0 F^\top) q_f. \tag{2.45}$$

According to Proposition 1(ii), the q_f-dynamics are shifted passive with (positive-definite) storage function $\mathbb{K} = \frac{1}{2}(q_f - \bar{q}_f)^\top (q_f - \bar{q}_f)$ and with respect to the supply rate $(q_f - \bar{q}_f)^\top (w_f - \bar{w}_f)$. This implies that $\dot{\mathbb{K}} \leq (q_f - \bar{q}_f)^\top (w_f - \bar{w}_f)$. Then, the first two terms of the controller (2.41) guarantee that $\dot{\mathbb{K}} \leq -k_f (q_f - \bar{q}_f)^\top (q_f - \bar{q}_f)$. The third term of the controller ensures a negative feedback interconnection [45] between the q_f-dynamics and the V_{sh}-dynamics. Note that $(\bar{q}_f, \bar{V}_{sh})$ is an equilibrium point of (2.44) in closed-loop with (2.41). For analyzing the asymptotic stability of this equilibrium point, consider the following Lyapunov function candidate:

$$\mathbb{V}_{f,sh} = \mathbb{K} + \frac{1}{2}(V_{sh} - \bar{V}_{sh})^\top (V_{sh} - \bar{V}_{sh}), \tag{2.46}$$

which to $\mathbb{K}$ adds a quadratic function on $(V_{sh} - \bar{V}_{sh})$. Note that $\mathbb{V}_{f,sh}$ is always positive, except at $(\bar{q}_f, \bar{V}_{sh})$ where is nullified. Moreover, the time-derivative of $\mathbb{V}_{f,sh}$ along solutions of the closed-loop system satisfies the following:

$$\dot{\mathbb{V}}_{f,sh} = (q_f - \bar{q}_f)^\top \mathscr{J}_f \dot{q}_f + (V_{sh} - \bar{V}_{sh})^\top \dot{V}_{sh}$$
$$\leq -k_f (q_f - \bar{q}_f)^\top (q_f - \bar{q}_f), \tag{2.47}$$

where to obtain the last inequality we have used the fact that $(W' \mathscr{B}_0 F^\top) \bar{q}_f = 0$ and the fact that f_f satisfies $(q_f - \bar{q}_f)^\top (f_f(q_f) - f_f(\bar{q}_f)) \geq 0$, $\forall q_f, \bar{q}_f \in \mathbb{R}^{n_f}$, as seen from the developments that lead to (2.18).

From (2.47), it is clear that $\mathbb{V}_{f,sh}$ is a Lyapunov function. Therefore, $(\bar{q}_f, \bar{V}_{sh})$ is stable. To see that this equilibrium point is in fact asymptotically stable, we can resort to LaSalle's invariance theorem (see, e.g., [45, Theorem 3.5]). Let us define the set of points in which the time-derivative of $\mathbb{V}_{f,sh}$ is nullified:

$$\mathscr{Z}_{f,sh} = \{(q_f, V_{sh}) : \dot{\mathbb{V}}_{f,sh} = 0 \Leftrightarrow q_f = \bar{q}_f\}. \tag{2.48}$$

Let (q_f, V_{sh}) be a solution of the closed-loop (q_f, V_{sh})-dynamics restricted to evolve in the set $\mathscr{Z}_{f,sh}$ in (2.48). Since $q_f = \bar{q}_f$ all the time, then $\dot{V}_{sh} = 0$; see (2.44). This implies that V_{sh} is a constant vector. To see that in fact $V_{sh} = \bar{V}_{sh}$, we note that the closed-loop q_f-dynamics evolving in $\mathscr{Z}_{f,sh}$ is reduced to the identity $(W' \mathbb{B}_0 F^\top)^\top (V_{sh} - \bar{V}_{sh}) = 0$. By assumption $W' \mathbb{B}_0 F^\top$ has full row rank, then the latter identity holds only if $V_{sh} = \bar{V}_{sh}$. Then, $(\bar{q}_f, \bar{V}_{sh})$ is the largest invariant set of the closed-loop (q_f, V_{sh})-dynamics contained in $\mathscr{Z}_{f,sh}$. Then, invoking [45, Theorem 3.5], we can conclude that this equilibrium is asymptotically stable.

Moving on to the second claim, let us assume that $q_f = \bar{q}_f$ all the time. For ease of reference we write the (T_E, T_N)-dynamics in open-loop as follows:

$$\text{diag}(V_E, V_N) \begin{bmatrix} \dot{T}_E \\ \dot{T}_N \end{bmatrix} = \mathscr{A}(\bar{q}_E) \begin{bmatrix} T_E \\ T_N \end{bmatrix} + B_E \text{diag}(\bar{q}_E) f_E(\bar{q}_E) + B_{pr} P_{pr} - B_c P_c + d_{env}. \tag{2.49}$$

Since $q_f = \bar{q}_f$ is fixed, then V_N is a constant vector, making the system (2.49) linear. Moreover, according to Proposition 1(iii), the dynamics (2.49) are shifted passive with (positive-definite) storage function $\mathbb{E}$ in (2.36) and with respect to the supply rate $(P_{pr} - \bar{P}_{pr})^\top (T_{pr} - \bar{T}_{pr})$, where we have incorporated the assumption that P_c is a constant disturbance vector. Then, the following holds:

$$\dot{\mathbb{E}} \leq -(T_{\mathscr{P}} - \bar{T}_{\mathscr{P}})^\top \text{diag}(\frac{\ell_{E,i}}{R_{env,i}})_{i \in \mathscr{P}} (T_{\mathscr{P}} - \bar{T}_{\mathscr{P}}) + (P_{pr} - \bar{P}_{pr})^\top (T_{pr} - \bar{T}_{pr})$$

$$= -(T_{\mathscr{P}} - \bar{T}_{\mathscr{P}})^\top \text{diag}(\frac{\ell_{E,i}}{R_{env,i}})_{i \in \mathscr{P}} (T_{\mathscr{P}} - \bar{T}_{\mathscr{P}}) - k_{pr}(T_{pr} - \bar{T}_{pr})^\top (T_{pr} - \bar{T}_{pr}),$$

$$(2.50)$$

where we have used $T_{\mathscr{P}} = \text{col}(T_{E,i})_{i \in \mathscr{P}}$ and have substituted the control law for P_{pr} in (2.43). The inequality (2.50) implies that the equilibrium $(\bar{T}_E, \bar{T}_N)$ is stable. To see that in fact this is an asymptotically stable equilibrium point, we follow steps analogous to the ones we used to establish the proposition's first claim. Consider the following set:

$$\mathscr{Z}_{E,N} = \{(T_E, T_N) : \dot{\mathbb{E}} = 0 \Leftrightarrow T_{pr} = \bar{T}_{pr}, \quad T_{E,i} = \bar{T}_{E,i}, \forall i \in \mathscr{P}\}, \qquad (2.51)$$

where we recall that for any pipe $i \in \mathscr{P}$, it holds that $\ell_i, R_{env,i} > 0$.

Let (T_E, T_N) be a solution of the temperature dynamics in (2.49). We will establish that that $T_{E,i} = \bar{T}_{E,i}$ and $T_{N,k} = \bar{T}_{N,k}$ for all $i \in \mathscr{E}$ and for all $k \in \mathscr{N}$. An immediate implication that (T_E, T_N) evolves in (2.51) is that the temperature of the source node of any pipe or any producer's heat exchanger is at equilibrium, i.e.,

$$T_{N,k} = \bar{T}_{N,k}, \forall k \in \hat{\mathscr{N}}_i^-, \forall i \in \mathscr{P} \cup \mathscr{H}_{pr}. \qquad (2.52)$$

To see this, consider the equilibrium equation associated to (2.23) for any $i \in \mathscr{P} \cup \mathscr{H}_{pr}$:

$$0 = \bar{q}_{E,i}(T_{E,i}^{in} - \bar{T}_{E,i}) + K_{E,i}|\bar{q}_{E,i}|\bar{q}_{E,i}^2 + \frac{\ell_{E,i}}{R_{env,i}}(T_{env} - \bar{T}_{E,i}) + \bar{P}_{pr}, \ i \in \mathscr{P} \cup \mathscr{H}_{pr}. \qquad (2.53)$$

This equation holds if and only if $T_{E,i}^{in} = \bar{T}_{E,i}^{in}$. Considering that $T_{E,i}^{in} = T_{N,k}$ for all $k \in \hat{\mathscr{N}}_i^-$ and all $i \in \mathscr{E}$ (see (2.25)), it follows that (2.52) holds.

One consequence of (2.52) is that the temperature $T_{E,j}$ of all the edges in series preceding any pipe or producer's heat exchanger, with respect to the direction of the water's stream, is at equilibrium, i.e.,

$$T_{E,j} = \bar{T}_{E,j}, \forall j \in \Gamma_i \subset \mathscr{E}, \ i \in \mathscr{P} \cup \mathscr{H}_{pr}, \qquad (2.54)$$

where Γ_i is the sequence formed by all the edges in series preceding $i \in \mathscr{P} \cup \mathscr{H}_{pr}$ (with respect to the direction of the flow). Indeed, since $T_{E,i} = \bar{T}_{E,i}$ for any $i \in \mathscr{P} \cup \mathscr{H}_{pr}$ it follows from the equilibrium condition $\dot{T}_{E,i} = 0$ that the temperature of the source node of i is at equilibrium, i.e., $T_{E,i}^{in} = T_{N,k} = \bar{T}_{N,k}$, where $k = \hat{\mathscr{N}}_i^-$;

see (2.23) and (2.25). Now, the equilibrium equation associated to $\dot{T}_{N,k} = 0$ is as follows (see (2.26)):

$$0 = \sum_{\mu \in \hat{\mathscr{E}}_k^+} \bar{q}_{E,\mu} T_{E,\mu} - \left(\sum_{\mu \in \hat{\mathscr{E}}_k^+} \bar{q}_{E,j} \right) \bar{T}_{N,k}. \tag{2.55}$$

Since i is the only edge sourcing from the node k (due to the considered sequence of series connected edges), it means that the preceding edge (with respect to the direction of the flow), say $\mu \in \mathscr{E}$, in series with i, is the only edge targeting the node k. Then, (2.55) holds if and only if $T_{E,\mu} = \bar{T}_{E,\mu}$. From here the procedure can be repeated to show that temperature of the source node of μ is at equilibrium and that consequently the temperature of any other edge in series preceding μ will be at equilibrium.

We move on to show that $T_{E,i} = \bar{T}_{E,i}$ for any consumer's heat exchanger $i \in \mathscr{H}_c$. Let $i \in \mathscr{H}_c$ be arbitrary. Since (2.42) holds, it follows that the target node of i, given by $k = \mathscr{N}_i^+$, has a positive outflow (and inflow). Then, only the following cases can occur, in all of which we will be able to show that $T_{E,i} = \bar{T}_{E,i}$.

(a) k is the source of a given pipe $j \in \mathscr{P}$. In this case $T_{N,k} = \bar{T}_{N,k}$ due to (2.52). From the equilibrium equation $\dot{T}_{N,k} = 0$ (see (2.26)), it holds that

$$0 = \sum_{\mu \in \hat{\mathscr{E}}_k^+} \bar{q}_{E,\mu} T_{E,\mu} - \left(\sum_{\mu \in \hat{\mathscr{E}}_k^+} \bar{q}_{E,j} \right) \bar{T}_{N,k}. \tag{2.56}$$

The term $\sum_{\mu \in \hat{\mathscr{E}}_k^+} \bar{q}_{E,\mu} T_{E,\mu}$ can be decomposed as follows:

$$\sum_{\mu \in \hat{\mathscr{E}}_k^+} \bar{q}_{E,\mu} T_{E,\mu} = \sum_{\mu \in \hat{\mathscr{E}}_k^+ \cap \mathscr{P}} \bar{q}_{E,\mu} \bar{T}_{E,\mu} + \sum_{\mu \in \hat{\mathscr{E}}_k^+ \cap \mathscr{H}_{pr}} \bar{q}_{E,\mu} \bar{T}_{E,\mu}$$

$$+ \sum_{\mu \in \hat{\mathscr{E}}_k^+ \cap \mathscr{E}_{valves}} \bar{q}_{E,\mu} \bar{T}_{E,\mu} + \sum_{\mu \in \hat{\mathscr{E}}_k^+ \cap \mathscr{E}_{pumps}} \bar{q}_{E,\mu} \bar{T}_{E,\mu}$$

$$+ \sum_{\mu \in \hat{\mathscr{E}}_k^+ \cap \mathscr{H}_c} \bar{q}_{E,\mu} T_{E,\mu}. \tag{2.57}$$

For writing the first two terms in the right-hand side we have used the fact that $T_{E,\mu} = \bar{T}_{E,\mu}, \forall \mu \in \mathscr{P} \cap \mathscr{H}_{pr}$. For the third term we note any valve $\mu \in \hat{\mathscr{E}}_k^+ \cap \mathscr{E}_{valves}$ is in series either with a pipe or with a heat exchanger of a producer, for which $T_{E,\mu} = \bar{T}_{E,\mu}$ in view of (2.54). A similar reasoning can be followed to see that $T_{E,\mu} = \bar{T}_{E,\mu}$ for any $\mu \in \hat{\mathscr{E}}_k^+ \cap \mathscr{E}_{pumps}$, which explains the fourth term in (2.57). Since consumers are not connected in parallel, it follows that the fifth term in the right-hand side of (2.57) is equivalent to $\bar{q}_{E,i} T_{E,i}$. Therefore, (2.56) holds if and only if $T_{E,i} = \bar{T}_{E,i}$ (recall that $|\bar{q}_{E,i}| > 0$ for all $i \in \mathscr{E}$).

(b) k is the source of a given valve $j \in \mathscr{E}_{valves}$. By assumption, this valve is in series either with a pipe or with the heat exchanger of a producer. Then,

$j \in \Gamma_\mu$ for a given $\mu \in \mathscr{P} \cup \mathscr{H}_{\mathrm{pr}}$, where recall the definition of Γ_μ as it appears in (2.54). In view of (2.54), $T_{\mathrm{E},j} = \bar{T}_{\mathrm{E},j}$. Then, the equilibrium equation associated to $\dot{T}_{\mathrm{E},j} = 0$ can be written as follows (see (2.23)):

$$0 = \bar{q}_{\mathrm{E},j}(T_{\mathrm{E},j}^{\mathrm{in}} - \bar{T}_{\mathrm{E},j}). \tag{2.58}$$

This equation holds if and only if $T_{\mathrm{E},j}^{\mathrm{in}} = T_{\mathrm{N},k} = \bar{T}_{\mathrm{N},k}$, where k is the source node of j, i.e., $k = \hat{\mathscr{N}}_j^-$. From here, the same steps appearing in the case (a) can be followed to establish that $T_{\mathrm{E},i} = \bar{T}_{\mathrm{E},i}$.

(c) k is the source of a given pump $j \in \mathscr{E}_{\mathrm{pumps}}$. By assumption, this pump is in series either with a pipe or with the heat exchanger of a producer. Then, $j \in \Gamma_\mu$, for a given $\mu \in \mathscr{P} \cup \mathscr{H}_{\mathrm{pr}}$. In view of (2.54), $T_{\mathrm{E},j} = \bar{T}_{\mathrm{E},j}$. Then, the equilibrium equation associated to $\dot{T}_{\mathrm{E},j} = 0$ can be written as follows (see (2.23)):

$$0 = \bar{q}_{\mathrm{E},j}(T_{\mathrm{E},j}^{\mathrm{in}} - \bar{T}_{\mathrm{E},j}) + K_{\mathrm{E},j}|\bar{q}_{\mathrm{E},j}|\bar{q}_{\mathrm{E},j}^2. \tag{2.59}$$

This equation holds if and only if $T_{\mathrm{E},j}^{\mathrm{in}} = T_{\mathrm{N},k} = \bar{T}_{\mathrm{N},k}$, where k is the source node of j, i.e., $k = \hat{\mathscr{N}}_j^-$. From here, the same steps appearing in the case (a) can be followed to establish that $T_{\mathrm{E},i} = \bar{T}_{\mathrm{E},i}$.

(d) k is the source node of a producer's heat exchanger $j \in \mathscr{H}_{\mathrm{pr}}$. From (2.52) we have that $T_{\mathrm{N},k} = \bar{T}_{\mathrm{N},k}$. From here, the same steps appearing in the case (a) can be followed to establish that $T_{\mathrm{E},i} = \bar{T}_{\mathrm{E},i}$.

Considering the above cases, we conclude that $T_{\mathrm{E},i} = \bar{T}_{\mathrm{E},i}$ for all $i \in \mathscr{H}_{\mathrm{c}}$. It remains to see that $T_{\mathrm{E},i} = \bar{T}_{\mathrm{E},i}$ for all $i \in \mathscr{E}_{\mathrm{valves}} \cup \mathscr{E}_{\mathrm{pumps}}$ and that $T_{\mathrm{N},k} = \bar{T}_{\mathrm{N},k}$ for all $k \in \mathscr{N}$, not covered already by (2.52). Let $i \in \mathscr{E}_{\mathrm{valve}}$ be an arbitrary valve. By assumption this valve is in series with some $j \in \mathscr{P} \cup \mathscr{H}_{\mathrm{pr}} \cup \mathscr{H}_{\mathrm{c}}$. Since $T_{\mathrm{E},\mu} = \bar{T}_{\mathrm{E},\mu}$ for all $\mu \in \mathscr{P} \cup \mathscr{H}_{\mathrm{pr}} \cup \mathscr{H}_{\mathrm{c}}$, it is possible to use a reasoning analogous to the one used for claiming (2.54) to establish that $T_{\mathrm{E},i} = \bar{T}_{\mathrm{E},i}$. Using a similar reasoning we can also claim that $T_{\mathrm{E},i} = \bar{T}_{\mathrm{E},i}$ for any $i \in \mathscr{E}_{\mathrm{pumps}}$.

At this point we have established that $T_{\mathrm{E},i} = \bar{T}_{\mathrm{E},i}$ for any edge $i \in \mathscr{E}$. Let us move on to see that $T_{\mathrm{N},k} = \bar{T}_{\mathrm{N},k}$ for any node $k \in \mathscr{N}$. Let $k \in \mathscr{N}$ be arbitrary. Since (2.42) holds, then this node has a positive inflow (and outflow). Then, there is at least one edge $i \in \mathscr{E}$ whose source node is k. Since we have established that $T_{\mathrm{E},i} = \bar{T}_{\mathrm{E},i}$ for any edge $i \in \mathscr{E}$, then the equilibrium equation $\dot{T}_{\mathrm{E},i} = 0$ reveals that $T_{\mathrm{E},i}^{\mathrm{in}} = T_{\mathrm{N},k} = \bar{T}_{\mathrm{N},k}$. Since k was chosen arbitrarily, then this should hold for any $k \in \mathscr{N}$. In summary $T_{\mathrm{E},i} = \bar{T}_{\mathrm{E},i}$ for all $i \in \mathscr{E}$ and $T_{\mathrm{N},k} = \bar{T}_{\mathrm{N},k}$ for all $k \in \mathscr{N}$. Then, the only solution of the $(T_{\mathrm{E}}, T_{\mathrm{N}})$-dynamics that can identically stay in $\mathscr{Z}_{\mathrm{E},\mathrm{N}}$ (see (2.51)) is the equilibrium $(\bar{T}_{\mathrm{E}}, \bar{T}_{\mathrm{N}})$. Then, invoking [45, Theorem 3.5] we can conclude that this equilibrium is asymptotically stable.   ∎

Next we state and prove a corollary for Proposition 2. It establishes that local asymptotic stability of the overall thermo-hydraulic DHS is also achieved via the controllers (2.41) and (2.43).

Corollary 1. *Consider the overall thermo-hydraulic model (2.31), in closed-loop with (2.41) and (2.43). Assume that $(\bar{q}_f, \bar{V}_{sh}, \bar{T}_E, \bar{T}_N)$ is an equilibrium in satisfaction of (2.42), provided it exists. Then, the system is asymptotically stable at $(\bar{q}_f, \bar{V}_{sh}, \bar{T}_E, \bar{T}_N)$.*

   Proof sketch: The proof is a direct application of [46, Proposition 4.1]. Indeed, it is enough to note the following aspects. First, the (q_f, V_{sh})-dynamics in closed-loop with (2.41) are in cascade with the (T_E, T_N)-dynamics in closed-loop with (2.43): the former system is independent of the states of the latter. Second, the closed-loop (q_f, V_{sh})-dynamics are asymptotically stable at $(\bar{q}_f, \bar{V}_{sh})$. Third, when the vector q_f, which is can be seen as an exogenous disturbance to the (T_E, T_N)-dynamics, is fixed to the equilibrium value $\bar{q}_f$, then the (T_E, T_N)-dynamics are asymptotically stable at $(\bar{T}_E, \bar{T}_N)$, subject to the satisfaction of (2.42). Based on these facts, [46, Proposition 4.1] can be invoked to claim that $(\bar{q}_f, \bar{V}_{sh}, \bar{T}_E, \bar{T}_N)$ is an (at least locally) asymptotically stable equilibrium of the overall thermo-hydraulic model (2.31). ∎

2.6.1   Notes of the section

The following remarks are in order:

   (i)   The results in Proposition 2 and Corollary 1 are relevant for claiming local stabilizability of the overall thermo-hydraulic DHS model. This property may be exploited, for example, for designing optimal controllers ensuring state and input constraints, as explained for general systems in [47], whose control design methodology is employed in [48] for a variant of the DHS model (2.31) in which q_f is directly treated as a control input.

  (ii)   The controllers (2.41) and (2.43) depend on an exact knowledge of the equilibrium point to be stabilized. Hence, they are sensitive to disturbances or uncertaintities in the system parameters. In [41, Appendix I], [14] and [15] dissipativity-based proportional-integral controllers are designed to address similar control goals while overcoming the aforementioned limitations. The reader is referred to [49] for a graph-theoretic analysis on hydraulic actuator placement for guaranteeing the existence and uniqueness of DHSs' hydraulic equilibrium solutions.

 (iii)   A notable feature of the dissipativity-based controllers (2.41) and (2.43) is that each component of the control input vector can be made dependent only on locally available measurements, i.e., the controller can be made decentralized (see, e.g., [50]), which adds a degree of robustness to the system by not requiring the exchange of measurements among the control entities via a communication network.

2.7   Conclusion

Using first principles, we have proposed a model to describe the thermo-hydraulic behavior of a DHS containing multiple heat producers, storage devices, and consumers, all of them interconnected through a common distribution network of potentially meshed topology. Moreover, we have analyzed conditions under which the proposed model as a whole, or some components of it, exhibit various dissipativity properties. The usefulness of this analysis has been reflected in the design of dissipativity-based controllers of simple structure that offer closed-loop stability guarantees.

Acknowledgments

(i)   The core contents and structure of the present chapter pertaining to modeling and dissipativity analysis are based on our works [21,41] and the references therein: the system setup takes inspiration from [23]; the procedure to obtain the DAE and the ODE-based hydraulic dynamics is based on the analysis in [24] and [20]; the derivation of the thermal model takes great inspiration from [29] and [36]; and the dissipativity properties analyzed in Section 2.5 draw inspiration from the Hamiltonian formulations of district heating systems in [29]. Nonetheless, compared to [21,41], we expand on several descriptions throughout the modeling process, making more intuitive the interpretation of the physical phenomena represented by our model. The proposed model is in fact extended from [21,41] in view of the relaxation of a number modeling assumptions, such as the fact that we allow for the presence of stand-alone storage tanks, and the consideration of dissipation and wall heat conduction in pipes. In particular, the latter aspect has been instrumental for introducing Proposition 1(i), which was not yet reported in our previous publications. Additionally, most of the material in Section 2.6 is either new or substantially expanded from the discussion in [41, Appendix I].

(ii) The production of this chapter was supported by the German Federal Government, the Federal Ministry of Education and Research, and the State of Brandenburg within the framework of the joint project EIZ: Energy Innovation Center (project numbers 85056897 and 03SF0693A) with funds from the Structural Development Act (Strukturstärkungsgesetz) for coal-mining regions.

References

[1]   Wang R, Assenova VA, and Hertwich EG. Energy system decarbonization and productivity gains reduced the coupling of CO_2 emissions and economic growth in 73 countries between 1970 and 2016. *One Earth*. 2021;4(11): 1614–1624.

[2]   United Nation's Intergovernmental Panel on Climate Change: Sixth Assessment Report; 2021. Available from: https://www.ipcc.ch/report/ar6/wg2/.

[3]   Papadis E, and Tsatsaronis G. Challenges in the decarbonization of the energy sector. *Energy*. 2020;205:118025.

[4]   Ember Climate: Global Electricity Review; 2023. Available from: https://ember-climate.org/insights/research/global-electricity-review-2023/#supporting-material.

[5]   International Energy Agency: Heating; 2023. Available from: https://www.iea.org/energy-system/buildings/heating.

[6]   International Energy Agency: District Heating; 2023. Available from: https://www.iea.org/energy-system/buildings/district-heating.

[7]   Lund H, Werner S, Wiltshire R, *et al.* 4th Generation District Heating (4GDH): Integrating smart thermal grids into future sustainable energy systems. *Energy*. 2014;68:1–11.

[8]   Werner S. International review of district heating and cooling. *Energy*. 2017;137:617–631.

[9]   Vandermeulen A, van der Heijde B, and Helsen L. Controlling district heating and cooling networks to unlock flexibility: A review. *Energy*. 2018;151:103–115.

[10] Guelpa E, and Verda V. Thermal energy storage in district heating and cooling systems: A review. *Applied Energy*. 2019;252:113474.

[11] van der Schaft A. *L2-gain and passivity techniques in nonlinear control*. vol. 2. Springer; 2000.

[12] Cucuzzella M, Kosaraju KC, and Scherpen JM. Distributed passivity-based control of DC microgrids. In: *2019 American Control Conference (ACC)*. IEEE; 2019. p. 652–657.

[13] Kawano Y, Cucuzzella M, Feng S, *et al.* Krasovskii and shifted passivity based output consensus. *Automatica*. 2023;155:111167.

[14] Machado JE, Cucuzzella M, Pronk N, *et al.* Adaptive control for flow and volume regulation in multi-producer district heating systems. *IEEE Control Systems Letters*. 2022;6:794–799.

[15] Strehle F, Machado JE, Cucuzzella M, *et al.* A unifying passivity-based framework for pressure and volume flow rate control in district heating networks. *IEEE Transactions on Control Systems Technology*. 2024;32(4): 1323–1340.

[16] Lasseter RH. MicroGrids. In: *2002 IEEE Power Engineering Society Winter Meeting. Conference Proceedings (Cat. No.02CH37309)*. vol. 1; 2002. p. 305–308.

[17] Krishna A, and Schiffer J. A port-Hamiltonian approach to modeling and control of an electro-thermal microgrid. *IFAC-PapersOnLine*. 2021;54(19):287–293.

[18] Tõnso M, Kaparin V, and Belikov J. Port-Hamiltonian framework in power systems domain: A survey. *Energy Reports*. 2023;10:2918–2930.

[19] Malan AJ, Rausche L, Strehle F, *et al.* Port-Hamiltonian modelling for analysis and control of gas networks. *IFAC-PapersOnLine*. 2023;56(2):5431–5437.

[20] Wang Y, You S, Zhang H, *et al.* Hydraulic performance optimization of meshed district heating network with multiple heat sources. *Energy*. 2017;126:603–621.

[21]   Machado JE, Cucuzzella M, and Scherpen JMA. Modeling and passivity properties of multi-producer district heating systems. *Automatica.* 2022;142:110397.

[22]   Talebi B, Mirzaei P, Bastani A, *et al.* A review of district heating systems: Modeling and optimization. *Frontiers in Built Environment.* 2016;2:22.

[23]   Scholten T, De Persis C, and Tesi P. Modeling and control of heat networks with storage: The single-producer multiple-consumer case. *IEEE Transactions on Control Systems Technology.* 2015;25(2):414–427.

[24]   De Persis C, and Kallesøe CS. Pressure regulation in nonlinear hydraulic networks by positive and quantized controls. *IEEE Transactions on Control Systems Technology.* 2011;19(6):1371–1383.

[25]   Bendtsen J, Val J, Kallesøe C, *et al.* Control of district heating system with flow-dependent delays. *IFAC-PapersOnLine.* 2017;50(1):13612–13617.

[26]   Ahmed S, Machado JE, Cucuzzella M, *et al.* Control-oriented modeling and passivity analysis of thermal dynamics in a multi-producer district heating system. *IFAC-PapersOnLine.* 2023;56(1):175–180.

[27]   Krug R, Mehrmann V, and Schmidt M. Nonlinear optimization of district heating networks. *Optimization and Engineering.* 2021;22:783–819.

[28]   Ismail K, Leal J, and Zanardi M. Models of liquid storage tanks. *Energy.* 1997;22(8):805–815.

[29]   Hauschild SA, Marheineke N, Mehrmann V, *et al.* Port-Hamiltonian modeling of district heating networks. In: *Progress in Differential-Algebraic Equations II.* Springer, Cham; 2020. p. 333–355.

[30]   Grosswindhager S, Voigt A, and Kozek M. Efficient physical modelling of district heating networks. *Proceedings of the IASTED International Conference on Modelling and Simulation.* 2011;(2):41–48.

[31]   Vladimarsson P. District heat distribution networks. *United Nations University Geothermal Training Programme.* 2014;30(9):239–240.

[32]   Bollobás B. *Modern Graph Theory.* Springer; 1998.

[33]   Desoer C, and Kuh E. *Basic Circuit Theory.* McGraw-Hill Book Company; 1969.

[34]   Ryu E, and Boyd S. A primer on monotone operator methods. *Applied and Computational Mathematics.* 2016;15(1):3–43.

[35]   Gong E, Wang N, You S, *et al.* Optimal operation of novel hybrid district heating system driven by central and distributed variable speed pumps. *Energy Conversion and Management.* 2019;196:211–226.

[36]   Dänschel H, Mehrmann V, Roland M, *et al.* Adaptive nonlinear optimization of district heating networks based on model and discretization catalogs. *SeMA Journal.* 2024;81:81–112.

[37]   Sandou G, Font S, Tebbani S, *et al.* Predictive control of a complex district heating network. In: *IEEE Conference on Decision and Control (CDC).* vol. 44; 2005. p. 7372.

[38]   Scholten T, Trip S, and De Persis C. Pressure regulation in large scale hydraulic networks with input constraints. In: *Proc. of the 2017 IFAC World Congress.* vol. 50. Elsevier; 2017. p. 5367–5372.

[39]   Duquette J, Rowe A, and Wild P. Thermal performance of a steady state physical pipe model for simulating district heating grids with variable flow. *Applied Energy*. 2016;178:383–393.

[40]   Ortega R, Van Der Schaft AJ, Mareels I, *et al*. Putting energy back in control. *IEEE Control Systems Magazine*. 2001;21(2):18–33.

[41]   Machado JE, Cucuzzella M, and Scherpen J. Modeling and passivity properties of multi-producer district heating systems. *Automatica*. 2022;142:110397.

[42]   Van Der Schaft A. Classical thermodynamics revisited: A systems and control perspective. *IEEE Control Systems Magazine*. 2021;41(5):32–60.

[43]   Haddad WM. *A Dynamical Systems Theory of Thermodynamics*. 1st ed. Princeton University Press; 2019.

[44]   Dong Z, Li B, and Huang X. Passivity-based control of heat exchanger networks. *Proceedings of the 38th Chinese Control Conference*. 2019. p. 6531–6536.

[45]   Khalil H. *Nonlinear Control, Global Edition*. Pearson; 2015.

[46]   Sepulchre R, Janković M, and Kokotovic P. *Constructive Nonlinear Control*. Springer; 1997.

[47]   Chen H, and Allgöwer F. A quasi-infinite horizon nonlinear model predictive control scheme with guaranteed stability. *Automatica*. 1998;34(10):1205–1217.

[48]   Rose M, Gernandt H, Machado JE, *et al*. Model predictive control of district heating grids using stabilizing terminal ingredients. In: *2024 European Control Conference (ECC)*; 2024. pp. 1083–1089.

[49]   Jeeninga M, Machado JE, Cucuzzella M, *et al*. On the existence and uniqueness of steady state solutions of a class of dynamic hydraulic networks via actuator placement. In: *2023 62nd IEEE Conference on Decision and Control (CDC)*; 2023. p. 3652–3657.

[50]   Machado JE, Ferguson J, Cucuzzella M, *et al*. Decentralized temperature and storage volume control in multiproducer district heating. *IEEE Control Systems Letters*. 2023;7:413–418.

Part II

Frequency control

Chapter 3

Dynamics and control of grid-connected microgrids

Ekaterina Dudkina[1], Emanuele Crisostomi[1], Pietro Ferraro[2] and Federico Milano[3]

Societal and economic challenges are driving the transition toward a sustainable, resilient, communication-reliant, and decentralized power grid. Microgrids tick all the boxes to become the building blocks of these modern grids. However, there are also increasing concerns about how this transition may ultimately impact the operation and stability. This chapter investigates the impact of grid-connected microgrids on power system dynamics under various scenarios, and evaluates their role in the energy market, and their ability to provide frequency services. The scenarios consider greedy approaches, i.e., microgrids that operate their resources, in particular their energy storage system, to maximize their revenues based on the variation of the electricity price; altruistic approaches, i.e., microgrids that provide frequency control as ancillary service to the grid; and a stochastic control that aims at blending both the greedy and altruistic approaches. Extensive realistic simulations and sensitivity analysis are carried out in a modified IEEE 39-bus transmission grid and show that the biggest impact on the dynamic performance of the system is obtained when the microgrids are freely allowed to operate in the market to greedily maximize their revenues.

3.1 Introduction

The concept of microgrid (MG) in the modern sense was proposed at the very beginning of this century [1], as a novel approach to manage and operate the distributed energy resources (DERs): an MG is a "cluster of micro-sources, storage systems and loads which presents itself to the grid as a single entity that can respond to central control signals. The heart of the MG concept is the notion of a flexible, yet controllable interface between the microgrid and the wider power system." Indeed, the

[1]Department of Energy, Systems, Territory and Constructions Engineering, University of Pisa, Italy
[2]Dyson School of Design Engineering, Imperial College London, UK
[3]School of Electrical and Electronic Engineering, University College Dublin, Ireland

increasing penetration of small-sized distributed power plants, usually from renewable sources, has played a pivotal role in the definition of microgrids. As suggested in [1], clustering of such micro-generators and storage systems can help overcome obstacles of direct connection of small generators to the main electrical network. MGs were initially supposed to follow the signals of the main power system to preserve normal operation; however, MGs were also designed to improve power quality and to satisfy the needs of consumers: "such issues as local voltage, reliability, losses and quality of power should be those that support the customers' objectives" [1]. The concept of MGs has pushed the development of a number of projects and case studies in different countries [2]. These studies have shown that an MG may provide technical, economic and environmental advantages, such as increased energy efficiency, decrease of overall energy consumption, and reduced environmental impacts. In addition, the possibility to switch between grid-connected and islanded mode improves energy system reliability and power quality [3].

With the continuous innovations in power systems and the advancement of MGs, their purpose, role, and functionalities have evolved as well. Initially, MGs were considered as a solution to electrify remote or isolated areas, where the interconnection with the main grid was impossible—for instance, the term "microgrid" had been already used in 1986 for a hybrid wind/vapor turbine generator system, located at the remote earth station in Black Island, McMurdo Sound, Antarctica [4]. In Europe, the first microgrid is considered the one installed in 2001 in the Greek island Kythnos. It has been a testbed for the innovative MG technologies for more than 20 years and it is still operating [5]. In addition, MGs were seen as an instrument to integrate small-scale generators [1] in the power system, while avoiding their direct connection to the grid. In this case, MGs are operating in a grid-connected mode, i.e., connected through a point of common coupling to the main grid. Through this connection, the MG may ensure power flow to (from) the main grid in case of generation surplus (lack of generation). With the constant spread of renewable energy sources (RES), MGs are seen as an instrument to mitigate fluctuations on the intermittent local sources and reduce the burden of the distribution grid [6]. The concept of MG is continuously evolving and alternative ways to organize MGs are being proposed. For example, clustered MGs (or grids of MGs), are a group of MGs that are physically linked together and primarily trade the excess energy among each other to reduce imbalances which, otherwise, should be compensated via the upstream grid [7]. Another novel MG concept is the Microgrid Building Block, which was suggested in [8] as a modular and standardized structure to help reduce the cost and increase the reliability of MGs operation. The concept of MGs is also related to those of virtual power plants (where the commercial aspects of the plants prevail on the geographic and physical aspects) [9,10], and also to those of energy communities, where cooperative aspects of the components of the MGs are emphasized.

Nowadays MGs are often seen as active participants in the power system. This active role is essential, as, in the near future, the number of MGs is expected to grow rapidly, also due to the tightening environmental policies that are gradually restricting the use of fossil-fuel based generators (e.g., in the European Union (EU)), to the spread of plug-in electrical vehicles (PEVs), and to the development of more efficient storage systems, which are the key elements in the accommodation of inter-mittent RES. However, the expected MGs' roll-out may cause distortions in power grid operation.

Although there is a general implicit assumption that MGs do not affect the operation of the main grid, when providing or absorbing excess energy, as MGs are "orders of magnitude smaller than the bulk power system" [11], the increas-ing penetration of grid-connected MGs in the system may tangibly affect the grid (voltage or frequency) stability [12]. Accordingly, an upcoming massive deployment of grid-connected MGs will likely require a careful adjustment of the operation of the power system from different aspects, including low inertia systems, uncertainty; energy management; and communication systems [13,14]. These aspects are detailed below.

3.1.1 Low inertia

Microgrids are mainly based on inverter-based resources (IBRs), which obviously do not inherit the same characteristics of the traditional synchronous generators, such as inertia and high fault current [15]. If the grid is "strong" enough to pro-vide a stable and clean voltage, microgrids and RES can inject power into the grid without considering the impact on its operation. However, with the deployment of non-synchronous generators, the lack of inertia will lead to a weaker grid and may cause large frequency swings, instabilities, and even blackouts [15]. For example, it has been demonstrated that for the 36-bus test system the maximum share of MGs that does not cause challenges in frequency control is slightly higher than 7% [16], while for the IEEE 50-machine systems, the MGs penetration should not exceed 2% [17]. Of course, these values depend on the different structure and operation of the MGs, but the accepted level of MG roll-out is still relatively low.

3.1.2 Uncertainty

The economical and reliable operation of microgrids requires a certain level of coor-dination among different DERs. This coordination is challenging for MGs because they mainly rely on intermittent generation, and thus require solving a problem over an extended horizon, taking into account the uncertainty of parameters such as load profile and weather forecast. This uncertainty is higher than that in bulk power sys-tems, due to the reduced number of loads and highly correlated variations of available energy resources.

3.1.3 Energy management

With the competitive nature of MGs, the optimization becomes a multi-objective problem, which should consider technical, economic, and environmental aspects. While the centralized control architecture is the common choice, the increasing penetration of distributed generators (DGs) in the power system may entail high com-putational cost, poor system scalability, and instability in case of failures. In addition, the optimal solution for an MG must not disturb the work of upstream system and;

on the other side, it must consider the provision of auxiliary services to the main grid.

3.1.4 Communication system

Communication systems are essential elements in power systems with a high share of MGs due to the large number of elements and volatility of RES production. To ensure stability and reliability of the grid system with grown network complexity, a reliable and efficient communication system is needed. Poorly designed communication systems may fail to provide the necessary communication consistently and promptly, what entails failure to provide an optimal energy management and, in the worst case, may even cause grid damage.

3.2 Hierarchical control of microgrids

To mitigate some of the negative effects of the MGs' integration into the power system, specific management and control strategies can be used. In this chapter, we discuss in particular the control aspects of MGs.

The control functions of MGs' energy management systems (EMS) are addressed in the IEEE standard 2030.7 [18]. First, this standard widens the definition of an MG, which is "a group of interconnected loads and DERs with clearly defined electrical boundaries that acts as a single controllable entity with respect to the grid and can connect and disconnect from the grid to enable it to operate in both grid-connected or island modes." Then, the standard specifies that EMS and control systems should operate simultaneously to provide stable operation for the following scenarios:

- Operation of MGs in grid-connected and islanded modes.
- Transition from grid-connected to islanded mode.
- Resynchronization and reconnection from islanded to grid-connected mode.
- Optimization of real and reactive power generation and consumption.
- Participation in the energy market, also as a provider of ancillary services.

To provide stable functioning during the aforementioned scenarios and to mitigate the disturbances, created by a higher share of MGs connected to the grid, several control methods have been proposed. Following the architecture of conventional power systems, MG control is usually divided into three layers—primary, secondary, and tertiary [13,19,20]. This hierarchical structure allows managing stationary and dynamic performance, while considering market signals, even if, for MGs, the time period and functionality of each level are not strictly defined.

3.2.1 Primary control

The primary control operates on the level of devices of the MG. It provides the fastest response (usually less than seconds); it operates in the decentralized mode, thus the communications are minor. It usually includes control hardware—inner current and voltage control loops, a virtual impedance control, and a droop controller. Correspondingly, its main functionalities are to maintain voltage and frequency and power exchange between units.

The power sharing control may be categorized depending on whether it uses the droop control. This idea is taken from conventional power systems; however, due to the high level of power electronics and lack of synchronous machines inertia, it does not perform well during transient states. As alternatives, non-droop or communication-based control methods have been suggested [21], which provide good transient response and good power quality, but at the expense of increased costs and a complicated communication loop [22]. Among the solutions to compensate the lack of inertia in the grid, the most considered one is virtual inertia (also known as artificial inertia or synthetic inertia) [23,24].

3.2.2   Secondary control

The secondary control is often the actual EMS, it acts at a slower time scale and ensures reliable and economical operation of MGs. In particular, it mitigates voltage and frequency deviations that persist after the operation of the primary control. Also, it defines the optimal economic dispatch, aiming to minimize the operational costs of the MG and enhance their reliability, by ensuring generation and demand balance, respecting power line and energy storage capacity limits, and the power ratings of controllable generators.

Secondary control strategies may be classified in centralized, decentralized, and distributed architectures [3,25]. These architectures are shown in Figure 3.1.

3.2.2.1   Centralized control architecture

In this approach, a central controller coordinates the operations of the DGs (DERs) in an MG. It assumes that there is a single controller that collects all information and manages all other MG components. In this case, all control inputs are combined into a single optimization problem. This architecture allows finding global optimal solution. However, it has limited scalability and, with the growing number of DGs, the computational burden may become too high. In addition, an extensive communication infrastructure is needed. The centralized control strategy conventionally applies proportional-integral (PI) controllers.

3.2.2.2   Decentralized control architecture

The decentralized control architecture consists of several controllers, which operate autonomously and do not share any information. Each DG unit restores its voltage and frequency amplitudes to nominal values independently, based on the measurements done via the communication link. Although there are minor communication requirements (only for transferring data between levels [25]), the scalability is high and computational burden is low, there is no guarantee that the global optimal solution will be achieved, since only local information is available. Among the algorithms that are applied for the decentralized control we mention Washout-Filter-Based control [26], Local-Variable-Based control [27], and Estimation-Based control [28].

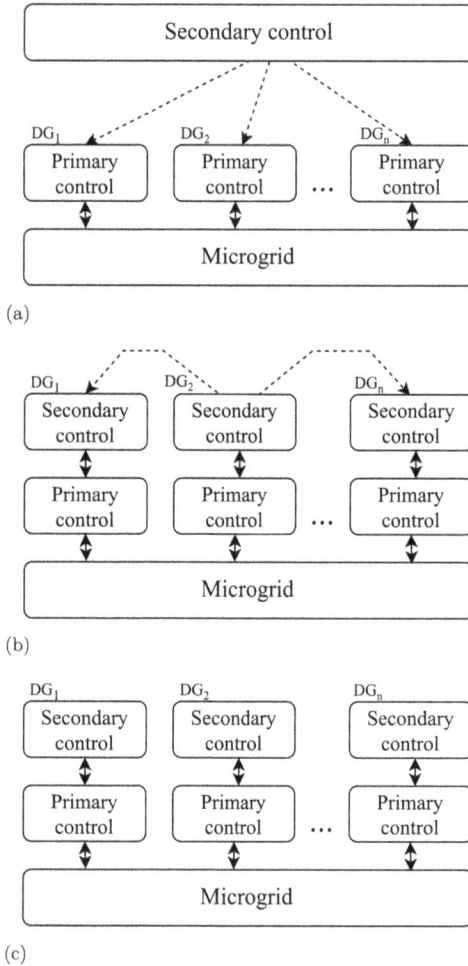

*Figure 3.1   Architectures of secondary controllers of MGs: (a) centralized;
(b) distributed; and (c) decentralized*

3.2.2.3   Distributed control architecture

The distributed control architecture assumes that some information is shared among neighboring controllers. This strategy better fits the concept of MG with spatially distributed DGs and has a range of advantages. Indeed, the distributed controllers require less computational power, are robust to a single point of failure (i.e., the failure of a controller will not cause a system blackout), and the communication requirements are limited due to local decision making. The distributed control framework is more flexible, and it is scalable for the future expansion of MGs. Considering that the energy system is moving toward a more distributed organization, due to the spread of DERs and appearance of novel organization of energy structures, such

as local communities, the distributed architecture is gaining more and more attention [29]. Among the algorithms that implement distributed control, the most popular
ones are the averaging [30], consensus algorithms [31], and event-triggered control
methods [32].

3.2.3   Tertiary control

Tertiary control refers to the grid level and its functionality is defined by the MG
interaction with neighboring MGs and upstream grid. Tertiary control operates at a
slower time scale, starting from few minutes. At this stage, economic, environmental,
and technological requirements from the external parties, such as electricity prices
or weather forecast data, are considered. This control level provides set points to the
secondary control, which further ensure the stability of system operation.

Tertiary control is usually considered as a functionality of the utility and not
as a part of MG system. However, with the further increase of MGs' number, the
centralized controller may be overloaded, and it is predicted that EMS of MG may
have a control counterpart in the tertiary control level.

3.3   Modeling

The hierarchical control of MGs described in the previous section is inspired by the
existing centralized control system of the grid. However, some particular features
of MG operation (see Section 3.1) can complicate the functioning of the control
system. Indeed, new aspects, which were not considered for the centralized system
operation, are often debated, such as consumer demand response programs and peer-
to-peer energy sharing. All of these are targeting consumers and producers behavior
and attempting to adapt it in favor of the grid needs.

A reason for these changes is that MGs compete to reach their own goals, such
as maximization of revenues, or cost reduction of energy bought on the external market applied [33,34]. In these studies, MGs are still considered an order of magnitude
smaller that the main grid, and therefore, they are allowed to purchase energy deficit
or inject the excess of production to the main grid without limitations and by the for-
merly set prices. However, with the gradual decentralization of generation capacities
and the expected growing share of energy supplied by MGs, their impact on the grid
should be reconsidered.

To avoid strict limitations of MGs operation and eliminate the need for excessive
reserves procurement, market-driven methods may be also applied to MG management, i.e., a price signal may be used to nudge the MGs' desirable energy exchanges
with the power grid. The electricity market price indeed may play an important role
in the control of power system operation. It has been shown, in fact, that appropriate
price signals may change the behavior of market players, affecting power grid energy
balance requirements and, therefore, stabilizing the system frequency, assuring active
and reactive power reserves, mitigating line congestion, etc. [35,36].

The remainder of this section discusses the impact of the price-signal on the
frequency stability of a grid with integrated MGs. With this aim, we describe the

model of the MG and their EMS. We also introduce the model of power system dynamics and describe a price setting mechanism.

3.3.1   MG model

As mentioned in Section 3.1, a microgrid includes loads, DERs, storage systems, and an EMS. The latter is the core of the MG, as it orchestrates the operations of each device that forms the MG. In particular, the EMS is in charge of the collection of information about consumed and generated power, the state-of-charge of the storage devices, the electricity price, possibly the frequency of the system (which we will represent as the frequency of the center of inertia), and smartly decides the most convenient reference power generation, say p^{ref}, according to the rules that we shall describe in Section 3.3.2.

3.3.1.1   Storage

We assume that all storage devices may be aggregated as a single virtual storage system, whose dynamics may be described by the following equation, which is a time-continuous equivalent of the model used in [37]:

$$K \frac{d}{dt} \text{SOC} = p_s = p_g - p_l - p_{\text{out}}, \tag{3.1}$$

where SOC is the level of charge of the MG's aggregated storage system, K is the parameter that can be used to vary the size of the storage system, p_s is the power generated ($p_s < 0$) or absorbed ($p_s > 0$) by the storage device, p_g and p_l are the active power generated by the MG and absorbed by its load, respectively, and p_{out} is the power output of the MG. Finally, the charging level is constrained to take into account if the storage is fully charged/discharged.

3.3.1.2   Distributed energy resource

The DER dynamic model included in the MG is an elaboration of the DER models discussed in [38,39]. This model is suitable for transient and voltage stability analysis and, hence, only current controllers of the power electronic converters included in the DER are modeled, not the converters themselves. The power injections into the AC bus are as follows:

$$p_g = v_d i_d + v_q i_q, \tag{3.2}$$
$$q_g = v_d i_d - v_d i_q, \tag{3.3}$$

where i_d and i_q are the AC-side dq-frame currents of the Voltage Sourced Converter (VSC), respectively, and v_d and v_q are the dq-frame components of the bus voltage phasor of the point of connection of the VSC with the AC grid. The reference currents i_d^{ref} and i_q^{ref} are obtained from active and reactive powers p_g and q_g, as follows:

$$\begin{bmatrix} i_d^{\text{ref}} \\ i_q^{\text{ref}} \end{bmatrix} = \begin{bmatrix} v_d & v_q \\ v_q & -v_d \end{bmatrix}^{-1} \begin{bmatrix} p_g \\ q_g \end{bmatrix}. \tag{3.4}$$

Both i_d^{ref} and i_q^{ref} are bounded by the converter thermal limits.

Uncertainty and volatility of both generators and loads are accounted for by modeling the net power produced by the MG as a stochastic process according to the following:

$$p_{\text{net}} = p_g - p_l = \bar{p}_{gT} - \bar{p}_{lT} + \eta_M, \tag{3.5}$$

where η_M is a white noise, and $\bar{p}_{gT}$ and $\bar{p}_{lT}$ are piece-wise constant functions that account for uncertainty and change randomly with a period T as discussed in [39].

It is also assumed that p^{ref} is defined by EMS system, according to the rules listed in 3.1 and imposed by p_s, which is the slack variable.

3.3.2 EMS rules

We assume that the EMS is based on simple *if-then* rules, on the basis of which the EMS decides the most convenient active power set-point. The input quantities of the EMS are the produced power p_g and load p_l, the price λ, the level of charge SOC of storage systems, and the frequency of the center of inertia when frequency services are provided to the grid. The rules are divided into two sets: the seller state, for which $p_{\text{net}} \geq 0$ (i.e., an MG is producing more than it is consuming and it will most likely sell energy) $p_{\text{net}} < 0$ (i.e., an MG is consuming more than it is producing and it will most likely buy energy). The thresholds ε_h and ε_l reflect the convenience of buying or selling energy:

$$\varepsilon_h = (1 + \rho)\bar{\lambda}, \tag{3.6}$$
$$\varepsilon_l = (1 - \rho)\bar{\lambda}, \tag{3.7}$$

where $\bar{\lambda}$ is the average price and ρ defines a price fluctuation, which makes the final price too low or too high for MGs to sell or buy energy.

EMS parameters are specific of each MG, i.e., they depend on the market strategy of the MG owner or on its marginal cost (all the EMS parameters are positive). A summary of the meaning of the parameters is given in Table 3.1. Such rules are however only one possible choice, and other rules or a more sophisticated logic may be taken into consideration, depending on the goals and priorities of MGs.

3.3.3 Power grid dynamics

The effect of the MGs on the power system dynamics is assessed by merging the described MG model into the power system model, which dynamics is described based on a system of Differential Algebraic Equation (DAE) as follows:

$$\frac{d}{dt}\mathbf{x} = \mathbf{f}(\mathbf{x}, \mathbf{y}, \mathbf{u}), \tag{3.8}$$
$$\mathbf{0} = \mathbf{g}(\mathbf{x}, \mathbf{y}, \mathbf{u}), \tag{3.9}$$

where $\mathbf{f}$ are the differential equations and $\mathbf{g}$ are the algebraic equations; $\mathbf{x}$ ($\mathbf{x} \in R^{n_x}$) are the state variables, $\mathbf{y}$ ($\mathbf{y} \in R^{n_y}$) are the algebraic variables, and $\mathbf{u}$ ($\mathbf{u} \in R^{n_u}$) are discrete events, which are used to model the logic of the EMS of MGs. The detailed description of this model is presented in [40].

Table 3.1   Microgrid EMS rules

Seller mode, $p_{net} > 0$		
Rule	**Action**	**Rationale**
If $SOC < SOC_{mid}$	$p^{ref} = 0$	Charge the storage with the production surplus
Else if $SOC \geq SOC_{high}$ and $\lambda \geq \varepsilon_h$	$p^{ref} = p_{net}(1 + k_s)$	The price of energy is high, and the battery is fully charged. It is convenient to sell more energy
Else if $SOC \geq SOC_{high}$ or $(\lambda \geq \varepsilon_h$ and $SOC \geq SOC_{mid})$	$p^{ref} = p_{net}$	Sell the surplus
Else if $\varepsilon_l \leq \lambda \leq \varepsilon_h$ or $(\lambda < \varepsilon_l$ and $SOC_{mid} \leq SOC \leq SOC_{high})$	$p^{ref} = 0$	Charge the storage with the production surplus
Else if $\lambda < \varepsilon_l$ and $SOC < SOC_{mid}$	$p^{ref} = -p_{buy} - K_{buy}\lambda$	Storage is low on charge and despite the surplus of production, it is convenient to buy energy proportionally to the price (i.e. the lower the price, the more EMS can buy)
Buyer mode, $p_{net} < 0$		
Rule	**Action**	**Rationale**
if $SOC \leq SOC_{low}$	$p^{ref} = p_{net} - p_{ch}$	Storage is very low on charge, buy the deficit of energy plus an extra amount to charge the storage
else if $\lambda \geq \varepsilon_h$ and $SOC \geq SOC_{high}$	$p^{ref} = p_{net}(1 + k_b)$	The price of energy is high and the storage is full, sell energy
else if $\lambda \geq \varepsilon_h$ and $SOC_{mid} \leq SOC < SOC_{high}$	$p^{ref} = 0$	The price of energy is very high and the storage has a middle charge, use it to compensate energy deficit
else if $\lambda \geq \varepsilon_h$ and $SOC < SOC_{mid}$	$p^{ref} = p_{net}$	The price of energy is high and the storage is medium-low on charge, buy the energy deficit
else if $\lambda \leq \varepsilon_l$ and $SOC < SOC_{high}$	$p^{ref} = p_{net} - p_{ch} - K_{ch}\lambda$	The price of energy is very low, and the storage is fully charged, buy an extra amount of energy to store, proportional to the price
else if $\lambda \leq \varepsilon_l$ and $SOC \geq SOC_{high}$	$p^{ref} = p_{net}$	The price of energy is very low, and the storage is fully charged, buy the energy deficit
else if $\lambda > \varepsilon_l$ and $SOC \geq SOC_{mid}$	$p^{ref} = 0$	The price of energy is not very low, and the storage is medium-high on charge, use it to compensate the energy deficit
else if $\lambda > \varepsilon_l$ and $SOC_{low} < SOC < SOC_{mid}$	$p^{ref} = p_{net}$	The price of energy is not very low, and the storage is medium-low on charge, buy the energy deficit

3.4   Case study

Following the model described in Section 3.3, we now investigate and evaluate the possible effect of integrating MGs into the power network. Specifically, we are interested in the effects of the specific strategy adopted by various MGs. With this purpose, we consider four possible alternative scenarios:

- *Price is an exogenous signal:* in this scenario, MGs are price takers and their strategies do not affect the price on the market. This means that the price signal is independent of the energy balance within the power system and MGs can take advantage of price fluctuations to maximize their revenues. This scenario corresponds to the case of a low penetration of MGs in the power system.
- *MGs are price makers:* the price signal is dependent of the power imbalance (i.e., difference between power demand and power generation), and demand/supply within MGs plays an important role in this context. For example, MGs that provide a large amount of power to the grid might offset the electricity price and make it decrease. MGs again take advantage of price fluctuations to maximize their revenues. This scenario corresponds to the case of a significant penetration of MGs in the power system.
- *MGs also provide ancillary services:* MGs are price makers but, if the power grid experiences relevant frequency deviations, most likely due to the market activities of MGs, then all, or some of the, MGs have to participate in the primary frequency regulation of the grid. Conversely, if the power frequency is within a safe range of values, MGs are still allowed to maximize their revenues by selling/buying power as in the other scenarios.
- *MGs only provide ancillary services:* MGs do not try to maximize their revenues, but they only provide frequency services to the power grid. This scenario corresponds to a case where MGs do not compete in the energy market, but they rather cooperate to improve the operation of the power grid.

The simulations of these scenarios are performed based on the IEEE 39-bus grid, which we have duly modified by allocating MGs on certain buses as described in Table 3.2 and adding uncertainties to the loads. The simulations are performed using Dome, a Python-based power system software tool [41]. Parameters of MGs are also presented in Table 3.2, where $\bar{p}_g$ and $\bar{p}_l$ are the average values of p_g and p_l, respectively, before adding uncertainties and perturbations; σ_{net} is the standard deviation of the net active power production p_{net}. The storage levels of SOC_{low}, SOC_{mid}, and SOC_{high} are equal to 20%, 50%, and 80%, respectively.

In the power system model (3.8), we consider lumped models of the transmission system and conventional dynamic models of synchronous machines (e.g., 6th order models) and their controllers, such as automatic voltage regulators, turbine governors, and power system stabilizers, as well as DERs, storage devices, and the controllers included in the MGs. The simplified scheme of the system is shown in Figure 3.2.

3.4.1 Scenario 1: exogenous price

In this first scenario, the price of electricity is imposed from the outer grid, i.e., MGs are considered as price takers. MGs try to profit from the price fluctuations to maximize their revenues (for instance, by charging batteries or directly consuming electricity when the price is low). The set of rules followed by the MGs is the

Table 3.2   Microgrid EMS rules

MG	Bus	p_g/p_l	K	σ_{net}
1	18	0.24	5.0	0.013
2	3	0.20	7.0	0.040
3	15	0.40	6.5	0.030
4	28	0.20	7.0	0.040
5	24	0.20	7.0	0.010
6	17	0.10	7.0	0.020

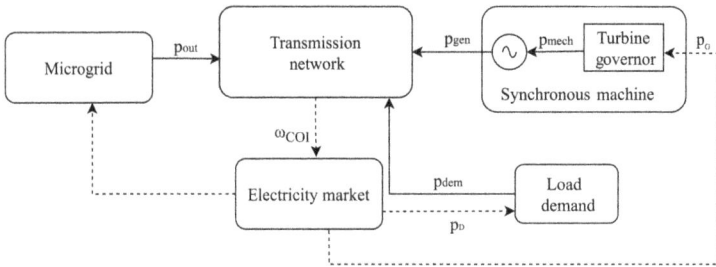

*Figure 3.2   Structure of the connections among MGs, loads, generators,
transmission network, and electricity market*

one presented in Table 3.1 and the conceptual structure of an MG is illustrated in
Figure 3.3. The variation of the price and the resulting frequency of the center of
inertia (ω_{COI}) of the system is presented in Figure 3.4, where:

$$\omega_{COI} = \frac{\sum_{i=1}^{n_g} H_i\omega_i}{\sum_{i=1}^{r} H_i}, \tag{3.10}$$

with ω_i and H_i the rotor angular speed and the inertia constant of the *i*-th synchronous machine, respectively, and n_g the total number of machines. For the sake
of readability, the average value μ_{COI} is shown with a dashed line in Figure 3.3,
together with two upper and lower dashed lines corresponding to $\mu_{COI} + 3\sigma_{COI}$ and
$\mu_{COI} + 3\sigma_{COI}$, respectively.

The standard deviation of the system frequency—σ_{COI}—is 0.001 pu(Hz) in this
scenario. Due to the low penetration of MGs, the independence of the price signal of
the power imbalance and due to the automatic generation control of the synchronous
generators, the fluctuation of system frequency is not critical and remains within
reasonable limits. It should be noted however that the fluctuations are larger than the
ones that would have been obtained if the MGs were not trying to maximize their
revenues: in that case the standard deviation of ω_{COI} is only 0.0006 pu(Hz). It is
also important to highlight that the share of MGs' average generation in the grid is
relatively low—only 2.9% of the overall grid generation.

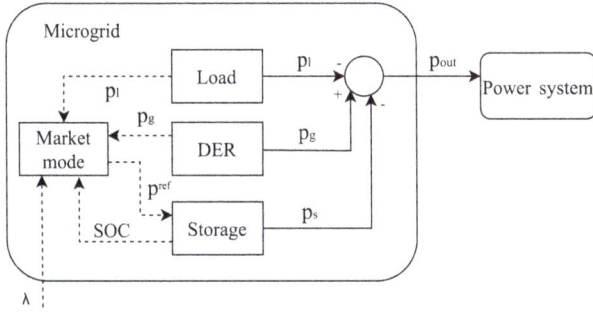

Figure 3.3   Scenario 1: Structure of the MG, when price is an exogenous signal (i.e., MGs are price-takers)

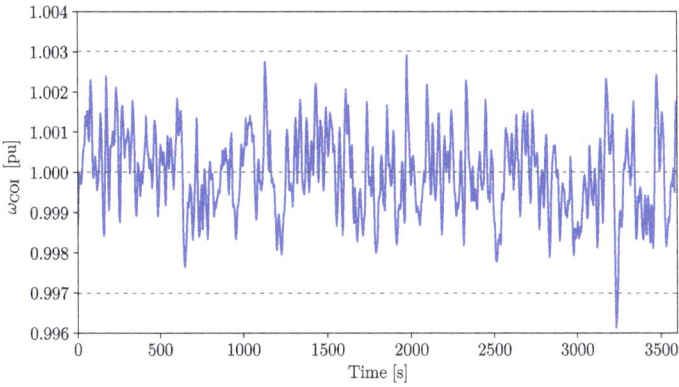

(a)

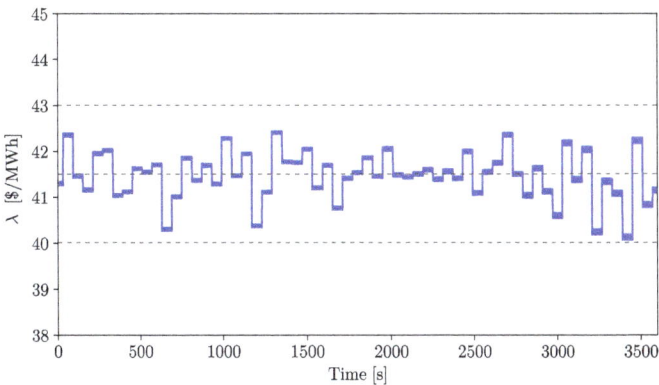

(b)

Figure 3.4   Scenario 1: Frequency ω_{COI} (a) and market clearing price λ (b) for a market-based frequency control and exogenous price

3.4.2   Scenario 2: price-makers MGs

With the increasing share of MGs connected to the grid, it can be reasonably expected that their influence on the power system balance will become more relevant. In that case, the energy price will be directly affected by the operation of MGs that will become price makers. We now consider this particular situation, and the main difference from before is that now the power system model needs to be coupled with a real-time electricity market model, in a way that the price variation helps to manage the balance of generation and consumption. Roughly speaking, this means that, if the electricity price is higher that its marginal cost, generators will increase their production until the cost becomes equal to the market price, and vice versa. Price is considered as a continuous state variable and it is computed and adjusted rapidly enough with respect to the dynamic response of the transmission system. The scheme of an MG in a "price-making" scenario is shown in Figure 3.5, and the main difference from Figure 3.3 is that now the value of λ is fed-back in a close-loop (i.e., to mimic the fact that the behavior of the MGs has a direct effect on the energy price).

In order to describe a dynamic market model with a single price of energy and with n power suppliers and m power consumers we consider the market equations proposed in [35]:

$$\frac{d}{dt}E = \sum_{i=1}^{n} p_{Gi} - \sum_{j=1}^{m} p_{Dj} - p_{\text{loss}}, \tag{3.11}$$

$$T_\lambda \frac{d}{dt}\lambda = -K_E E - \lambda, \tag{3.12}$$

where p_{Gi} are the generated active powers of the n suppliers connected to the grid; p_{Dj} are the active power consumption of m loads connected to the grid; p_{loss} are the active power losses in the transmission system; E and λ, are the energy imbalance and the electricity price, respectively; and K_E and T_λ are parameters that depend on the design of the market itself. Since in real-world systems it is hard to measure the power imbalance in (3.11), such an imbalance is deduced in [35] implicitly based

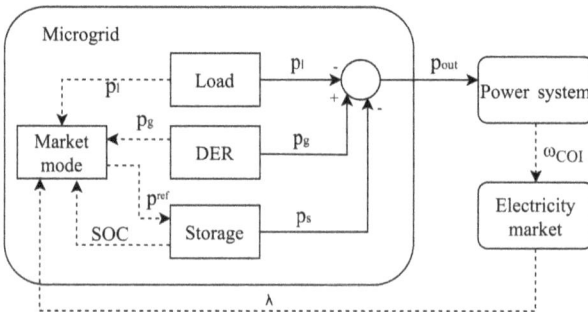

Figure 3.5   Scenario 2: Structure of an MG in a scenario where MGs are price-makers

on the frequency deviation of the center of inertia. Hence, (3.12) the market clearing price dynamic is expressed as

$$T_\lambda \frac{d}{dt} \lambda = K_E(1 - \omega_{\text{COI}}) - \lambda. \tag{3.13}$$

Finally, generator and load active powers are linked to the market clearing price λ via the following differential equations [35]:

$$T_{Gi} \frac{d}{dt} p_{Gi} = \lambda - c_{Gi} p_{Gi} - b_{Gi}, \tag{3.14}$$

$$T_{Di} \frac{d}{dt} p_{Di} = -\lambda - c_{Di} p_{Di} - b_{Di}, \tag{3.15}$$

where c_{Gi}, c_{Di}, and b_{Gi}, b_{Di} are the parameters of the marginal benefit of the generators, respectively, and T_{Gi} and T_{Di} are time constants modeling generator and demand, respectively, delayed response to variations of the market clearing price λ. Note that in [35], the equations for the load active powers are expressed with different signs.

For this scenario, the participation of MGs in the frequency control service has a positive effect on the standard deviation of ω_{COI} that is decreased down to 0.00073 pu(Hz), compared to the previous scenario, where the price was endogenously set by the outer system. Intuitively, we can explain the smaller variations as follows: as frequency increases price will decrease (as a consequence of the energy imbalance). Consequently the actors that try to maximize their revenues will start buying energy to store it in their storage, to take advantage of the low price. This result is consistent with the interpretation of (3.13) as a power price feedback, that tends to stabilize the system, as already pointed out in [35]. The variation of ω_{COI} and the resulting clearing price fluctuations are shown in Figure 3.6. Notice however that this result does not imply that MGs that try to maximize their revenues lead to an improvement in frequency stability. For example, different parameter choices might easily lead to instability (see [35]). It does however suggest that, while MGs that try to maximize revenues have a negative impact on the grid stability, MGs as price makers might have less of an impact than MGs as price takers (as shown by the results of the previous scenario).

3.4.3   Scenario 3: MGs provide ancillary services when needed

To mitigate the possible negative effect of MGs' greedy operation (especially when MGs are not price makers) on the power grid, and at the same time, to allow MGs to participate in the market, when the grid conditions are not critical, the following stochastic scheme may be applied.

- MGs are operating in the market mode (M-mode), trying to maximize their profits, as long as the system frequency lies within an acceptable range, σ_w is lower than the threshold value σ_{w_m}. In this case, MGs are following the rules described in Section 3.4.2.
- When the frequency deviation exceeds the maximum allowed value, some MGs switch to the frequency regulation (F-mode) and follow the droop control equations that are conventionally employed in the primary frequency regulation of

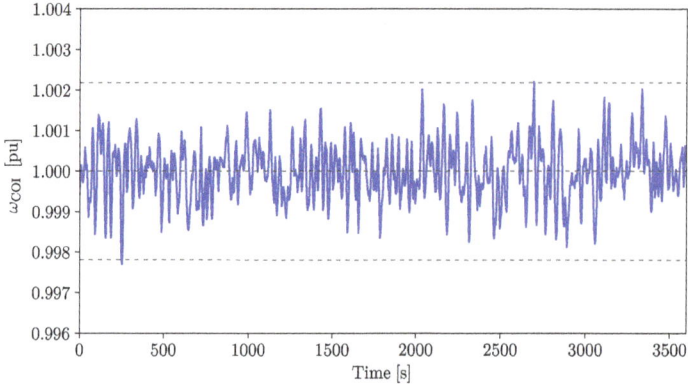

(a)

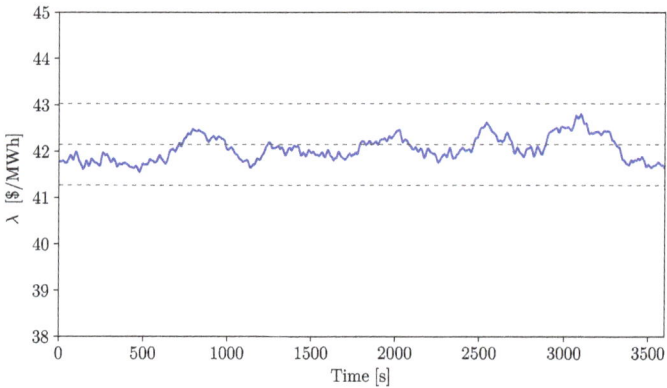

(b)

Figure 3.6   Frequency ω_{COI} (a) and market clearing price λ (b) for a market-based frequency control

the power system. To control the switching between two modes (M-mode and F-mode), every period T, a probability $\pi_s \in [0, 1]$ defines if the MG will operate in F-mode during the next time window.

The structure of MG in this scenario is shown in Figure 3.7. The MG control strategy leads to a reduction of ω_{COI} fluctuation to 0.00069 pu(Hz). The market participation of MGs in this scenario is also reduced, which results in less fluctuation of the market clearing price. The results can be seen in Figure 3.8(a) and (b).

3.4.4   Scenario 4: MGs provide only ancillary services

In this scenario, we assume that MGs do not compete in a market context in order to maximize their revenues, but we assume that the MGs are willing to cooperate

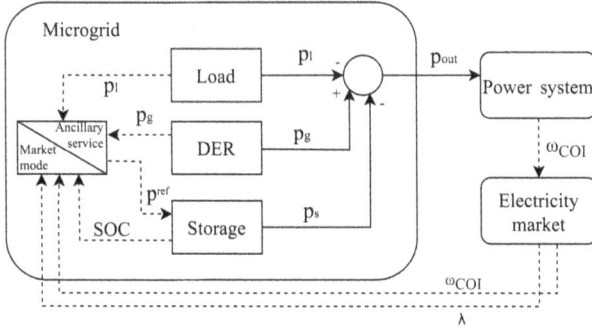

*Figure 3.7   Scenario 3: Structure of an MG in the scenario where MGs switch
between the greedy mode (when they try to maximize their revenues),
and the mode where they provide ancillary services (when required by
the grid)*

to improve the operation of the overall power grid. Note that such a new scenario
may be easily obtained as a special case of the third scenario when the value of
σ_{ω_m} is set to a very small one (i.e., MGs are always forced to provide frequency
services to the grid). In this new case, the price does not affect the operation of the
system anymore, as shown in Figure 3.9. The impact of such a control on the system
frequency is presented in Figure 3.10. In particular, the standard deviation of ω_{COI}
has reached 0.00062 pu(Hz), which is very close to the case where the power grid
does not contain MGs, and, as one could easily expect, is also the best situation, from
a frequency stability point of view, among all the investigated scenarios.

3.4.5   On the impact of the storage capacity and price deviations

The four presented scenarios show that depending on the chosen strategy, MGs may
either pose stability issues to the power grid (as the system frequency oscillates
severely), or they may improve the stability of grid, as in the case where MGs provide
frequency services.

In the remainder of this section, we now further investigate the influence of the
MGs on the system in case of changing conditions. In particular, we now develop a
simple sensitivity analysis to evaluate the impact of the availability of storage devices
(and their capacity), and of price deviations on the obtained results.

3.4.5.1   Storage capacity

To assess the effect of the storage capacity on the performance of the described scenarios, parameter K (see (3.1)) is multiplied (or divided) by 2, 10, and 100, to mimic
a longer (or shorter) time to charge the devices, which corresponds to increased (or
decreased) capacities. The resulting ω_{COI} for the market-based frequency regulation (the second scenario described in Section 3.4.2) is presented in Figure 3.11.
Initially a greater storage capacity allows to slightly reduce the frequency standard
deviation, as it reaches values of 0.00071 pu(Hz) and 0.00067 pu(Hz), for 2 and 10

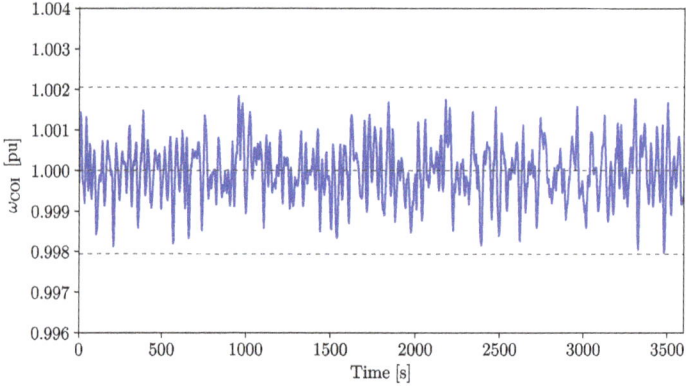

(a)

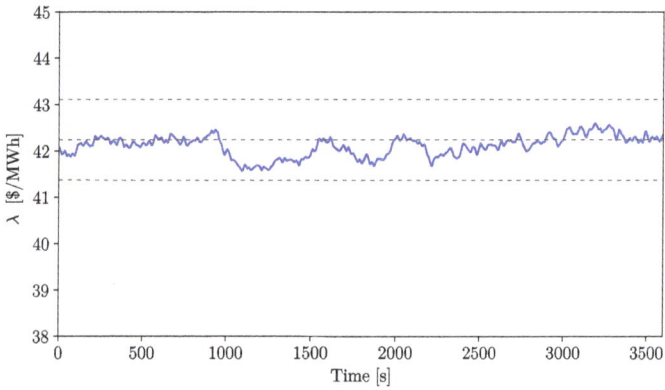

(b)

Figure 3.8   Scenario 4: Frequency ω_{COI} for a stochastic frequency control with MGs

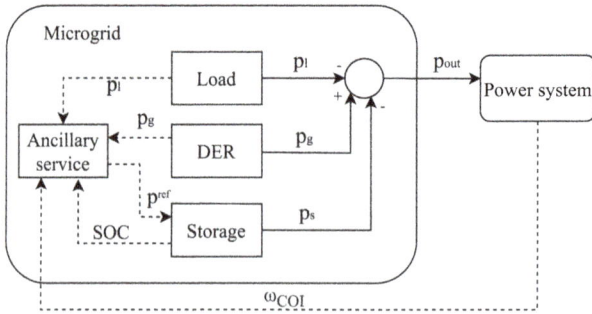

Figure 3.9   Scenario 4: Structure of an MG in the scenario where MGs provide only ancillary services

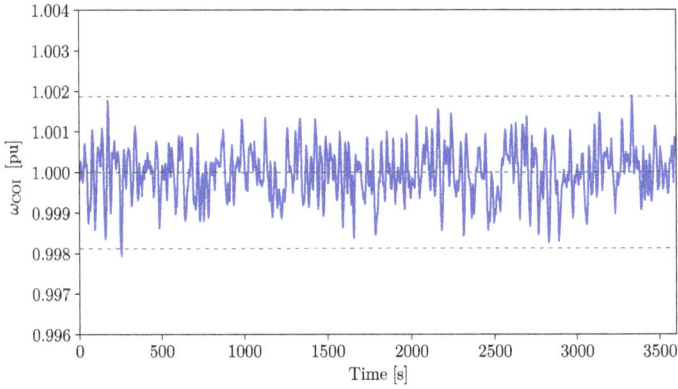

Figure 3.10   Frequency ω_{COI} for a frequency control with MGs

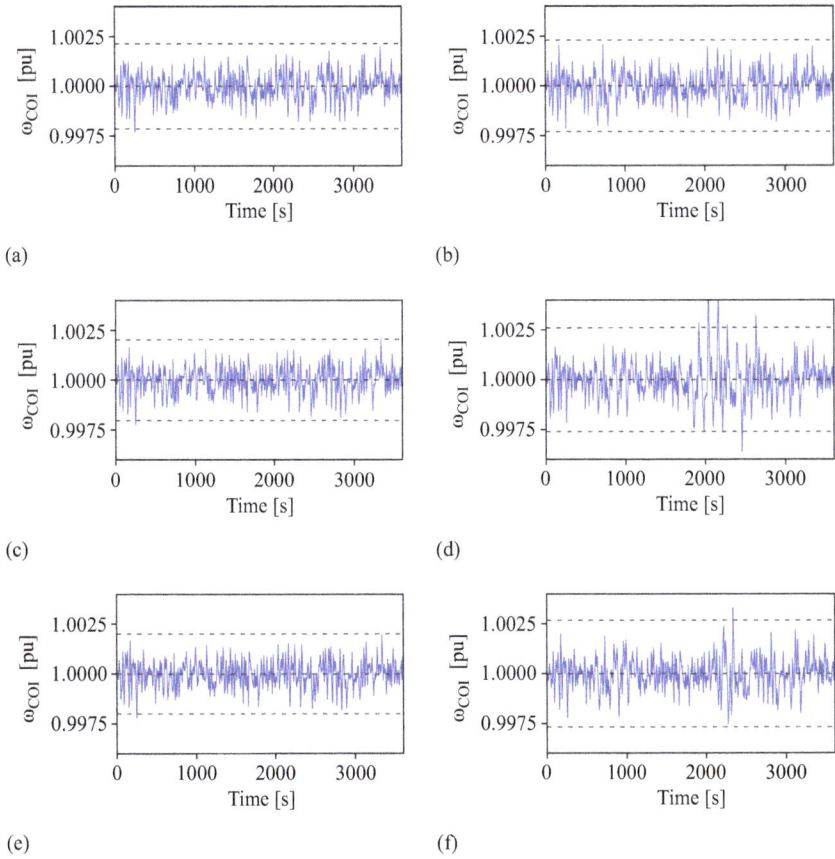

(a)

(b)

(c)

(d)

(e)

(f)

Figure 3.11   Frequency ω_{COI} for K multiplied/divided by 2 (a) and (b), 10 (c) and (d) and 100 (e) and (f) for market-based control

times increase of K, respectively. However, further increases of the storage capacity (including $K * 100$) do not lead to further improvements of σ_{COI}. Conversely, ω_{COI} increases with the reduction of the storage system, reaching a maximum value of 0.00087 pu(Hz) for the reduction of the storage capacity by 100. A possible explanation of this is that the storage capacities allow the MGs to increase their flexibility and to provide more convenient services to the grid. However, when the storage capacity is too large (i.e., around 10 times the initial value), then it is never fully charged and no further improvements can be noticed.

3.4.5.2 Price sensitivity

We now vary the price threshold ρ and the corresponding deviations of ω_{COI} are presented in Figure 3.12. A smaller threshold encourages MGs to change their strategy more often, which leads to more severe frequency fluctuations. Besides, the doubling of ρ causes a slightly lower frequency standard deviation of the system. If ρ is further increased, then MGs never change their behavior, and therefore the system frequency does not change. The same effect (increasing ω_{COI} fluctuations with the

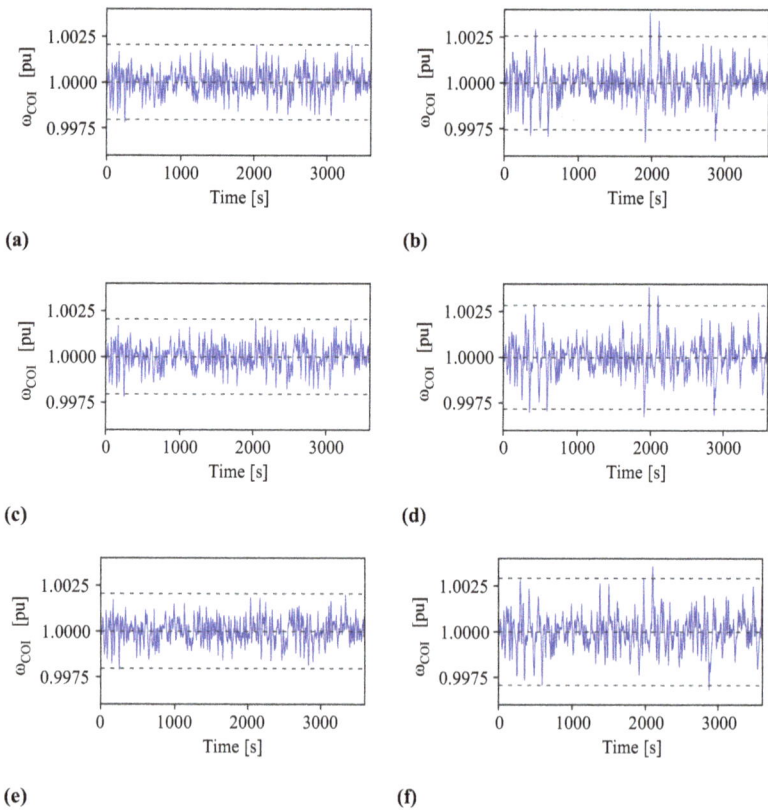

Figure 3.12   Price fluctuations threshold for EMS, ρ is multiplied/divided by 2 (a) and (b), 10 (c) and (d), and 100 (e) and (f) for market-based control

Table 3.3   Variation of ω_{COI} depending on the storage size and price fluctuation threshold

K	$\omega_{\text{COI}} * 10^{-3}$	ρ	$\omega_{\text{COI}} * 10^{-3}$
	Exogenous price		
	Bigger storage		**Higher price threshold**
K	1.00	ρ	1.0
$K \cdot 2$	0.97	$\rho \cdot 2$	0.98
$K \cdot 10$	0.88	$\rho \cdot 10$	0.88
$K \cdot 100$	0.86	$\rho \cdot 100$	0.85
	Smaller storage		**Lower price threshold**
$K/2$	1.02	$\rho/2$	1.00
$K/10$	1.01	$\rho/10$	1.01
$K/100$	1.01	$\rho/100$	1.01
	MGs are price makers		
	Bigger storage		**Higher price threshold**
K	0.73	ρ	0.73
$K \cdot 2$	0.71	$\rho \cdot 2$	0.68
$K \cdot 10$	0.67	$\rho \cdot 10$	0.68
$K \cdot 100$	0.67	$\rho \cdot 100$	0.68
	Smaller storage		**Lower price threshold**
$K/2$	0.77	$\rho/2$	0.85
$K/10$	0.85	$\rho/10$	0.93
$K/100$	0.87	$\rho/100$	0.98
	MGs provide ancillary services		
	Bigger storage		**Higher price threshold**
K	0.69	ρ	0.69
$K \cdot 2$	0.67	$\rho \cdot 2$	0.67
$K \cdot 10$	0.66	$\rho \cdot 10$	0.67
$K \cdot 100$	0.65	$\rho \cdot 100$	0.66
	Smaller storage		**Lower price threshold**
$K/2$	0.70	$\rho/2$	0.73
$K/10$	0.71	$\rho/10$	0.76
$K/100$	0.71	$\rho/100$	0.77

increase of ρ) is observed in the scenarios corresponding to the stochastic and exogenous price cases. Conversely, price does not affect the behavior of the MGs in the last scenario, which is thus omitted here.

The results of the sensitivity analysis for the first three scenarios are summarized in Table 3.3. The patterns appear to be similar in all the three cases (i.e., ω_{COI}

increases when the size of the storage decreases, and it increases when ρ decreases). While little differences can be noted in the first and in the third scenarios, the second scenario, where MGs are price-makers, shows more evident frequency fluctuations for different values of the storage capacity or of the price threshold. The reason is that the effect of the greedy behavior of the MGs is amplified by their larger penetration in the power grid.

3.5 Conclusions

As there is an increased interest in MGs, there are also increased concerns about whether a larger penetration could affect the operation and/or the stability of the power grid. This chapter shows that, indeed, the presence of MGs does have an impact on the frequency of the power grid, in terms of ω_{COI}, and that the impact is more significant when MGs are allowed to greedily buy/sell energy in the market with the objective to maximize their revenues, without caring about stability issues.

However, there also are several ways to mitigate such an impact. An option is to correct the behavior of MGs with a price signal, which corresponds to the second scenario we have examined. Another option is to force MGs to cooperate, and not compete, in the energy market, and to make them contribute to provide frequency services to the power grid (fourth scenario). A trade-off, examined in the third scenario, is to allow MGs to run their business in the energy market when frequency oscillations are minor, and ask them to provide frequency regulation when frequency oscillations increase.

As the interest in new MGs is gathering increasing social and economic consensus, this chapter is a first step into the exploration of new business models and more advanced logic of EMS of MGs, and also toward new energy systems with very few large-size power generators, and a multitude of little *prosumers*—that is, a combination of producer and consumer—entering the energy market.

References

[1] Lasseter B. Microgrids [distributed power generation]. In: *2001 IEEE Power Engineering Society Winter Meeting. Conference Proceedings (Cat. No.01CH37194)*. vol. 1; 2001-01. p. 146–149.
[2] Hatziargyriou N, Hiroshi A, Reza I, *et al. Microgrids: An Overview of Ongoing Research, Development, and Demonstration Projects | Energy Technologies Area*; 2007. Available from: https://eta.lbl.gov/publications/microgrids-overview-ongoing-research.
[3] Vasilakis A, Zafeiratou I, Lagos DT, *et al.* The evolution of research in microgrids control. *IEEE Open Access Journal of Power and Energy*. 2020;7:331–343.

[4]   Kueffner JH. Wind hybrid power system for Antarctica Inmarsat link. In: *INT-ELEC '86 – International Telecommunications Energy Conference*; 1986. p. 297–298.

[5]   Hatziargyriou N, Dimeas A, Vasilakis N, *et al.* The Kythnos microgrid: a 20-year history. *IEEE Electrification Magazine*. 2020;8(4):46–54.

[6]   Liang X, Saaklayen MA, Igder MA, *et al.* Planning and service restoration through microgrid formation and soft open points for distribution network modernization: a review. *IEEE Transactions on Industry Applications*. 2022;58(2):1843–1857.

[7]   Osama RA, Zobaa AF, and Abdelaziz AY. A planning framework for optimal partitioning of distribution networks into microgrids. *IEEE* Systems Journal. 2020;14(1):916–926.

[8]   Liu CC, Jain AK, Boroyevich D, *et al.* Microgrid building blocks: concept and feasibility. *IEEE Open Access Journal of Power and Energy*. 2023;10:463–476.

[9]   Khan R, Islam N, Das SK, *et al.* Energy sustainability–Survey on technology and control of microgrid, smart grid and virtual power plant. *IEEE Access*. 2021;9:104663–104694.

[10] Wang Y, Li Y, Cao Y, *et al.* Optimal operation strategy for multi-energy microgrid participating in auxiliary service. *IEEE Transactions on Smart Grid*. 2023;14(5):3523–3534.

[11] Awal MA, Yu H, Tu H, *et al.* Hierarchical control for virtual oscillator based grid-connected and islanded microgrids. *IEEE Transactions on Power Electronics*. 2020;35(1):988–1001.

[12] Ferraro P, Crisostomi E, Raugi M, *et al.* Analysis of the impact of microgrid penetration on power system dynamics. *IEEE Transactions on Power Systems*. 2017;32(5):4101–4109.

[13] Olivares DE, Mehrizi-Sani A, Etemadi AH, *et al.* Trends in microgrid control. *IEEE Transactions on Smart Grid*. 2014;5(4):1905–1919.

[14] Roslan MF, Hannan MA, Ker PJ, *et al.* Microgrid control methods toward achieving sustainable energy management: a bibliometric analysis for future directions. *Journal of Cleaner Production*. 2022;348:131340.

[15] Tuckey A, and Round S. Grid-forming inverters for grid-connected microgrids: Developing "good citizens" to ensure the continued flow of stable, reliable power. *IEEE Electrification Magazine*. 2022;10(1):39–51.

[16] Zhang J, Su S, Chen J, *et al.* Stability analysis of the power system with the large penetration ratios of microgrids. In: *2009 International Conference on Sustainable Power Generation and Supply*; 2009. p. 1–5.

[17] Golpîra H, Seifi H, Messina AR, *et al.* Maximum penetration level of microgrids in large-scale power systems: frequency stability viewpoint. *IEEE Transactions on Power Systems*. 2016;31(6):5163–5171.

[18] IEEE Std 2030.7-2017 IEEE Standard for the Specification of Microgrid Controllers. Piscataway, NJ: IEEE; 2018. p. 1–43.

[19] Feng W, Sun K, Guan Y, *et al.* Active power quality improvement strategy for grid-connected microgrid based on hierarchical Control. *IEEE Transactions on Smart Grid*. 2018;9(4):3486–3495.

[20]   Bidram A, and Davoudi A. Hierarchical structure of microgrids control system. *IEEE Transactions on Smart Grid.* 2012;3(4):1963–1976.

[21]   Nasirian V, Shafiee Q, Guerrero JM, *et al.* Droop-free distributed control for AC microgrids. *IEEE Transactions on Power Electronics.* 2016;31(2):1600–1617.

[22]   Ahmed K, Seyedmahmoudian M, Mekhilef S, *et al.* A review on primary and secondary controls of inverter-interfaced microgrid. *Journal of Modern Power Systems and Clean Energy.* 2021;9(5):969–985.

[23]   Zhao J, Lyu X, Fu Y, *et al.* Coordinated microgrid frequency regulation based on DFIG variable coefficient using virtual inertia and primary frequency control. *IEEE Transactions on Energy Conversion.* 2016;31(3):833–845.

[24]   Kerdphol T, Watanabe M, Hongesombut K, *et al.* Self-adaptive virtual inertia control-based fuzzy logic to improve frequency stability of microgrid with high renewable penetration. *IEEE Access.* 2019;7:76071–76083.

[25]   Khayat Y, Shafiee Q, Heydari R, *et al.* On the secondary control architectures of AC microgrids: an overview. *IEEE Transactions on Power Electronics.* 2020;35(6):6482–6500.

[26]   Sun Y, Hou X, Yang J, *et al.* New perspectives on droop control in AC microgrid. *IEEE Transactions on Industrial Electronics.* 2017;64(7):5741–5745.

[27]   Rey JM, Martí P, Velasco M, *et al.* Secondary switched control with no communications for islanded microgrids. *IEEE Transactions on Industrial Electronics.* 2017;64(11):8534–8545.

[28]   Lou G, Gu W, Wang L, *et al.* Decentralised secondary voltage and frequency control scheme for islanded microgrid based on adaptive state estimator. *IET Generation, Transmission & Distribution.* 2017;11(15):3683–3693.

[29]   Cheng Z, Duan J, and Chow MY. To centralize or to distribute: that is the question: a comparison of advanced microgrid management systems. *IEEE Industrial Electronics Magazine.* 2018;12(1):6–24.

[30]   Simpson-Porco JW, Shafiee Q, Dörfler F, *et al.* Secondary frequency and voltage control of islanded microgrids via distributed averaging. *IEEE Transactions on Industrial Electronics.* 2015;62(11):7025–7038.

[31]   Lu X, Yu X, Lai J, *et al.* Distributed secondary voltage and frequency control for islanded microgrids with uncertain communication links. *IEEE Transactions on Industrial Informatics.* 2016;13(2):448–460.

[32]   Weng S, Yue D, Dou C, *et al.* Distributed event-triggered cooperative control for frequency and voltage stability and power sharing in isolated inverter-based microgrid. *IEEE Transactions on Cybernetics.* 2018;49(4):1427–1439.

[33]   Yuan G, Gao Y, Ye B, *et al.* Real-time pricing for smart grid with multi-energy microgrids and uncertain loads: a bilevel programming method. *International Journal of Electrical Power & Energy Systems.* 2020;123:106206.

[34]   Dong X, Li X, and Cheng S. Energy management optimization of microgrid cluster based on multi-agent-system and hierarchical Stackelberg game theory. *IEEE Access.* 2020;8:206183–206197.

[35]   Alvarado FL, Meng J, DeMarco CL, *et al.* Stability analysis of interconnected power systems coupled with market dynamics. *IEEE Transactions on Power Systems.* 2001;16(4):695–701.

[36]   Alvarado FL. Is system control entirely by price feasible? In: *Proceedings of the 2003 36th Annual Hawaii International Conference on System Sciences.* Piscataway, NJ: IEEE; 2003. p. 6.

[37]   Chiu WY, Sun H, and Poor HV. A multiobjective approach to multimicrogrid system design. *IEEE Transactions on Smart Grid*. 2015;6(5):2263–2272.

[38]   Tamimi B, Cañizares C, Bhattacharya K. Modeling and performance analysis of large solar photo-voltaic generation on voltage stability and inter-area oscillations. In: *2011 IEEE Power and Energy Society General Meeting*. Piscataway, NJ: IEEE; 2011. p. 1–6.

[39]   Milano F. Control and stability of future transmission networks. *Handbook of Clean Energy Systems*. John Wiley & Sons, Ltd; 2015. p. 1–15.

[40]   Milano F. Power System Modelling and Scripting. Berlin: Springer Science & Business Media; 2010.

[41]   Milano F. A Python-based software tool for power system analysis. In: *2013 IEEE Power & Energy Society General Meeting*. Piscataway, NJ: IEEE; 2013. p. 1–5.

Chapter 4
Fully distributed and economic frequency regulation solutions for autonomous microgrids

Wenchuan Wu[1] and Zhongguan Wang[1]

Since the prediction results of volatile renewables involve significant errors, the economic dispatch is impractical for the operation of autonomous microgrids. Therefore, the frequency control schemes should consider the operating costs. This chapter introduces a fully distributed and economic frequency regulation solution for autonomous microgrids, which includes secondary and primary frequency control strategies to both stabilize frequency and minimize operating costs. For the secondary frequency control, a two-stage fully distributed regulation method is proposed. In the first stage, a subgradient-based consensus algorithm incorporated with equal increment rate criteria is used to recover frequency with minimal operating costs. In the second stage, an average consensus algorithm is adopted to compress frequency oscillations caused by measurement errors. For the primary frequency control, a distributed quasi-Newton method is outlined. Each distributed resource shares limited information with its neighboring resources across a sparse communication network. By applying the equal increment rate criteria, the frequency control method simultaneously reduces generation costs and optimizes renewable energy utilization. Simulation results demonstrate that the proposed method is robust against errors and time delays, effectively preventing overshooting.

4.1 Introduction

A microgrid is a cluster of distributed generators (DGs), loads, energy storage systems (ESSs), and control devices. Organized as an autonomous system with advanced control and management strategies, a microgrid can improve energy efficiency, make use of a high penetration level of renewable energy generation, provide ancillary services, and enhance power reliability and quality for customers [1–5].

Autonomy of a microgrid means that it is able to operate in isolated mode [6]. However, intermittency of renewable energy generation, such as wind turbines (WTs) and photovoltaic systems (PV), poses technical challenges, especially when microgrids are operating in isolation, wherein a microgrid must maintain its own active

[1]State Key Laboratory of Power Systems, Department of Electrical Engineering, Tsinghua University, PR China

power supply-demand balance [7]. Uncertainty in the distributed generation and load demand can cause significant frequency fluctuations because of the low inertia of a microgrid. Therefore, effective power balance and frequency control schemes are essential.

A three-layer hierarchical control structure is commonly employed [8–10]. The primary control, typically implemented through a droop mechanism, operates on a short timescale to maintain power balance and ensure proportional load sharing among inverters [11]. This is achieved by using a predetermined droop curve to distribute the total load demand, thereby preventing overstressing of power sources. The secondary control, functioning on an intermediate timescale, regulates the power set point to correct frequency deviations introduced by the primary control [6,12]. Lastly, the tertiary control manages the generation schedule over a longer timescale to achieve economic dispatch [13].

The droop-based primary control is fully decentralized and implemented locally. Despite its simplicity, droop control has several limitations due to its lack of coordination. First, since predetermined droop curves conflict with real-time optimization goals, power sharing among sources is often neither efficient nor economical. Second, the droop mechanism's poor dynamic performance can cause inconsistencies between controls and frequency fluctuations [14], leading to prolonged frequency stabilization times. Third, the droop mechanism is highly sensitive to measurement errors, necessitating very precise measurements [15]. Specifically, a flat droop curve can result in significant power deviations from small frequency errors, while a steep curve can cause severe steady-state frequency deviations when loads change. Furthermore, droop control struggles with managing nonlinear loads [16]. Although many improved and adaptive droop-based methods have been developed [17–19], these drawbacks have not been fully resolved. As a result, novel droop-free control schemes have been proposed to replace local decision-making with cooperative control [20,21].

Cooperative control in autonomous microgrids can be categorized into centralized and distributed approaches. In a centralized setup, a microgrid central controller (MGCC) communicates with all DGs and sends operational commands to the inverters. However, the robustness of the cyber-physical system is a concern, as the MGCC is susceptible to targeted cyber and physical attacks. Centralized schemes also face challenges with time delays [22], especially in complex communication networks, making the MGCC a potential single point of failure and rendering the centralized approach unreliable [23]. Additionally, any expansion of the microgrid requires reconfiguration of the control model, which is particularly problematic for remote microgrids, such as those on islands or in rural areas, where repair and system restart may take several days following an MGCC failure.

Distributed control, often based on the multi-agent system (MAS) theory [24], offers a more resilient and efficient alternative for microgrid operation. In fully distributed control architectures, the central controller is eliminated, and each agent communicates with others using a limited communication protocol [25]. Some methods require a communication network where each agent is connected to all others [26,27]. However, the reliability of such a system can be compromised if any communication link fails, affecting the overall system performance. In systems with sparse communication networks, such as those proposed in [28–30], the master node in a master–slave architecture can become a single point of failure. Fully distributed cooperative control schemes, which rely on peer-to-peer information sharing over sparse communication networks, have been applied to microgrids as proposed in [31–33]. However, many of these approaches are still based on the droop mechanism, carrying the inherent drawbacks previously discussed.

A previous study proposes a distributed control architecture that maintains power balance, regulates frequency, and minimizes power generation costs [33]. This event-triggered scheme separates frequency control and cost optimization into two stages. However, due to the unpredictability of renewable energy and rapid changes in load demand, significant errors in predicted values can degrade the results of prediction-based economic dispatch. Therefore, real-time control schemes should incorporate optimal operation [33]. In other words, frequency control and economic dispatch objectives, such as power balance, minimizing generation costs, and maximizing renewable energy utilization, should be achieved through a single control process.

This chapter introduces fully distributed and economically efficient frequency regulation solutions for autonomous microgrids, enabling rapid frequency recovery and minimal generation costs using distributed schemes.

4.2 Distributed control framework for frequency regulation

Microgrids establish a direct physical link between power sources, storage devices, and loads. This physical infrastructure is complemented by a cyber network that enables the implementation of various control schemes. In this context, a microgrid is considered a cyber-physical system, with a communication network that facilitates data exchange between sources for monitoring and control.

In centralized control schemes, each source communicates with the MGCC, where all data are gathered, and control signals are sent to each controller after complex calculations. In contrast, distributed control schemes do not need the MGCC, with each source exchanging information only with a few neighboring sources over a sparse communication network. This distributed communication network in a microgrid can be represented by a directed graph (digraph), where the nodes correspond to distributed sources, and the edges represent communication links.

The digraph is expressed as $G = (\mathbf{v}, \varepsilon, \mathbf{A})$, where $\mathbf{v} = \{v_1, v_2, \cdots, v_m\}$ is a nonempty finite set of m nodes, $\varepsilon \subseteq \mathbf{v} \times \mathbf{v}$ is the set of edges and the matrix $\mathbf{A} = [\mu_{ij}]_{m \times m}$ is a weighted adjacent matrix with nonnegative adjacency elements μ_{ij}. The set of neighbors of node v_i is described by the following:

$$N_i = \{v_j \in \mathbf{v} : (v_i, v_j) \in \varepsilon\} \tag{4.1}$$

Let $x_i \in \mathbb{R}$ denote the value of node v_i, and let the consensus variable vector be given by $\mathbf{x} = [x_1, x_2, \cdots, x_m]^T$. The value of a node may represent physical quantities such as voltage, and frequency. We say nodes v_i and v_j agree if and only if $x_i = x_j$, and all nodes in a network reach consensus if and only if $x_i = x_j$ for all

$i, j \in (1, 2, \cdots, m), i \neq j$. In other words, a consensus scheme establishes a map $\chi : \mathbb{R}^n \to \mathbb{R}$ converging all elements of vector $\mathbf{x}$ to a common value x^*.

A consensus protocol is used to solve agreement problems in a network of agents with a discrete-time (DT) model. We have

$$x_i(k + 1) = x_i(k) + u_i(k) \tag{4.2}$$

where $x_i(k)$ is the value of x_i at step k, and $u_i(k)$ is the adjustment amount at step k, whose information is obtained only from node i and its neighbors.

In autonomous microgrids, CGs, RGs, and ESSs are designated as frequency regulation components, each managed by an intelligent agent. Frequency regulation and cost optimization are achieved through local iterations of these agents, coupled with information sharing among neighboring agents. Each agent measures the performance of its associated source, communicates with neighboring agents, conducts local calculations, and sends control signals to its respective source.

4.3 Optimization model for frequency regulation

4.3.1 Generation cost and the equal increment rate criteria

The general form of the generation cost associated with different sources can be expressed as a quadratic function of the active power output:

$$C_i(P_i) = a_i P_i^2 + b_i P_i + c_i \tag{4.3}$$

where P_i denotes the active power output of DG i, and a_i, b_i, c_i are coefficients of DG i.

By taking the derivative of $C_i(P_i)$, the incremental cost rate (ICR) of DG i or ESS i has the following form:

$$\text{ICR}_i(P_i) = \frac{\mathrm{d}C_i(P_i)}{\mathrm{d}P_i} = 2a_i P_i + b_i \tag{4.4}$$

For an RG, we define a "virtual" generation cost. The objective of power dispatch for RG is to minimize the curtailment of renewable energy. Therefore, a virtual generation cost for RG i can be formulated as

$$C_i(P_i) = \frac{(P_i - P_i^{\max})^2}{P_i^{\max}} = \frac{P_i^2}{P_i^{\max}} - 2P_i + P_i^{\max} \tag{4.5}$$

where $P_i^{\max}$ denotes the predicted maximum generation capacity of RG i.

By introducing

$$a_i = \frac{1}{P_i^{\max}} \tag{4.6}$$

$$b_i = -2 \tag{4.7}$$

$$c_i = P_i^{\max} \tag{4.8}$$

the generation cost function of an RG has the same form as that of a CG. From (4.5), we can find that the closer is the actual output P_i to the predicted maximum generation capacity $P_i^{\max}$, the lower is the generation cost. Thus, renewable energy curtailment will be minimized when the minimum generation cost is obtained.

For an ESS, the generation cost function is defined as

$$C_i(P_i) = \begin{cases} \bar{a}_i P_i^2 + c_i, & P_i \geq 0 \\ \underline{a}_i P_i^2 + c_i, & P_i < 0 \end{cases} \tag{4.9}$$

where $\bar{a}_i$ and $\underline{a}_i$ denote the different coefficients in charging and discharging state. This is of the same form as the cost function of a CG if $b_i = 0$.

According to the equal ICR criterion, the total generation cost is minimized when the ICRs of different DGs are equal, or they reach their limits if transmission capacity allows (refer to Chapter 9 in [34]).

Thus, the generation cost can be minimized by distributing power according to the equal ICR criterion during the frequency control process. It is important to note that the ICRs of CGs and RGs must meet the following conditions:

$$ICR_i = \frac{\mathrm{d}C_i(P_i)}{\mathrm{d}P_i} = 2a_i P_i + b_i \geq 0 \tag{4.10}$$

$$ICR_j = \frac{\mathrm{d}C_j(P_j)}{\mathrm{d}P_j} = 2\frac{P_j}{P_j^{\max}} - 2 \leq 0 \tag{4.11}$$

This means that the ICRs of CGs are always positive, whereas the ICRs of RGs are non-positive. In other words, ICRs of CGs are always greater than those of RGs. Therefore,

1. When load is light, DGs reach lower bounds, and RGs distribute power by the equal ICR criterion;
2. When load is heavy, RGs reach upper bounds, and CGs distribute power according to the equal ICR criterion.

Additionally, because the cost ICRs of RGs are always less than those of CGs, RGs have higher power generation priority, leading to an increased utilization of renewable energy. When the ICRs of RGs are equal, the power outputs of RG i and RG j meet the following equation:

$$\frac{P_i}{P_i^{\max}} = \frac{P_j}{P_j^{\max}} \tag{4.12}$$

When the system is unable to accommodate all renewable power generation, (4.12) indicates that the power generation for each RG is proportional to their predicted maximum generation capacity.

4.3.2   Optimization model

Traditionally, frequency control is achieved using a droop control scheme. The basic principle involves increasing active output when the frequency drops and decreasing it when the frequency rises. Distributed DGs under droop control share the power imbalance based on predetermined droop coefficients.

The droop control is highly sensitive to frequency measurement errors. With a flat droop curve, even a small error can lead to significant power deviations. Conversely, when the curve is steep, the system's ability to regulate frequency diminishes. Other drawbacks of droop control include poor dynamic performance, such as overshooting and contention. Since the curve is predetermined, it is not possible to achieve optimal power sharing or minimize generation costs using the droop control scheme.

To reduce power generation costs during frequency regulation, an optimization model must be established for the frequency regulation problem. The goal of frequency regulation is to maintain active power balance in an islanded microgrid. To minimize generation costs, the equal ICR criterion must also be satisfied. Therefore, the frequency control problem can be formulated as the following optimization model:

$$\min \frac{1}{2}\left(\sum_{i=1}^{m} P_i - P_D - P_{\text{loss}}\right)^2 \tag{4.13}$$

$$\text{s.t. } ICR_i = ICR_j, \text{ for all } i, j \in \{1, 2, \cdots, m\} \tag{4.14}$$

where P_D denotes the total load demand, and P_{loss} denotes the total power loss.

Power output constraints of DGs and ESSs should also be satisfied. These constraints can be expressed as

$$P_i^{\min} \leq P_i \leq P_i^{\max}, \quad i \in G_{CG} \tag{4.15}$$

$$0 \leq P_j \leq P_j^{\max}, \quad j \in G_{RG} \tag{4.16}$$

$$P_k^{\min} \leq P_k \leq P_k^{\max}, \quad k \in G_{ESS} \tag{4.17}$$

where $P_i^{\min}$ and $P_i^{\max}$ denote the lower and upper bounds of output, and G_{CG}, G_{RG}, and G_{ESS} denote the index sets of CG, RG, and ESS, respectively.

It is important to note that the equality constraint (4.14) may conflict with the inequality constraints (4.15)–(4.17). According to the equal ICR criterion, when the total cost is minimized, some sources will reach their output limits, while the ICRs of the remaining sources will be equal. In the following discussion, the inequality constraints are initially disregarded, focusing only on solving the convex problem defined by (4.13)–(4.14). A method for addressing the inequality constraints will be introduced later.

Use x to denote the common ICR when the generation cost reaches its minimum. According to (4.4), the original model (4.13)–(4.14) can be rearranged to obtain the following unconstrained optimization problem:

$$\min \Phi(x) = \frac{1}{2}\left(\sum_{i=1}^{m} \frac{x - b_i}{2a_i} - P_D - P_{\text{loss}}\right)^2 \tag{4.18}$$

4.4 Secondary frequency regulation based on subgradient method

4.4.1 Subgradient method for secondary frequency regulation

First, we introduce a fully distributed power dispatch and frequency recovery method [35], which combines frequency regulation and economic dispatch in one process.

The objective (4.18) is a convex function, which can be solved efficiently in a distributed manner by using the subgradient algorithm [36–38].

Specifically, according to [37], if $ICR_i(k)$ represents the increment rate of DG i at step k, then $ICR_i(k)$ can be updated according to:

$$ICR_i(k+1) = \sum_{j=1}^{m} \mu_{ij} ICR_j(k) - \alpha \cdot d_i(k) \qquad (4.19)$$

where μ_{ij} is a communication coefficient, α is the step size, and $d_i(k)$ is the subgradient of DG i objective function at step k.

According to (4.18), $d_i(k)$ can be calculated using

$$
\begin{aligned}
d_i(k) &= K \cdot \left(\sum_{i=1}^{m} \frac{ICR_i(k) - b_i}{2a_i} - P_D - P_{\text{loss}} \right) \\
&= K \cdot \left(\sum_{i=1}^{m} P_i(k) - P_D - P_{\text{loss}} \right)
\end{aligned}
\qquad (4.20)
$$

where K is a constant and $P_i(k)$ represents the power output of DG i at step k.

Substituting (4.20) into (4.19) yields

$$ICR_i(k+1) = \sum_{j=1}^{m} \mu_{ij} ICR_j(k) - \alpha \cdot K \cdot \left(\sum_{i=1}^{m} P_i(k) - P_D - P_{loss} \right) \qquad (4.21)$$

Since the mismatch between power supply and demand ($\sum_{i=1}^{m} P_i(k) - P_D - P_{\text{loss}}$) can be represented by the frequency deviation, we have

$$ICR_i(k+1) = \sum_{j=1}^{m} \mu_{ij} ICR_j(k) - \lambda_i \left(f_i^m(k) - f^* \right) \qquad (4.22)$$

where λ_i is determined by α, K, and the inertia of system, $f_i^m(k)$ is the measured frequency and f^* is the nominal frequency value.

In the deviation process of (4.20), the change of P_{loss} is neglected. However, in the final updating formula (4.22), the power mismatch is replaced by frequency deviation. Thus, the change of P_{loss} has been taken into account during the iteration process because it is reflected in frequency deviation.

The power generation adjustment for DG i at step $k+1$ is

$$\Delta P_i(k+1) = \frac{ICR_i(k+1) - b_i}{2a_i} - P_i^m(k+1) \qquad (4.23)$$

where $P_i^m(k+1)$ is the latest measurement value for the power generation of DG i.

Therefore, the power generation adjustment for a renewable energy generator can be rearranged as

$$\Delta P_i(k+1) = P_i^{\max} \frac{ICR_i(k+1)+2}{2} - P_i^m(k+1) \tag{4.24}$$

As discussed earlier, the actual ICRs cannot reach consensus among all DGs, so we introduce "virtual" ICR. During the control procedure, once certain DG reaches its limiting bound, its output will be fixed to that value. However, those fixed DGs will still update their "virtual" ICRs and pass their new "virtual" ICR to neighbors in subsequent iterations. At this time, the "virtual" ICRs of those fixed DGs are not associated with their actual output anymore. This means that the "virtual" ICR is just used to keep message passing procedure uninterrupted and guarantee the subsequent iteration in whole system can be continued. The advantage of "virtual" ICR is that it does not influence convergence because power unbalance is reflected by the measured frequency deviation.

According to [37], the convergence of the distributed subgradient algorithm is affected by following two conditions:

The first condition is that the matrix $\mathbf{A}$ satisfies:

$$\lim_{k \to \infty} \mathbf{A}^k = \begin{pmatrix} \frac{1}{n} & \cdots & \frac{1}{n} \\ \vdots & \ddots & \vdots \\ \frac{1}{n} & \cdots & \frac{1}{n} \end{pmatrix} \tag{4.25}$$

Thus, the coefficient μ_{ij} in matrix $\mathbf{A}$ is critical and requires proper design. The metropolis algorithm is used to set the elements of matrix $\mathbf{A}$, given by the following equation:

$$\mu_{ij} = \begin{cases} \frac{2}{n_i + n_j + 1}, & j \in N_i \\ 1 - \sum_{j \in N_i} \frac{2}{n_i + n_j + 1}, & i = j \\ 0, & \text{otherwise} \end{cases} \tag{4.26}$$

In (4.26), n_i and n_j are the number of agents connected to agent i and agent j, respectively. N_i denotes the set of agents connected to agent i. The matrix $\mathbf{A}$ is a doubly stochastic matrix, so it has the property that all eigenvalues are less or equal to 1, and matrix $\mathbf{A}$ satisfies:

$$\mathbf{A} \cdot \begin{bmatrix} \frac{1}{\sqrt{n}} \\ \vdots \\ \frac{1}{\sqrt{n}} \end{bmatrix}_{n \times 1} = \begin{bmatrix} \frac{1}{\sqrt{n}} \\ \vdots \\ \frac{1}{\sqrt{n}} \end{bmatrix}_{n \times 1} \tag{4.27}$$

According to the property of (4.27), the matrix $\mathbf{A}$ satisfies condition (4.25).

This algorithm has the adaptability to changes in network topology and is fully distributed. It also has satisfactory convergence speed because condition (4.25) is satisfied.

The second condition is that λ_i should be appropriately tuned to guarantee performance. Since the total number of iterations required to reach the final result is inversely proportional to λ_i, λ_i cannot be excessively small. In contrast, too large λ_i

may result in no convergence. Thus, λ_i should be tuned artificially to balance precision of final result with convergence speed. Thus, the convergence of the proposed algorithm is guaranteed.

The frequency of each DG is restored to the nominal value f^* and the ICR of each DG output reaches a common value when the control process converges. As discussed earlier, this control method can achieve minimal total generation cost. Meanwhile, power generation for each RG is proportional to their predicted maximum generation capacity when the system cannot accommodate all renewable power.

4.4.2   Frequency regulation performance

This section presents numerical simulations to illustrate the results of the proposed method. First, a small islanded microgrid system is used to test the method under different scenarios including heavy load, light load, and existing communication delays. For comparison, the performance of control of other methods proposed in by other also tested. Moreover, numerical tests on an actual and a more complicated system are also given.

4.4.2.1   A small-scale islanded microgrid

The simulation system for a microgrid is developed in Simulink. As shown in Figure 4.1, the microgrid contains two CGs, three PVs, one ESS, and two loads. The DGs and ESS are modeled as individual agents who participate in frequency control. The topology of the communication network between agents is the same as the electrical network, meaning that neighbors can participate in peer-to-peer (P2P) communication.

In the simulation system, the nominal value for frequency is 50 Hz and the time interval between two adjacent iterations is 0.1 s, which can be obtained in practice. Important microgrid parameters are listed in Table 4.1.

Three test scenarios are designed to demonstrate the performance of the proposed method. The first two scenarios assume that there is no communication time

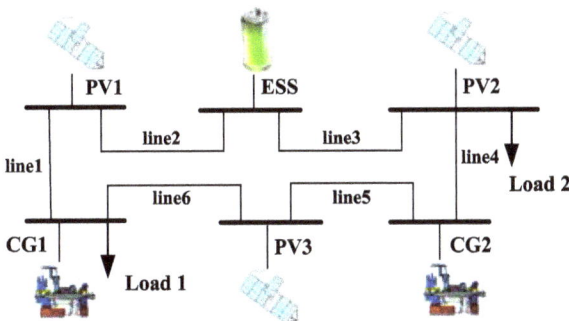

Figure 4.1   The simulation system for microgrid

Table 4.1   Simulation parameters

CG		a_i	b_i	c_i	Minimal output limit	Initial active power	Initial reactive power
	SG1	0.3	5.5	5	0.5 MW	4.2 MW	4Mvar
	SG2	0.2	6.7	5	0.5 MW	3.3 MW	3Mvar
ESS		$\bar{a}=2$	$\underline{a}=1$	5	–	2.5 MW	0
		Predicted maximum generation	Initial active power	Initial reactive power	Load	Initial active power	Initial reactive power
PV	PV1	3 MW	3 MW	1Mvar			
	PV2	3 MW	3 MW	1Mvar	Load1	10 MW	6Mvar
	PV3	2 MW	2 MW	1Mvar	Load2	8 MW	4Mvar
		Line1	Line2	Line3	Line4	Line5	Line6
Line	Resistance	0.23Ω	0.35Ω	0.35Ω	0.23Ω	0.35Ω	0.35Ω
	Inductance	0.318 mH	0.847 mH	0.847 mH	0.318 mH	0.847 mH	0.847 mH

delays. The third scenario considers the influence of communication time delays on the performance of the results.

1. Heavy load demand scenario
    In this case, the load demand is increased suddenly by 3.5 MW at $t = 7s$ from 17.5 MW to 21 MW. The frequency response curve for the proposed control method is shown in Figure 4.2. For comparison, the frequency response curves of two other methods are also shown in Figure 4.2. It can be concluded that the currently proposed method performs better than others at fast frequency recovery.

    As seen from Figure 4.2, following the sudden load power increase, the frequency of the system drops and the distributed frequency control is triggered. The system frequency is then restored to its nominal value (50 Hz) in 2 s after 20 iterations. Comparing the performance of the three methods, it can be demonstrated that the proposed scheme has better performance of frequency recovery speed. Moreover, the maximum deviation of the proposed scheme during the control process is significantly less than those resulting from the other methods. In the proposed method, control signals are executed in each iteration step instead of achieving consensus. This demonstrates that the proposed scheme can achieve fast frequency recovery.

    The DG power generation curves are shown in Figure 4.3. It can be seen that the PV power outputs maintain maximum capacity both before and after the sudden load demand increase. The power imbalance is compensated for by increasing power output of CGs and ESS only. As shown in Figure 4.4, the

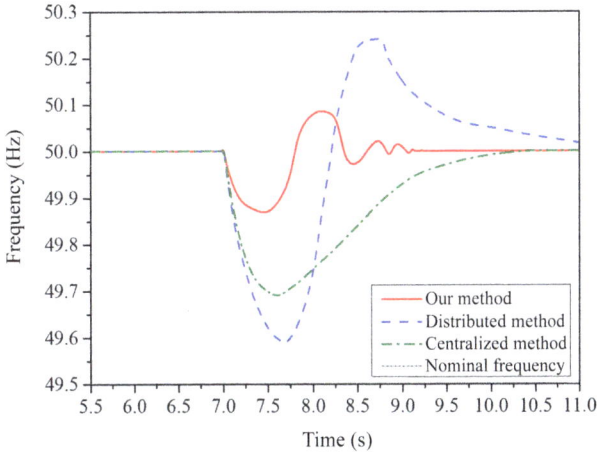

Figure 4.2   Frequency response curve after a sudden increase of load demand for case 1

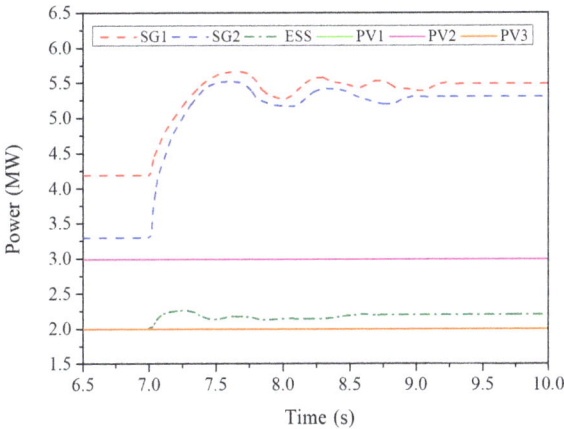

Figure 4.3   Active power outputs of DGs in case 1

incremental rates of the two CGs and ESS reach an identical value eventually, satisfying the equal increment rate criteria.

In this scenario, the total load demand is higher than the sum of the maximum generation capacities of three PVs, implying that all generated power from PVs can be absorbed. According to the analysis above, PVs should reach their maximum output limits, while CGs and ESS are supposed to share active power according to the equal increment rate criteria. The simulation results are consistent with these assumptions.

This generation cost minimization problem can also be formulated as an AC power flow-based optimal economic dispatch (A-ED), in which network losses are precisely incorporated. The final results for distributed frequency control

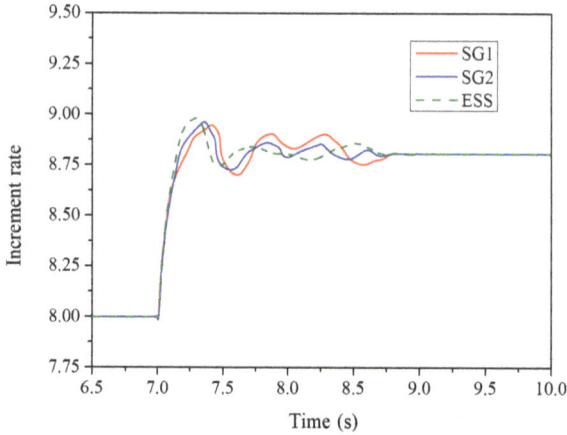

Figure 4.4   Increment rates of CGs and ESS in case 1

Table 4.2   Generation costs under proposed method and A-ED

Case	Actual generation costs	Theoretical optimized generation costs for A-ED
1	50.12	50.12
2	72.04	72.03
3	105.13	105.11

and A-ED are listed in Table 4.2. It can be seen that they are in good agreement with each other. Actual generation costs are slightly larger than those of A-ED, since changing of load demand caused by frequency variation is neglected in A-ED. In conclusion, the proposed distributed control method can achieve minimal generation cost as well as frequency recovery.

2. Light load demand scenario with curtailment of PV generation

In this scenario, the load demand is decreased by 13 MW from 17.5 MW to 4.5 MW suddenly at $t = 7$ s. Under these conditions, the load demand is lower than the total predicted maximum power generation capacities of three PVs and some solar energy should be curtailed.

Figure 4.5 shows the frequency response curves using three different control methods. Caused by the sudden load decrease, the system frequency increases. Similar to case 1, the system frequency is restored to its nominal value (50 Hz) in 3 sec with 30 iterations under proposed method in this paper. By contrast, it takes more time to restore frequency under the distributed method and the centralized method. Furthermore, the maximum frequency deviations in the results of these two methods are significantly larger than that of the proposed scheme during the

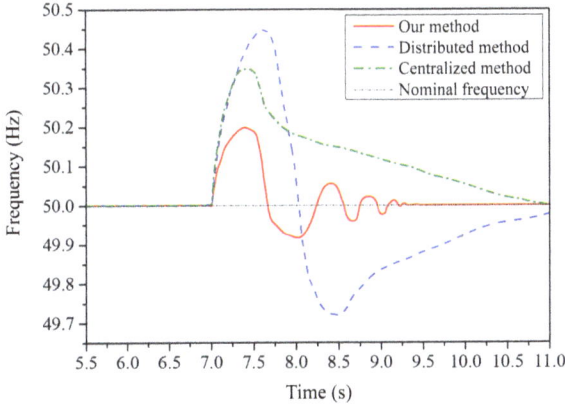

Figure 4.5   *Frequency response curve after a sudden load demand decrease for case 2*

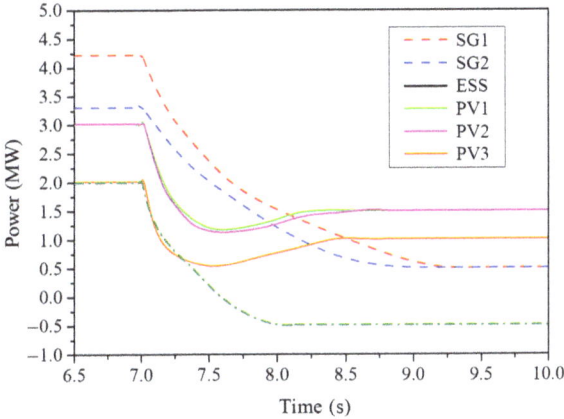

Figure 4.6   Active DG power outputs for case 2

control process. Similar to case 1, the result also demonstrates that the proposed scheme can achieve faster frequency recovery.

The active DG power outputs are shown in Figure 4.6. It can be seen that SG outputs drop to their minimal output limit −0.5 MW, and the ESS absorbs 0.5 MW from the system. In the steady state, the power outputs of the three PVs are 1.5 MW, 1.5 MW, and 1 MW, which are proportional to their maximum power generation capacities.

3.   Operational scenario with communication delays

Communication delay between agents is an inevitable limitation, because it is likely to affect the dynamics of the system. In this test case, three different P2P communication delays are simulated: 0.03 s, 0.07 s, and 0.1 s. Other conditions are identical to those in case 1. Similar to other test scenarios, the time interval between adjacent controls is set to be 0.1 s.

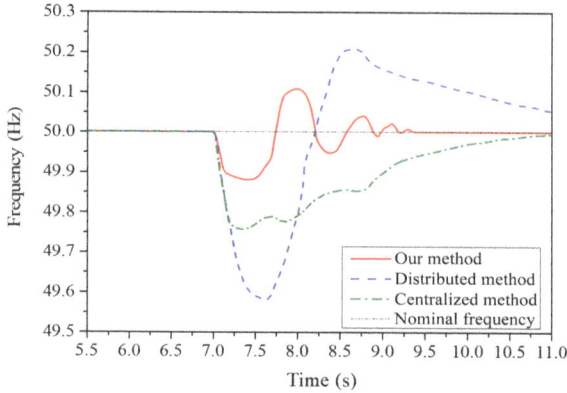

Figure 4.7   Frequency responses for 0.03 s communication delay

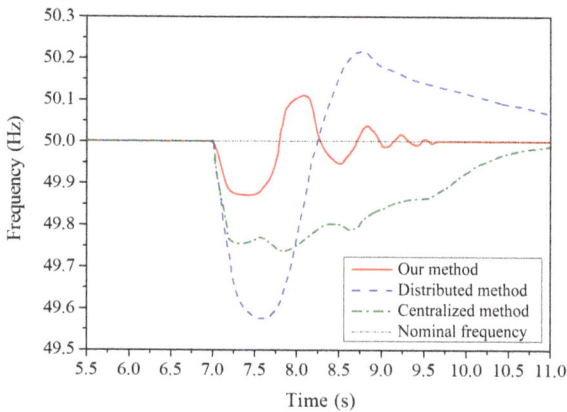

Figure 4.8   Frequency responses for 0.07 s communication delay

For comparison, the frequency response curves for the three methods under different communication time delays are shown in Figure 4.7 to Figure 4.9. As seen in the figures, the proposed method has much better performance on frequency recovery than the other two methods. Especially, when the delay increases to 0.1 s, our proposed method can still recover the frequency in 3 s, while the other two methods need more time. In conclusion, the proposed algorithm shows promising performance in the scenario with communication delays.

4.4.2.2   An actual islanded system

This study case is from an actual islanded industrial power system located in Inner Mongolia of China. The single line diagram of this system is shown in Figure 4.10.

This system consists of eight synchronous generators (thermal power units), two wind power plants, and three aluminum smelting loads.

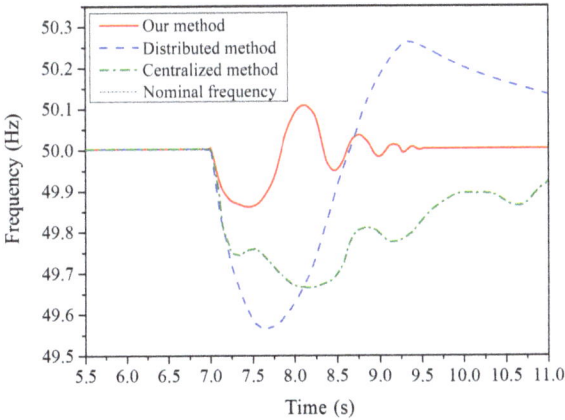

Figure 4.9   Frequency responses for 0.1 s communication delay

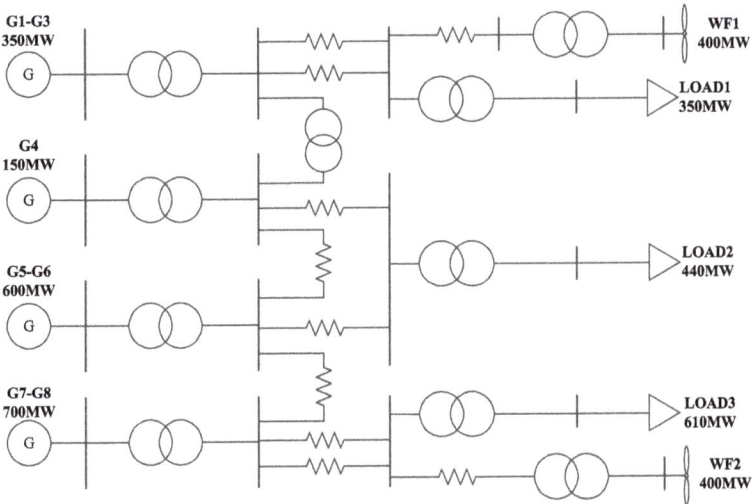

Figure 4.10   Diagram of an actual islanded system in Inner Mongolia

In this system, wind penetration level is as high as 30%, so frequency stability issue is very important. Stochastic wind power outputs may cause significant over/under frequency deviations. Therefore, fast generation control and frequency regulation are vitally important.

The simulation model for this islanded system is developed in Simulink. Synchronous generators are modeled in the same way as those in earlier microgrid cases, but with far larger capacity. As a renewable resource, wind power generation should be utilized as much as possible. For convenience, we assume that wind speed is appropriate so that the maximum available outputs of the two wind power plants are their generation capacity. Moreover, the aluminum smelting loads are regarded as constant loads (350 MW, 440 MW, 610 MW, respectively). The communication network consists of fiber-optic cables and its topology is the same as the electrical network.

The generation costs of distributed frequency control and A-ED for the three test cases with different load demand are listed in Table 4.3. It can be seen that these two generation costs are identical and that the proposed distributed control method can achieve minimal generation cost.

Three test cases are designed to demonstrate the performance of the proposed method in this system.

1. Scenario of A sudden load reduction
    In this scenario, the load demand of Load1 is decreased suddenly by 200 MW from 660 MW to 460 MW suddenly at $t = 5$ s. The frequency response curve for the proposed control method is shown in Figure 4.11. For comparison, the frequency response curves for local droop control and another distributed control are also shown in Figure 4.11.

Table 4.3   Generation costs under proposed method and A-ED

Case	Actual generation costs	Theoretical optimized generation costs for A-ED
1	172.46	172.46
2	188.44	188.44
3	206.31	206.31
4	237.59	237.59

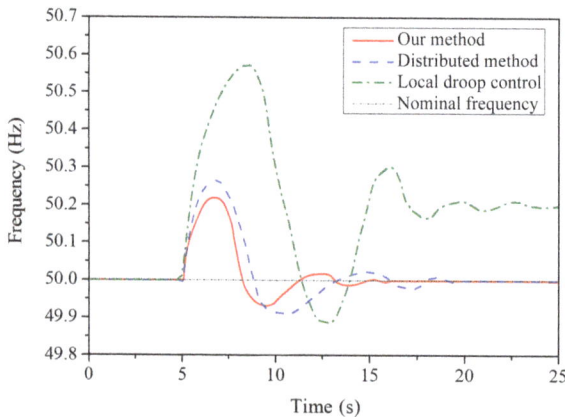

Figure 4.11   Frequency response curve after a sudden decrease of load demand

As can be seen from Figure 4.11, after the sudden decrease of load demand, the frequency of the system increases. The system frequency is then restored to its nominal value (50 Hz) in 10s. Note that frequency cannot be restored to 50 Hz by only local control. And not only that, the maximum frequency deviation under local control is about 0.6 Hz, which cannot be accepted and might cause tripping of devices and cascading failures. By comparing the performance of the three methods, the conclusion is similar to that of the small microgrid. It can be demonstrated that the proposed scheme has better frequency recovery speed performance than the other methods. Moreover, the maximum deviation of the proposed scheme during the control process is significantly less than other methods. This demonstrates that the proposed scheme can achieve faster frequency recovery.

2.   Scenario of a sudden generator tripping
In this scenario, G3 suddenly trips at $t = 5$ sec. To maintain the balance of power supply and demand, outputs of the other generators must increase.

Figure 4.12 shows the frequency response curves using three different control methods. Caused by the sudden load decreasing, the system frequency increases. Similar to case 1, the system frequency is restored to its nominal value (50 Hz) in 10 sec under proposed method in this chapter. In contrast, it takes more time to restore frequency under the distributed method and local control. Furthermore, the maximum frequency deviations in the results of these two methods are significantly larger than that of the proposed scheme during the control process. Similar to that in the small microgrid, the result also demonstrates that the proposed scheme can achieve faster frequency recovery.

3.   Operational scenario with communication delays
In this test case, the communication delay between two neighbors is assumed to be 0.1 s. Other conditions are identical to those in case 1. For comparison, the

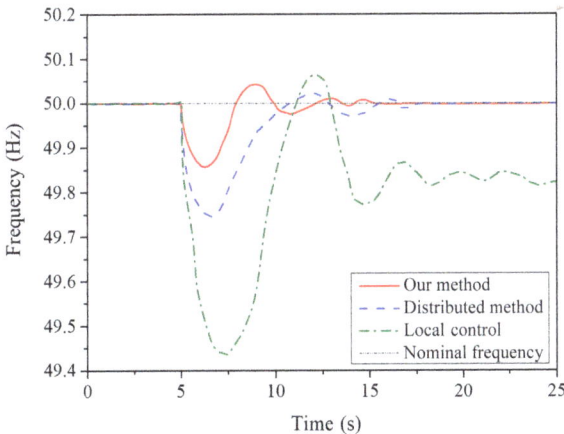

Figure 4.12   Frequency response curve after a sudden tripping

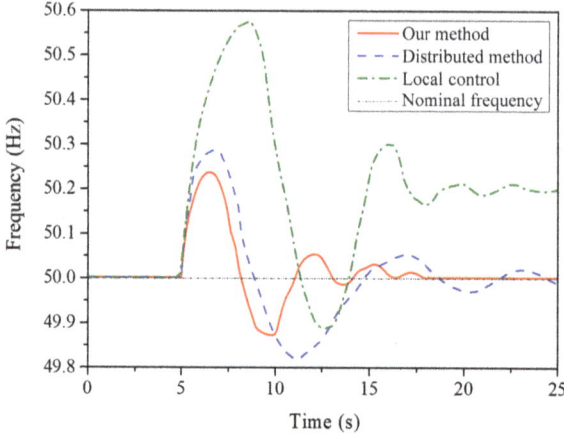

Figure 4.13   Frequency response curve under communication delay

frequency response curves of the three methods under different communication time delays are shown in Figure 4.13.

From Figure 4.13, we can see that the proposed algorithm achieves promising performance in the scenario with communication delays. This conclusion is the same as that attained in the small microgrid.

4.5   Primary frequency regulation based on Newton method

This section provides a summary of an effective method for primary frequency regulation based on the Newton method, developed in [39]. This summary includes key technical details on the optimization model, on Newton method, as well as a discussion about relevant challenges and solutions. The interested readers are referred to [39] for details and an extensive simulation study.

Transformation of the optimization model:
The optimization problem for primary frequency regulation is formulated as minimizing the cost function

$$\sum_{i=1}^{m} \Phi_i(x),$$

where each $\Phi_i(x)$ represents a cost function associated with each agent. Initially, the problem is simplified by ignoring inequality constraints and solving a convex problem involving an equality constraint. The problem can be solved in a distributed manner by using a distributed subgradient method, where each agent updates its estimate of the optimal solution based on its neighbors' estimates and its own gradient information, thus using a formula that involves gradients of local cost functions and a matrix **A** representing network connections. A penalty term for discrepancies among agents' estimates is included. However, this method exhibits slow linear convergence, making it less suitable for fast primary control applications but appropriate for secondary control.

Therefore, a more advanced technique based on distributed Newton method is proposed. This method is known for its second-order convergence properties, making it suitable for faster and more efficient optimization in the context of primary frequency control.

Newton method for primary frequency regulation:
The Newton method is applied to improve primary frequency regulation. The Newton method's iteration formula is introduced, where the Newton direction is computed using the inverse of the Hessian matrix

$$\mathbf{x}(k + 1) = \mathbf{x}(k) - \lambda \mathbf{d}_k,$$

where $\mathbf{d}_k$ is the Newton direction, computed as

$$\mathbf{d}_k = \mathbf{H}_k^{-1} \mathbf{g}_k.$$

Here, $\mathbf{H}_k$ is the Hessian matrix, and $\mathbf{g}_k$ is the gradient of the objective function.

The challenge lies in calculating the Hessian matrix's inverse in a distributed manner due to its dependency on global information. To overcome this, the matrix is decomposed using a matrix splitting technique, allowing for local iterations.

Challenges and Solutions:
As mentioned above, the main challenge with distributed Newton methods is computing $\mathbf{H}_k^{-1}$ locally, which requires global information. To address this, the Hessian matrix $\mathbf{H}_k$ is split into two matrices ($\mathbf{D}_k$ and $\mathbf{B}$) with a similar sparsity pattern as $\mathbf{A}$. This allows the inversion to be approximated using a Taylor series expansion and implemented in a distributed manner, enabling the computation of the Newton direction iteratively. The recursion formula for updating the Newton direction, $\mathbf{d}_k^{(t+1)}$, is used and can be computed locally.

The steps an agent follows during each iteration are outlined next.

Local computation steps:

1. **Receive neighbor information:** Each agent shares and receives data with its neighbors.
2. **Calculate local parameters:** Compute local components like $D_{ii,k}^{-1}$ and gradient $g_{i,k}$.
3. **Iterate to compute direction:** Update the Newton direction using a recursive formula.
4. **Update solution:** Apply the computed direction to update the local estimate of the optimal solution.

Practical implementation and results:
The simulation study considers a low-voltage AC islanded microgrid model with multiple power sources and loads, along with the communication links between intelligent agents [39]. Results show that the developed Newton method stabilizes frequency fast and more accurately, even with sudden load increases or communication delays. It also handles communication failures effectively, maintaining control

performance by relying on available neighbor information. The overall method shows a super-linear convergence rate, making it a robust and fast solution for droop-free frequency control.

Furthermore, the proposed Newton method is shown to perform well compared to other methods, including droop control and the subgradient method:

- **Speed and accuracy:** The Newton method outperforms droop control and subgradient methods in terms of speed and accuracy for primary frequency regulation.
- **Generation costs:** The costs associated with the Newton method are close to optimal values, demonstrating its effectiveness in cost minimization.

To achieve optimal frequency control, the Newton method can be combined with a secondary control method. The Newton method provides rapid stabilization, while the secondary control, such as the subgradient method, can restore the frequency to its nominal value.

In conclusion, the developed Newton method offers a robust and efficient solution for primary frequency regulation, addressing the limitations of slower methods and handling practical challenges such as communication delays and failures effectively.

4.6 Conclusions

For bulk power system, a three-level hierarchical control scheme is widely used, in which the tertiary control (called economic dispatch) is to minimize the operating cost and the others maintain system safety. However, the economic dispatch is impractical for microgrid due to the significant prediction errors of renewable energy. Therefore, the secondary and primary frequency control should consider both the operating cost and the safety. In this chapter, we presented a fully distributed and economic secondary and primary frequency control schemes for islanded microgrids. For the proposed secondary frequency control solution, it can recover frequency and achieve minimal generation costs by the distributed subgradient-based consensus algorithm in combination with the equal increment rate criteria. Meanwhile, the power generation of RGs is proportional to their maximal generation capacity when the system cannot absorb all generated renewable power. Moreover, the control method can suppress the oscillation phenomena caused by measurement errors by iterative switching of the subgradient-based consensus algorithm and the classical consensus algorithm in different stages. For the primary frequency control, we developed a distributed Newton method as a substitute for droop control. Due to the super-linear convergence, faster power balance is achieved when using our method. The simulation results indicate that the proposed method has very good dynamic performance in a variety of conditions. In particular, the method is insensitive to measurement errors and time delays, and it effectively avoids overshooting.

References

[1]   Lasseter RH. MicroGrids. In: *2002 IEEE Power Engineering Society Winter Meeting*. vol. 1; 2002. p. 305–308.

[2]   Lasseter RH. Smart distribution: coupled microgrids. *Proceedings of the IEEE*. 2011;99(6):1074–1082.

[3]   Hatziargyriou N, Asano H, Iravani R, *et al.* Microgrids. *IEEE Power and Energy Magazine*. 2007;5(4):78–94.

[4]   Beer S, Gomez T, Dallinger D, *et al.* An economic analysis of used electric vehicle batteries integrated into commercial building microgrids. *IEEE Transactions on Smart Grid*. 2012;3(1):517–525.

[5]   Venkataramanan G, and Marnay C. A larger role for microgrids. *IEEE Power and Energy Magazine*. 2008;6(3):78–82.

[6]   Lopes JAP, Moreira CL, and Madureira AG. Defining control strategies for microGrids islanded operation. *IEEE Transactions on Power Systems*. 2006;21(2):916–924.

[7]   Skea J, Anderson D, Green T, *et al.* Intermittent renewable generation and the cost of maintaining power system reliability. *IET Generation, Transmission & Distribution*. 2008;2(1):82–89.

[8]   Bidram A, and Davoudi A. Hierarchical structure of microgrids control system. *IEEE Transactions on Smart Grid*. 2012;3(4):1963–1976.

[9]   Vasquez JC, Guerrero JM, Miret J, *et al.* Hierarchical control of intelligent microgrids. *IEEE Industrial Electronics Magazine*. 2010;4(4):23–29.

[10] Olivares DE, Mehrizi-Sani A, Etemadi AH, *et al.* Trends in microgrid control. *IEEE Transactions on Smart Grid*. 2014;5(4):1905–1919.

[11] Rocabert J, Luna A, Blaabjerg F, *et al.* Control of power converters in AC microgrids. *IEEE Transactions on Power Electronics*. 2012;27(11): 4734–4749.

[12] Wu D, Tang F, Dragicevic T, *et al.* Autonomous active power control for islanded AC microgrids with photovoltaic generation and energy storage system. *IEEE Transactions on Energy Conversion*. 2014;29(4):882–892.

[13] Kanchev H, Lu D, Colas F, *et al.* Energy management and operational planning of a microgrid with a PV-based active generator for smart grid applications. *IEEE Transactions on Industrial Electronics*. 2011;58(10):4583–4592.

[14] Haddadi A, Yazdani A, Joós G, *et al.* A gain-scheduled decoupling control strategy for enhanced transient performance and stability of an islanded active distribution network. *IEEE Transactions on Power Delivery*. 2014;29(2): 560–569.

[15] Kim JW, Choi HS, and Cho BH. A novel droop method for converter parallel operation. *IEEE Transactions on Power Electronics*. 2002;17(1):25–32.

[16] Etemadi AH, Davison EJ, and Iravani R. A decentralized robust control strategy for multi-DER microgrids—Part I: Fundamental concepts. *IEEE Transactions on Power Delivery*. 2012;27(4):1843–1853.

[17] Mohamed YARI, and El-Saadany EF. Adaptive decentralized droop controller to preserve power sharing stability of paralleled inverters in

distributed generation microgrids. *IEEE Transactions on Power Electronics*. 2008;23(6):2806–2816.

[18]   Ashabani SM, and Mohamed YARI. General interface for power management of micro-grids using nonlinear cooperative droop control. *IEEE Transactions on Power Systems*. 2013;28(3):2929–2941.

[19]   Hu J, Zhu J, Dorrell DG, *et al.* Virtual flux droop method–A new control strategy of inverters in microgrids. *IEEE Transactions on Power Electronics*. 2014;29(9):4704–4711.

[20]   Nasirian V, Shafiee Q, Guerrero JM, *et al.* Droop-free team-oriented control for AC distribution systems. In: *2015 IEEE Applied Power Electronics Conference and Exposition (APEC)*; 2015. p. 2911–2918.

[21]   Nasirian V, Shafiee Q, Guerrero JM, *et al.* Droop-free distributed control for AC microgrids. *IEEE Transactions on Power Electronics*. 2016;31(2): 1600–1617.

[22]   Tsikalakis AG, and Hatziargyriou ND. Centralized control for optimizing microgrids operation. *IEEE Transactions on Energy Conversion*. 2008;23(1):241–248.

[23]   Amin SM. Smart grid security, privacy, and resilient architectures: Opportunities and challenges. In: *2012 IEEE Power and Energy Society General Meeting*; 2012. p. 1–2.

[24]   Dimeas A, and Hatziargyriou N. A multiagent system for microgrids, *IEEE Power Engineering Society General Meeting*. 2004;1:55–58.

[25]   Loh PC, Li D, Chai YK, *et al.* Autonomous operation of hybrid microgrid with AC and DC subgrids. *IEEE Transactions on Power Electronics*. 2013;28(5):2214–2223.

[26]   Shafiee Q, Vasquez JC, and Guerrero JM. Distributed secondary control for islanded MicroGrids – a networked control systems approach. In: *IECON 2012 – 38th Annual Conference on IEEE Industrial Electronics Society*; 2012. p. 5637–5642.

[27]   Shafiee Q, Guerrero JM, and Vasquez JC. Distributed secondary control for islanded microgrids—a novel approach. *IEEE Transactions on Power Electronics*. 2014;29(2):1018–1031.

[28]   Marwali MN, and Keyhani A. *Control of distributed generation systems— Part I: Voltages and currents control. IEEE Transactions on Power Electronics*. 2004;19(6):1541–1550.

[29]   Zhang Y, and Ma H. Theoretical and experimental investigation of networked control for parallel operation of inverters. *IEEE Transactions on Industrial Electronics*. 2012;59(4):1961–1970.

[30]   Bidram A, Davoudi A, Lewis FL, *et al.* Secondary control of microgrids based on distributed cooperative control of multi-agent systems. *IET Generation, Transmission & Distribution*. 2013;7(8):822–831.

[31]   Dominguez-Garcia AD, Hadjicostis CN, and Vaidya NH. Resilient networked control of distributed energy resources. *IEEE Journal on Selected Areas in Communications*. 2012;30(6):1137–1148.

[32]   Simpson-Porco JW, Dörfler F, and Bullo F. Synchronization and power sharing for droop-controlled inverters in islanded microgrids. *Automatica.* 2013;49(9):2603–2611.

[33]   Dörfler F, Simpson-Porco JW, and Bullo F. Breaking the hierarchy: distributed control and economic optimality in microgrids. *IEEE Transactions on Control of Network Systems.* 2016;3(3):241–253.

[34]   Stevenson WD. *Elements of power system analysis.* 4th ed. McGraw-Hill; 1982.

[35]   Wang Z, Wu W, and Zhang B. A fully distributed power dispatch method for fast frequency recovery and minimal generation cost in autonomous microgrids. *IEEE Transactions on Smart Grid.* 2016;7(1):19–31.

[36]   Xiao L, and Boyd S. Optimal scaling of a gradient method for distributed resource allocation. *Journal of Optimization Theory and Applications.* 2006;129:469–488.

[37]   Nedic A, and Ozdaglar A. Distributed subgradient methods for multi-agent optimization. *IEEE Transactions on Automatic Control.* 2009;54(1):48–61.

[38]   Lobel I, and Ozdaglar A. Distributed subgradient methods for convex optimization over random networks. *IEEE Transactions on Automatic Control.* 2011;56(6):1291–1306.

[39]   Wang Z, Wu W, and Zhang B. A distributed quasi-Newton method for droop-free primary frequency control in autonomous microgrids. *IEEE Transactions on Smart Grid.* 2018;9(3):2214–2223.

Part III

Electric energy management

Chapter 5

Distributed optimization for energy grids: a tutorial on ADMM and ALADIN

Lukas Lanza[1], Timm Faulwasser[2] and Karl Worthmann[1]

The ongoing transition toward energy and power systems dominated by renewable power injections to the distribution grid poses severe challenges for system operation, coordination, and control. Optimization-based methods for coordination and control are of substantial research interest in this context. In this regard, we provide a tutorial introduction of distributed optimization algorithms for energy systems with a large share of renewables, such as, e.g., microgrids in residential areas. Specifically, we focus on the Alternating Direction Method of Multipliers (ADMM) and on the Augmented Lagrangian Alternating Direction Inexact Newton (ALADIN) method as both algorithms are frequently considered for coordination and control of power and energy systems. Exemplarily, we discuss the application of ALADIN and ADMM to AC optimal power flow and to energy management in microgrids. Moreover, we give an outlook on open problems.

5.1 Introduction

The ongoing energy transition toward power and energy systems involving a large number of distributed resources and a rapidly growing share of volatile renewable energy production induces multiple challenges for operation, coordination, and control of such systems. Traditionally, the flow of electrical power is mostly unidirectional, i.e., from the large fossil power plants feeding into transmission systems to the customers supplied by the distribution grids. Networks with increasing numbers of distributed resources (prosumers, battery storage, and controllable renewable generation) exhibit bidirectional flow of power and the established coordination mechanisms between transmission and distribution systems have to be rethought, see, e.g., [1,2] and references therein. We discuss solution concepts for the described situation from a mathematical point of view. In particular, we focus on coordination and control via numerical optimization.

[1]Optimization-based Control Group, Institute of Mathematics, Technische Universität Ilmenau, Germany
[2]Institute of Control Systems, Hamburg University of Technology, Germany This work has been conducted while Timm Faulwasser was with Institute of Energy Systems, Energy Efficiency and Energy Economics, Technical University Dortmund, Germany.

We consider *distributed optimization algorithms* as these methods allow to tackle the problem of coordinating a large number of individual systems while avoiding the clustering of all information in a single centralized entity. Specifically, we discuss the Alternating Direction Method of Multipliers (ADMM) and the Augmented Lagrangian Alternating Direction Inexact Newton (ALADIN) method as both are frequently considered for power and energy systems. An alternative route would be complexity reduction via model reduction, see, e.g., [3,4]. However, this typically results in sub-optimal performance. Moreover, in the context of frequency and voltage stabilization, also the design of distributed sub-optimal feedback laws is considered [5–7].

The remainder is structured as follows: In Section 5.2, we explain the conceptual idea of distributed optimization including a brief exposition of ADMM and ALADIN. The focus of Section 5.3 is put on two illustrative use cases for distributed optimization in energy systems: energy management and AC Optimal Power Flow (AC-OPF) before we conclude with an outlook on open problems in Section 5.4.

Notation: We use $\langle \cdot, \cdot \rangle$ as the scalar product on $\mathbb{R}^n$; $\| \cdot \|_p$ denotes the p-norm in $\mathbb{R}^n$, and for brevity, we use $\| \cdot \| := \| \cdot \|_2$. Moreover, $\| \cdot \|_Q^2 := \langle \cdot, Q \cdot \rangle$ for a symmetric positive-definite matrix Q. Further, $I_n := \mathrm{diag}(1, \ldots, 1) \in \mathbb{R}^{n \times n}$ denotes the identity matrix of dimension $n \in \mathbb{N}$. In the optimization problems, we write the Lagrangian multiplier after the constraints, i.e., "S $|\lambda$" means that λ is the Lagrange multiplier for the constraint S. For $p \in \mathbb{N}$ the space of p-times continuously differentiable functions on $V \subset \mathbb{R}^m$ with image in $\mathbb{R}^n$ is denoted by $C^p(V, \mathbb{R}^n)$.

5.2   Distributed optimization

Optimization aims at finding solutions which are superior to other solutions in a particular sense. The notion of optimality is defined by asking a pre-assigned performance criterion to be minimal (or maximal). In the context of energy systems and power supply, typical optimality criteria are related to transmission losses [8,9], generation costs [10], efficiency [11], incorporation of small-scale renewables and storage devices [12–14], building control [15], or voltage stability [16,17], see also [18] as well as the surveys [19,20] and references therein. A common requirement for power grid operation is that line losses are minimized and, simultaneously, as much as possible energy from residential photovoltaic generation is used while local storage devices are limited by capacity and charging constraints. Therefore, the resulting optimization problems are large scale and of high complexity with many individual, potentially conflicting cost functions and constraints [21] and the surveys [22,23] for multi-objective optimization in general. To solve these problems efficiently, it is necessary to investigate the structure and to consider, e.g., whether the problems can be decomposed into subproblems to be solved in parallel. For the sake of clarity, we briefly state our notion of decentralized and distributed optimization. Following [24], in decentralized optimization, the overall problem can be split into subproblems, which are coupled such that each subproblem can be solved separately while only communicating its solution to the *coupled* subproblems, see, e.g., Problem (5.2) below. The subproblems communicate with each other directly, in particular, no extra communication entity is involved. In distributed optimization, the overall problem can be split into subproblems, which are coupled with each other via constraints or via the cost functions, see e.g., Problem (5.3). Each subproblem subject to the connecting constraints is solved individually. A central entity, e.g., a market maker [10,25], communicates the local solutions and solves, e.g., a consensus optimization problem based on the local solutions. With reference to [20, Section I], we emphasize that the above terms are used ambiguously.

We highlight two main aspects leading to distribution of optimization problems. First, the computational tasks to be solved can be distributed. This means that a large-scale optimization problem is split into several small-scale subproblems, where each subproblem is amenable to optimization algorithms individually. Therefore, the computation task is parallelizable, which may save computation time. The parallelization may even be necessary to render the problem solvable at all, as the overall problem would have, e.g., required too much memory. A concise overview of distributed optimization with regard to energy systems, can be found, e.g., in [20,26]. Since the overall optimization problem typically does not consist of completely decoupled subproblems, the artificially split subproblems have to be equipped with coupling constraints to obtain an equivalent problem, cf. [27–30]. This idea is discussed in Section 5.2.1. The second main aspect of distributing and, thus, decomposing the optimization problem is linked to data security and sovereignty. Solving the overall problem as a whole, the computation entity must have access to all constraints, all objectives, and all states; in particular, all local information has to be shared. If, however, the overall problem is split into several subproblems, which can be solved individually in parallel, only little information, e.g., local optimal values, has to be shared with a central entity since the latter only has to solve a consensus problem, see, for example, the works [20,31,32]. This means that privacy is maintained to a much greater extent. Another possible option to take care of data security is to encrypt local information, share it with the computation entity which operates on the encrypted data, receive a still encrypted solution, locally decrypt the data and apply the solution, cf. [33,34]. Although these techniques are subject of ongoing research, they are still quite limited in terms of the complexity of the computational tasks.

Besides parallel computation and higher level of data privacy, a distributed optimization problem should have the following properties: The problem allows plug-and-play, i.e., including new components (e.g., PV or storage devices) does not change the overall problem, but only few modifications are required, cf. [35–38]. Moreover, the solution techniques are scalable in the sense that the same algorithms are applicable if the problem is extended, e.g. by medium-scale energy generation. Furthermore, simple manipulations such as utilizing averaged quantities allow to keep the number of unknowns [39] and the communication effort [40] independent of the number of subsystems.

Distributed optimization aims at separating an large-scale and/or structured problem into several small-scale subproblems connected via coupling constraints. This typically leads to partially separable problems of the form

$$\min_{x} \sum_{i=1}^{N} k_i(x_i) \text{ subject to } \begin{cases} \sum_{i=1}^{N} A_i x_i = b, \\ h_i(x_i) \le 0, \qquad i \in \{1, \dots, N\}, \end{cases} \tag{5.1}$$

where $k_i : \mathbb{R}^n \rightarrow \mathbb{R}$, $h_i : \mathbb{R}^n \rightarrow \mathbb{R}^{n_h}$, $i \in \{1, \ldots, N\}$ are to be specified in the respective context, e.g., their regularity and convexity properties. The matrices $A_i \in \mathbb{R}^{m \times n}$ and the vector $b \in \mathbb{R}^m$ couple the subproblems.

Before we present two approaches to solve optimization problems (5.1) in Section 5.2.2, namely the algorithms ADMM and ALADIN, we first illustrate simple and briefly the idea of decomposing optimization problems.

5.2.1 Decomposition techniques

In this section, we briefly discuss two decomposition strategies, namely decomposition in primal variables, and dual decomposition. The underlying idea is to split a complex optimization problem into more simple subproblems and a connecting master problem. Roughly speaking the main difference is that decomposition in primal variables leads to subproblems where the connecting variables are primal variables, while dual decomposition yields subproblems with duals as connecting variables. The difference can be illustrated as follows: Consider the case where usage of several resources is coupled in a complex way. Decomposition in primal variables yields a set of subproblems where the connecting master problem directly controls the resources. In contrast, via dual decomposition the master problem controls the resources' prices, which correspond to the Lagrange multipliers.

5.2.1.1 Decomposition in primal variables

Consider a structured optimization problem

$$\min_{x,y,z} K(x, y, z) = k_1(x, z) + k_2(y, z), \tag{5.2}$$

where x, y are local variables, and z are coupling variables. To solve (5.2), fix z and simultaneously solve the subproblems

$$\min_x k_1(x, z) \quad \text{and} \quad \min_y k_2(y, z),$$

and obtain optimal $k_1^\star(z)$, $k_2^\star(z)$. Problem (5.2) is then equivalent to the *primal master problem*

$$\min_z k_1^\star(z) + k_2^\star(z),$$

with coupling variable z. Each iteration of solving the primal master problem requires solving the subproblems. Note that, if the original problem (5.2) is convex, then the primal master problem is convex.

The decomposition in primal variables is reminiscent of Bellman's Dynamic Programming. Indeed, structured decomposition of path and tree-structured problems can be achieved via decomposition in primal variables and recursion (which alleviates the need for a master problem), see, e.g., [41] and [42,43] and references therein.

5.2.1.2 Dual decomposition

If the optimization problem is subject to further constraints, the decomposition can be applied to the dual problem. Originally proposed in [44], the idea is to include the constraints via Lagrange multipliers and to solve a dual ascent problem.

Consider

$$\min_{x,y,z_1,z_2} K(x, y, z_1, z_2) = k_1(x, z_1) + k_2(y, z_2), \quad \text{subject to } z_1 = z_2, \tag{5.3}$$

with local variables x, y, and coupling variables z_1, z_2 to satisfy so-called consensus. We define the corresponding Lagrangian function

$$L(x, y, z_1, z_2, \lambda) := K(x, y, z_1, z_2) + \lambda(z_1 - z_2),$$

where λ is the so-called Lagrange multiplier. Denoting $p^\star := \inf_{x,y,z} k_1(x, z) + k_2(y, z)$ we may estimate with the dual function

$$D(\lambda) := \min_{x,z} (k_1(x, z) - \lambda z) + \min_{y,z} (k_2(y, z) + \lambda z) \leq p^\star$$

a lower bound for $p^\star$ for all λ. Then, the dual problem

$$\max_{\lambda} D(\lambda) = \max_{\lambda} \left\{ \min_{x,z} (k_1(x, z) + \lambda z) + \min_{y,z} (k_2(y, z) - \lambda z) \right\}$$

provides the tightest lower bound for $p^\star$. Note that if the primal problem (5.3) is convex and feasible, then $D(\lambda^\star) = p^\star$. However, usually in practice the so-called duality gap $D(\lambda^\star) < p^\star$ exists, even for optimal dual $\lambda^\star$, that is, sub-optimality of solutions. Moreover, if $z_1^\star \neq z_2^\star$, then it is not straightforward to recover the primal optimal $(x^\star, y^\star, z_1^\star, z_2^\star)$.

The method of dual decomposition has been applied to various optimization problems in systems and control, see, e.g., [27,30,45,46].

5.2.2 ADMM and ALADIN

In this section, we present and discuss two algorithms, namely ADMM and ALADIN, to solve distributed optimization problems (5.1), where k_i, h_i are local costs and constraints, i.e., dependence is only on the local variables x_i. The matrices $A_i \in \mathbb{R}^{m \times n}$ and the vector $b \in \mathbb{R}^m$ are used to model the coupling between subsystems. Regularity and convexity requirements of k_i, h_i for each algorithm will be discussed in the next sections.

5.2.2.1 ADMM

The acronym ADMM refers to the Alternating Direction Method of Multipliers, originally proposed in [47,48]. It is a further development of dual decomposition. The idea can be illustrated as follows. For $k_i : \mathscr{D} \to \mathbb{R} \cup \{+\infty\}$, $i = 1, 2$ consider the simplified problem with consensus constraints

$$\min_{z_1, z_2 \in \mathscr{D}} K(z_1, z_2) = k_1(z_1) + k_2(z_2), \quad \text{subject to } z_1 = z_2. \tag{5.4}$$

This problem is solved approximately by alternating computing z_1 for z_2 fixed, and then calculating z_2 for z_1 fixed. For a regularization (augmentation) parameter $\mu > 0$ we define the augmented Lagrangian

$$L_\mu(z_1, z_2, \lambda) := k_1(z_1) + k_2(z_2) + \langle z_1 - z_2, \lambda \rangle + \tfrac{\mu}{2}\|z_1 - z_2\|^2. \tag{5.5}$$

Then the ADMM algorithm essentially repeats the following steps

$$z_1^k = \arg\min_{z_1} L(z_1, z_2^{k-1}, \lambda^{k-1}),$$

$$z_2^k = \arg\min_{z_2} L(z_1^k, z_2, \lambda^{k-1}),$$

$$\lambda^k = \lambda^{k-1} + \mu(z_1^k - z_2^k),$$

and so approaches the optimal solution, see also [49, Algorithm 2] for a concise presentation of the algorithm. To solve the more general problem (5.1), we recall in Algorithm 1 below the version of ADMM from [29]. In [50, Section 3.2, Appendix A] it has been proven for separable problems (5.4) with closed, proper, and convex objective functions k_1, k_2, under a saddle-point condition on the unaugmented Lagrangian ($\mu = 0$), that the iterates are feasible (($z_1^n - z_2^n)_{n\in\mathbb{N}} \to 0$ as $n \to \infty$), the costs converge to an optimal value (($K(z_1^n, z_2^n, \lambda^n))_{n\in\mathbb{N}} \to K^\star$ as $n \to \infty$), and that the duals are optimal too ($\lambda^n \to \lambda^\star$ as $n \to \infty$). ADMM (Algorithm 1) has been successfully applied to optimization problems arising in energy systems, see, e.g., [39,51,52]. Moreover, one should note that ADMM as such is rather a family of algorithms than a single rule, cf. the different expositions and variants in [50,53,54].

Algorithm 1: ADMM in virtue of [29]

Input: Initial guess $x_i \in \mathbb{R}^n$ and $\lambda_i \in \mathbb{R}^m$, penalty parameter $\rho > 0$, and numerical tolerance $\varepsilon > 0$

Repeat:

1 Solve for $i \in \{1, \dots, N\}$ the decoupled problems

$$\min_{y_i} k_i(y_i) + \langle A_i y_i, \lambda \rangle + \tfrac{\rho}{2}\|A(y_i - x_i)\|^2 \text{ subject to } h_i(y_i) \leq 0. \quad \text{(NLP)}$$

2 If $\|\sum_{i=1}^N A_i y_i - b\|_1 \leq \varepsilon$, then terminate with $x^\star = y$.

3 Implement the dual gradient steps $\lambda_i^+ = \lambda_i + \rho A_i(y_i - x_i)$.

4 Solve the coupled QP

$$\min_{x^+} \sum_{i=1}^N \frac{\rho}{2}\|A_i(y_i - x_i^+)\|^2 + \langle A_i x_i^+, \lambda_i^+ \rangle \text{ subject to } \sum_{i=1}^N A_i x_i^+ = b. \quad \text{(QP)}$$

5 Update the iterates $x \leftarrow x^+$ and $\lambda \leftarrow \lambda^+$, and go to Step 1.

Output: Numerical optimal solution $x^\star$

5.2.2.2 ALADIN

The acronym ALADIN stands for Augmented Lagrangian Alternating Direction Inexact Newton. First proposed in [29], this algorithm is designed to solve non-convex problems (5.1), where we assume regularity of the costs $k_i \in C^2(\mathbb{R}^n; \mathbb{R})$ as well as for the constraints $h_i \in C^2(\mathbb{R}^n; \mathbb{R}^{n_h})$ for all $i \in \{1, \dots, N\}$. Furthermore, we assume that (5.1) is feasible and all local minimizers are regular KKT points. Note that, in contrast to ADMM, convexity of the functions k_i, h_i is not required. Importantly, the functions k_i, h_i take only x_i, this means, the problem is separable into N subproblems, which are coupled via the consensus constraint $[A_1, \dots, A_N]x = b$. Similar to ADMM, the idea is to decompose (5.1) into subproblems, which can be solved in parallel, where the coupling constraints are formulated as costs. The first two steps of Algorithm 2 (ALADIN) solve the decoupled subproblems in primal variables analogous to Algorithm 1 (ADMM). In Step 3, an approximation of the constraint Jacobian is obtained. The latter is similar to inexact SQP methods and it allows to cope with inexact constraint Jacobian and resulting errors in the step direction. Moreover, in contrast to ADMM, an approximation of the Hessian of $k_i(x_i) + \langle h_i(x_i), \kappa_i \rangle$ is required to solve the coupling problem (QP) in Step 4. For the update of x and λ in Step 5 the line search parameters can be set $\alpha_1 = \alpha_2 = \alpha_3 = 1$, or can be computed according to [29, Algorithm 3] to obtain global convergence guarantees. Algorithm 2 summarizes the ALADIN as proposed by [29].

5.2.2.3 Comparing ADMM and ALADIN

In both algorithms, Algorithm 1 (ADMM) and Algorithm 2 (ALADIN), the N subproblems (NLP) are nonlinear, but compared to the overall problem of small scale. The coupling problems are of large scale, but only linear equality constraints are involved. Moreover, due to the separated structure of (NLP), no detailed information about the costs k_i, the constraints h_i and the coupling A_i have to be shared to the central entity. For $\rho = 0$ the method in Algorithm 2 is equivalent to dual decomposition, see [29, Appendix A] for detailed analysis. For $H_i = \rho A_i^\top A_i$, $\Sigma_i = A_i^\top A_i$, $C_i = 0$, and large values of μ ($\mu \to \infty$) similarity between Algorithm 1 and Algorithm 2 is in detail discussed in [29, Section 5].

Unlike ADMM, ALADIN can cope with non-convex costs k_i as well as with non-convex constraints h_i, cf. [29]. Invoking an L_1-penalty function, it is possible to choose $\alpha_1, \alpha_2, \alpha_3$, cf. [29, Algorithm 3], such that Algorithm 2 converges globally to local minimizers of (5.1), cf. [29, Lemma 3]. It turns out that in each iteration either $\alpha_1 = \alpha_2 = \alpha_3 = 1$, or $\alpha_1 = \alpha_2 = 0$ and $\alpha_3 = 1$, cf. [29, Theorem 4]. If (5.1) is feasible and bounded from below, $\alpha_1, \alpha_2, \alpha_3$ computed by [29, Algorithm 3], and ρ large enough, then Algorithm 2 terminates after finite steps, cf. [29, Theorem 2]. As elaborated in [55, Section III] the approximation of the Hessian in Step 3 can be replaced by the damped Boyden–Fletcher–Goldfarb–Shanno (BFGS) update, which only involves previous gradient updates of the Lagrangian. Using this update also reduces the overall communication effort.

In Table 5.1 we summarize a qualitative comparison of ADMM and ALADIN algorithms. The communication effort is derived in the context of optimal power flow (OPF), where $\mathscr{S}$ is the set of all sub-networks, cf. Section 5.3.2 and [55, Table II]. The difference in communication effort is mainly due to the second-order

---

Algorithm 2: ALADIN [29]

Input: Initial guess for $x_i \in \mathbb{R}^n$ and $\lambda \in \mathbb{R}^m$, and numerical tolerance $\varepsilon > 0$

Repeat :

1 Choose penalty parameter $\rho \geq 0$ and $0 \leq \Sigma_i \in \mathbb{R}^{n \times n}$, and solve for $i \in \{1, \ldots, N\}$ the decoupled problems

$$\min_{y_i} \ k_i(y_i) + \langle A_i y_i, \lambda \rangle + \tfrac{\rho}{2} \|y_i - x_i\|^2_{\Sigma_i}, \quad \text{subject to } h_i(y_i) \leq 0 \quad | \kappa_i \quad \text{(NLP)}$$

to either local or global optimality.

2 If $\| \sum_{i=1}^{N} A_i y_i - b \|_1 \leq \varepsilon$ and $\| \Sigma_i (y_i - x_i) \|_1 \leq \varepsilon$, then terminate with $x^\star = y$.

3 Choose for $i \in \{1, \ldots, N\}$ an approximation of the Jacobian $C_i \approx C_i^\star$

$$\forall j = 1, \ldots, n_h : \quad C_{ij}^\star = \begin{cases} \nabla_x(h_i(x))_j|_{x=y_i}, & \text{if } (h_i(y_i))_j = 0, \\ 0, & \text{otherwise,} \end{cases}$$

and compute $g_i = \nabla k_i(y_i) + \langle (C_i^\star - C_i), \kappa_i \rangle$, and choose a symmetric positive-definite approximation of the Hessian

$$H_i \approx \nabla^2 (k_i(x_i) + \langle h_i(y_i), \kappa_i \rangle) \in \mathbb{R}^{n \times n}.$$

4 For $\mu > 0$ solve the coupled QP

$$\min_{\Delta y, s} \sum_{i=1}^{N} \tfrac{1}{2} \|\Delta y_i\|^2_{H_i} + \langle \Delta y_i, g_i \rangle + \langle s, \lambda \rangle + \tfrac{\mu}{2} \|s\|^2_2$$

$$\text{subject to} \begin{cases} \sum_{i=1}^{N} A_i(y_i + \Delta y_i) = b + s \quad | \lambda_{\text{QP}}, \\ C_i \Delta y_i = 0, \quad i \in \{1, \ldots, N\}. \end{cases} \quad \text{(QP)}$$

5 For step-sizes $\alpha_1, \alpha_2, \alpha_3 \geq 0$ set

$$x^+ = x + \alpha_1(y - x) + \alpha_2 \Delta y, \quad \lambda^+ = \lambda + \alpha_3(\lambda_{\text{QP}} - \lambda).$$

6 Update the iterates $x \leftarrow x^+$ and $\lambda \leftarrow \lambda^+$, and go to Step 1.

Output: Numerical optimal solution $x^\star$

---

information used in ALADIN. While ADMM only communicates local minimizers in both directions (local $\to$ CE and CE $\to$ local), the ALADIN algorithm also has to share information of second-order derivatives (local $\to$ CE), and minimizers of (QP) (CE $\to$ local).

Due to its flexibility and convergence properties, Algorithm 2 is widely considered in the literature, see [57–59]. We refer to [60] for applications in the energy management context, wherein also a comparison between ADMM and ALADIN is presented. Moreover, in [61,62] ALADIN is utilized for vehicle coordination at traffic intersections; to name but two exemplary instances.

Table 5.1 Comparison of Algorithm 1 (ADMM) and Algorithm 2 (ALADIN) based on the findings in [29,40,55]. Here, CE = central entity

	ADMM	ALADIN	ALADIN + BFGS
Objective function	convex	non-convex	non-convex
Constraints	convex	non-convex	non-convex
Convergence guarantee	no	yes	yes
Convergence rate	(sub-)linear	quadratic	superlinear
Information local $\rightarrow$ CE	$\sum_{i \in \mathscr{S}} n_i$	$\sum_{i \in \mathscr{S}} \frac{n_i(2n_i+3)}{2}$	$\sum_{i \in \mathscr{S}} \frac{n_i(n_i+4)}{2}$
Information CE $\rightarrow$ local	$\sum_{i \in \mathscr{S}} n_i$	$\sum_{i \in \mathscr{S}} 2n_i$	$\sum_{i \in \mathscr{S}} 2n_i$
Applied to energy systems	[35,39,51]	[40,55]	[55,56]

A potential bottleneck of ALADIN (Algorithm 2) is its coupling QP. It involves all decision variables, an active set detection, and a central coordinator. Hence the original ALADIN variant presented in Algorithm 2 has seen a number of modifications. A trick to overcome the issue of active set detection is to formulate the coupling QP with all inequality constraints, see [46] for a case study on EV charging. This latter trick, however, does not resolve the scaling issue.

To avoid the need for a central coordinator, one can exploit the observation that the coupling QP has the same partially separable structure as the original problem. In [63] a second inner layer of problem decomposition is applied to the convex coupling QP. On this inner level, one may apply ADMM or distributed variants of the conjugate gradient method [64]. Moreover, one can project the coupling QP onto the consensus constraints, which ensures that the inner problem scales in the number of coupling variables. This later aspect is of interest for energy grids, which are typically coupled only in a few variables describing line flows or virtual nodes. We refer to [65] for details.

Similar to ADMM, by now several algorithmic variants have been proposed. Indeed there is a trend of ALADIN becoming a family of algorithms, cf., e.g., [29, 46,66–68].

5.3 Applications

Next, we present and discuss two prominent applications of the previously introduced optimization methods, namely Nonlinear Model Predictive Control of energy distribution in Section 5.3.1, and Optimal Power Flow in Section 5.3.2. While the latter can be considered as steady-state problem, the first application explicitly considers dynamical components.

5.3.1 Nonlinear model predictive energy management

The underlying idea of MPC is the repeated solution of optimal control problems to determine the control input. To this end, a model to predict the near future behavior of the dynamical system in consideration is required to find controls with minimal costs with respect to a given cost function on the prediction horizon, see [69] for an introduction. Due to its capability to take state and input constraints into account, MPC has been successfully applied in many applications, cf. [70–74].

In the context of power grids, a typical objective is to avoid rapid changes in the power demand, i.e., the grid operator aims at a flat aggregated power demand profile. However, since more and more renewable energy sources become part of the overall power network, the overall grid cannot be modeled accurately, and hence, the idea of MPC is not directly applicable. In particular, due to many small-scale power generation devices, e.g., private PV, the generation and demand profile strongly depends on the local conditions, e.g., sun and wind. Here, recent progress in time-series prediction may be used to generate day-ahead predictions, see, e.g., [75–77]. The idea is to consider small-scale semi-autonomous subsystems of the overall power net, and to control the generation-demand balance based on the local sources and conditions, see for example [21,39,40,54,78,79] and references therein, respectively.

In virtue of [40], we consider the following situation as an example. A set of $N \in \mathbb{N}$ residential energy systems or microgrids (private or small-scale commercial) are coupled via a grid operator. The grid operator (the central entity CE) has to balance demand and surplus of energy in the smart grid. A particular component in the smart grid is energy storage devices, e.g., batteries. In [40], the following charging and discharging dynamics for a single energy storage device $i \in \{1, \ldots, N\}$ are stated

$$x_i(t+1) = \alpha_i x_i(t) + \tau(\beta_i u_i^+(t) + u_i^-(t)), \tag{5.6a}$$
$$z(t) = w_i(t) + u_i^+(t) - \gamma_i u_i^-(t), \tag{5.6b}$$

where $t \in \mathbb{N}$ is the current time instant, and $\alpha, \beta, \gamma \in (0, 1]$ model efficiency of self-discharge and conversion. In (5.6a) the state $x_i(t) \in \mathbb{R}$ is the state of charge of the ith battery at time $t \in \mathbb{N}$, $z_i(t) \in \mathbb{R}$ is the power demand in (5.6b), $w_i(t) \in \mathbb{R}$ is the net energy consumption disregarding the battery, and $\tau > 0$ is the sampling time. The state of charge x_i is influenced by the charging u_i^+ and discharging u_i^- rate, which is considered to be independently controllable, respectively. Hence, we may control the battery by

$$u_i(t) = \begin{pmatrix} u_i^+(t) \\ u_i^-(t) \end{pmatrix} \in \mathbb{R}^2.$$

The states x_i and the charging/discharging rate are subject to constraints. For battery capacity $C_i \geq 0$, and bounds on the charging $\bar{u}_i \geq 0$ and discharging rate $\underline{u}_i \leq 0$ typical constraints read, for $i = 1, \ldots, N$,

$$
\begin{aligned}
0 &\leq & x_i(t) & \leq C_i, \\
0 &\leq & u_i^+(t) & \leq \bar{u}_i, \\
\underline{u}_i &\leq & u_i^-(t) & \leq 0, \\
0 &\leq & \frac{u_i^-(t)}{\underline{u}_i} + \frac{u_i^+(t)}{\bar{u}_i} & \leq 1.
\end{aligned}
\tag{5.7}
$$

Invoking linearity in (5.6), the state x_i at time instant $t + k$ reads

$$x_i(t + k) = \alpha_i^t x_i(k) + \tau \sum_{j=0}^{t-1} \alpha_i^{t-i-j} \begin{bmatrix} \beta_i & 1 \end{bmatrix} u_i(j).$$

Hence, on the prediction horizon $0 < T \in \mathbb{N}$, for appropriate expressions $D_i \in \mathbb{R}^{8T \times 2T}$ and $d_i \in \mathbb{R}^{8T}$, the constraints on both, the state and the control (5.7), can be compactly written as

$$D_i u_i \leq d_i, \quad i \in \{1, \ldots, N\},$$

which in particular means that on the prediction horizon T we have polyhedral constraints on the states and controls. The overall net consumption is given by

$$W(t) = \frac{1}{T} \sum_{j=t-T+1}^{t} \sum_{i=1}^{N} w_i(j)$$

for $t \in \{k, \ldots, k + T - 1\}$ with current time step $k \geq T - 1$. Setting $\bar{z}(t) = \sum_{i=1}^{N} z_i(t)$, in virtue of MPC, for $\sigma_0 > 0$ we may formulate the global consensus objective function $k_0 : \mathbb{R}^T \to \mathbb{R}_{\geq 0}$

$$k_0(\bar{z}) := \frac{\sigma_0}{TN^2} \sum_{j=k}^{k+T-1} (\bar{z}(j) - W(j))^2 = \frac{\sigma_0}{TN^2} \|\bar{z} - W\|^2, \tag{5.8a}$$

which describes the deviation between the aggregated power demand and the overall net consumption on the time horizon T. In the following, we use the predicted demand $z_i := (z_i(t), \ldots, z_i(t + T - 1))^\top \in \mathbb{R}^T$ and the expected net consumption $w_i := (w_i(t), \ldots, w_i(t + T - 1))^\top \in \mathbb{R}^T$ on the prediction horizon T.

To address local balance between consumption w_i and demand z_i, and to account for charging/discharging costs, we introduce the local objective functions $k_i : \mathbb{R}^T \times \mathbb{R}^{2T} \to \mathbb{R}_{\geq 0}, i \in \{1, \ldots, N\}$ as

$$k_i(u_i) := \frac{\sigma_i}{2} (\|z_i - w_i\|^2 + \|u_i\|^2) = \frac{\sigma_i}{2} \|u_i\|_{Q_i}^2, \tag{5.8b}$$

where we used (5.6b) to rewrite $z_i - w_i$. We set $Q_i := \sigma_i I_{2T} + \sigma_i I_T \otimes \begin{bmatrix} 1 & \gamma_i \\ \gamma_i & \gamma_i^2 \end{bmatrix}$. The factor σ_i penalizes the use of the ith battery. With the costs introduced in (5.8) and setting $\bar{w} := \sum_{i=1}^{N} w_i$ we formulate the optimal control problem

$$\min_{\bar{z}, u} k_0(\bar{z}) + \sum_{i=1}^{N} k_i(u_i)$$

subject to

$$\bar{z} = \bar{w} + \sum_{i=1}^{N} A_i u_i, \quad |\lambda_i$$

$$d_i \geq D_i u_i, \qquad\qquad i \in \{1, \ldots, N\}.$$

(5.9)

To efficiently solve the optimal control problem (5.9), in [40] ALADIN is utilized. In each of the N microgrids the following is solved locally and in parallel

$$\min_{v_i} \ k_i(v_i) - \langle A_i v_i, \lambda \rangle + \frac{1}{2} \|v_i - u_i\|_{Q_i}^2 \ \text{ subject to } d_i \geq D_i v_i \,|\, \kappa_i, \qquad (5.10a)$$

according to Algorithm 2 Step 1, where we use the current primal u_i and dual λ iterate. First-order optimality $\nabla k_i(v_i) + \langle D_i, \kappa_i \rangle - \langle A_i, \lambda \rangle + Q_i(v_i - u_i) = 0$ allows evaluation of the modified gradient as $g_i = \langle A_i, \lambda \rangle - Q_i(v_i - u_i)$. The local solutions v_i are communicated to the CE, where the equality constrained QP

$$\min_{\bar{z},u,s} k_0(\bar{z}) + \sum_{i=1}^{N} \frac{1}{2}\|u_i - v_i\|_Q^2 + \langle g_i, u_i \rangle + \frac{\mu_i}{2}\|s_i\|^2$$

subject to
$$\hspace{9cm} (5.10b)$$

$$\bar{z} = \bar{w} + \sum_{i=1}^{N} A_i u_i \quad |\lambda^+$$

$$s_i = D_i^{\text{act}}(u_i - v_i) \quad |\kappa_i^{\text{QP}} \ i \in \{1, \dots, N\}$$

is solved. Here, $D_i^{\text{act}} \in \mathbb{R}^{8T \times 2T}$ is the Jacobian of the active constraints at local solutions, meaning $D_i^{\text{act}} v_i = d_i$ for all $i = 1, \dots, N$.

   For the above problem it is proven in [40, Proposition 1] that the optimization problem (5.9) has a unique solution, and further that for suitable updates of the parameters μ_i in (5.10b) the solution locally converges to the optimum with quadratic rate, cf. [40, Proposition 2]. Moreover, under regularity assumptions on the solution and a certain Armijo step-size condition [40, Theorem 1] ensures global convergence of (5.9) to the optimal solution $u^\star$.

5.3.2   Optimal power flow

The notion of OPF refers to one of the most important optimization problems in power systems, where the power grid is described via stationary power flow equations. These equations model routing and distribution as well as loss of power within electricity networks. OPF computations and other optimization problems involving the power flow equations as constraints are applied for a number of different problems in power systems, ranging from the computation of generator setpoints via grid planning to state estimation and others. We refer to [10,80] for tutorial introductions.

   One typical variant of OPF considers the computation of setpoints for grid-supporting generators, e.g., for so-called redispatch problems when the outcome of electricity markets is incompatible with physical requirements and operational constraints of the grid, cf. [10]. Since stability of the power grid is essential for availability of electrical power, OPF has been and is still a topic of research, see, e.g., [12,81] and references therein, respectively. Introductory overviews and reports on recent developments can be found in the recent work [82], and [83–85]. In this section, we consider the rather simple formulation OPF under consideration in [10].

   The stationary behavior of an AC electrical network can be modeled using the triple $(\mathcal{N}, \mathcal{G}, Y)$, where $\mathcal{N} = \{1, \dots, N\}$ is the set of nodes (also called buses), the (non-empty) set $\mathscr{G}$ describes the available generators, and $Y = G + jB \in \mathbb{C}^{N \times N}$ is the bus admittance matrix, where $j^2 = -1$. We introduce OPF for AC systems and then briefly show DC as a special case.

5.3.2.1 AC power flow

In a balanced symmetric three-phase AC system every bus $l \in \mathscr{N}$ can be described by its voltage magnitude v_l, the voltage phase θ_l, and the net active p_l and reactive power q_l. With these quantities we may recall from [10] the steady-state AC power flow equations for all $l \in \mathscr{N}$

$$p_l = v_l \sum_{k \in \mathscr{N}} v_k(G_{lk} \cos(\theta_{lk}) + B_{lk} \sin(\theta_{lk})), \tag{5.11a}$$

$$q_l = v_l \sum_{k \in \mathscr{N}} v_k(G_{lk} \sin(\theta_{lk}) - B_{lk} \cos(\theta_{lk})), \tag{5.11b}$$

where $\theta_{lk} := \theta_l - \theta_k$ is the phase angle difference, and G_{lk}, B_{lk} denote the admittance of the bus connecting $l \in \mathscr{N}$ and $l \neq k \in \mathscr{N}$, respectively. Since θ_{lk} is defined as differences, we set one bus as reference θ_{l_0}, w.l.o.g $1 = l_0 \in \mathscr{N}$. To simplify notation and exposition we assume $\mathscr{G} \subseteq \mathscr{N}$. For each bus $l \in \mathbb{N}$ the net apparent power is given by

$$s_l := p_l + jq_l = \begin{cases} (p_l^g - p_l^d) + j(q_l^g - q_l^d), & l \in \mathscr{G}, \\ -(p_l^d + jq_l^d), & \text{else}, \end{cases}$$

where the superscript "d" indicates demand, and "g" refers to generation. For generator nodes $l \in \mathscr{G}$ the generated power injections p_l^g, q_l^g are the control variables, and for $m \in \mathscr{N}$ the power sinks or sources p_m^d, q_m^d are uncontrollable.

5.3.2.2 DC power flow

Assuming a lossless line $r_{lk} \approx 0$ for the Ohmic resistance, a constant voltage magnitude $v_l \approx 1$, and small phase differences $\theta_{lk} \approx 0$ for all $l, k \in \mathscr{N}$, we obtain the so-called DC power flow equations as a linear approximation of (5.11)

$$p_l = \sum_{k \in \mathscr{N} \setminus \{l\}} B_{lk}(\theta_l - \theta_k) \iff p = -B\theta, \tag{5.12}$$

where $B = Im(Y)$, and no conductance is present.

5.3.2.3 Optimal power flow

To formulate an optimization problem in a usual way, we introduce the state $x \in \mathbb{R}^{n_x}$, the disturbance $d \in \mathbb{R}^{n_d}$, and the control input $u \in \mathbb{R}^{n_u}$ for AC and DC configuration, respectively, by

$$x^{AC} := (v_l, \theta_l)_{l \in \mathscr{N}} \in \mathbb{R}^{n_x^{AC}}, \quad d^{AC} := (p_l^d, q_l^d)_{l \in \mathscr{N}} \in \mathbb{R}^{n_d^{AC}},$$

$$u^{AC} := (p_l^g, q_l^g)_{l \in \mathscr{G}} \in \mathbb{R}^{n_u^{AC}}, \tag{5.13a}$$

$$x^{DC} := (\theta_l)_{l \in \mathscr{N}} \in \mathbb{R}^{n_x^{DC}}, \quad d^{DC} := (p_l^d)_{l \in \mathscr{N}} \in \mathbb{R}^{n_d^{DC}},$$

$$u^{DC} := (p_l^g)_{l \in \mathscr{G}} \in \mathbb{R}^{n_u^{DC}}. \tag{5.13b}$$

In the following $\sigma \in \{AC, DC\}$ indicates the configuration. Note that the dimensions differ for AC and DC cases as sketched above are as follows:

σ	AC	DC
n_x^σ	$2|\mathcal{N}|$	$|\mathcal{N}|$
n_d^σ	$2|\mathcal{N}|$	$|\mathcal{N}|$
n_u^σ	$2|\mathcal{G}|$	$|\mathcal{G}|$

Using the triple (x, d, u) from (5.13) we may write the power flow equations (5.11) compactly as

$$F^\sigma : \mathbb{R}^{n_x^\sigma} \times \mathbb{R}^{n_d^\sigma} \times \mathbb{R}^{n_u^\sigma} \to \mathbb{R}^{N^\sigma}, \quad F^\sigma(x, u; d) = 0, \tag{5.14}$$

where for AC we have a nonlinear map, and for DC it simplifies to

$$F^{DC}(x^{DC}, u^{DC}; d^{DC}) = u^{DC} - d^{DC} + Bx^{DC} = 0.$$

Note that in this case the dimension of the controls coincides with the dimension of sinks/sources. Having the abstract formulation (5.14) at hand, we introduce the power flow manifold

$$\mathscr{F}^\sigma(d) := \left\{ (x, u) \in \mathbb{R}^{n_x^\sigma} \times \mathbb{R}^{n_u^\sigma} \;\middle|\; F^\sigma(x, u; d) = 0 \right\},$$

which represents all possible solutions of the power flow equations (5.11), respectively (5.12) for a given disturbance d, i.e., the solutions (x, u) are parameterized by d. In the following, we assume that the input u belongs to the box constraints set

$$\mathscr{U}^\sigma := \left\{ (u_l^\sigma)_{l \in \mathcal{G}} \in \mathbb{R}^{n_u^\sigma} \;\middle|\; \begin{cases} p_l^g \in [\underline{p}_l^g, \bar{p}_l^g], \; q_l^g \in [\underline{q}_l^g, \bar{q}_l^g] \; \forall l \in \mathcal{G}, & \sigma = AC \\ p_l^g \in [\underline{p}_l^g, \bar{p}_l^g], \; \forall l \in \mathcal{G}, & \sigma = DC \end{cases} \right\},$$

and the state x is restricted to the set

$$\mathscr{X} := \left\{ (x_l^\sigma)_{l \in \mathcal{N}} \in \mathbb{R}^{n_x^\sigma} \;\middle|\; \begin{cases} v_l \in [\underline{v}_l, \bar{v}_l], \; \theta_1 = 0 \; \forall l \in \mathcal{N}, & \sigma = AC \\ \theta_1 = 0 \; \forall l \in \mathcal{N}, & \sigma = DC \end{cases} \right\}.$$

In real applications, there will be further restrictions on the line flows, which are typically stated as constraints on the power

$$|s_{lk}| = \sqrt{p_{lk}^2 + q_{lk}^2} \leq |\bar{s}_{lk}| \; \forall (l, k) \in \mathcal{L},$$

for AC, where the set $\mathcal{L} \subseteq \mathcal{N} \times \mathcal{N}$ represents all connecting lines, and the active p_{lk} and reactive power q_{lk} across the line from $l \in \mathcal{N}$ to $l \neq k \in \mathcal{N}$, i.e., $(l, k) \in \mathcal{L}$, is

$$p_{lk} = v_l^2 g_{lk} - v_l v_m (g_{lk} \cos(\theta_{lk}) + b_{lk} \sin(\theta_{lk})),$$
$$q_{lk} = -v_l^2 b_{lk} - v_l v_m (b_{lk} \cos(\theta_{lk}) - g_{lk} \sin(\theta_{lk})).$$

Since the constraints on line power flow depend on the state only, we formulate these compactly

$$\mathscr{C}^\sigma := \left\{ x \in \mathbb{R}^{n_x^\sigma} \;\middle|\; \begin{cases} \sqrt{p_{lk}^2 + q_{lk}^2} \leq |\bar{s}_{lk}| \; \forall (l, k) \in \mathcal{L}, & \sigma = AC, \\ -\operatorname{diag}((b_{lk})_{(l,k) \in \mathcal{L}}) Ax \leq (\bar{p}_{lk})_{(l,k) \in \mathcal{L}}^\top, & \sigma = DC \end{cases} \right\},$$

where $A \in \mathbb{R}^{|\mathscr{L}| \times N}$ is the graph incidence matrix in case of DC. Note that for DC the constraints $\mathscr{C}^{DC}$ are obtained from Kirchhoff's laws, and we have $\mathscr{C}^{DC} \subset \mathbb{R}^{n_x^{DC}}$.

With the notation above, we arrive at the optimization problem

$$\min_{(x^\sigma, u^\sigma) \in \mathbb{R}^{n_x^\sigma + n_u^\sigma}} J(u^\sigma)$$
$$\text{subject to} \quad (x^\sigma, u^\sigma) \in \mathscr{F}^\sigma(d), \tag{5.15}$$
$$u^\sigma \in \mathscr{U}^\sigma,$$
$$x^\sigma \in \mathscr{X}^\sigma \cap \mathscr{C}^\sigma.$$

5.3.2.4 AC-OPF

Common control objectives in the AC setting are to reduce transmission losses, to avoid large injections of reactive power, and to minimize costs of active power generation. The latter is considered by [55], which leads to the following AC-OPF problem

$$\min_{\theta, v, p, q} \sum_{i \in \mathscr{G}} \alpha_i p_i^2 + \beta_i p_i + \gamma_i, \tag{5.16a}$$

subject to (5.11) and for all $k \in \mathscr{N}$

$$\left. \begin{array}{l} \underline{p}_i \leq p_i \leq \bar{p}_i, \ \forall i \in \mathscr{G}, \\ \underline{q}_i \leq q_i \leq \bar{q}_i, \ \forall i \in \mathscr{G}, \\ \underline{v}_i \leq v_i \leq \bar{v}_i, \ \forall i \in \mathscr{N}, \end{array} \right\} \tag{5.16b}$$

$$v_1 = 1, \ \theta_1 = 0, \tag{5.16c}$$

where $\alpha_i, \beta_i, \gamma_i > 0$. The terms in (5.16a) penalize generation of power, the constraints (5.16b) are explicit versions of $u \in \mathscr{U}^{AC}$ and $x \in \mathscr{X}^{AC} \cap \mathscr{C}^{AC}$ in (5.15), and (5.16c) defines the reference node. Note that (5.16) contains all nodes, all lines, and all constraints in one optimization problem.

In real applications it is often reasonable to consider geographically separated areas as semi-autonomous grids, in the sense that they are only connected to other areas (grids) by single lines. In this case the power flow can be considered to be controlled within each grid separately, respectively; while the connecting lines encode constraints on some buses. With this idea, we may formulate the OPF problem as a task of distributed optimization.

Aiming at representing the AC-OPF (5.16) in the form of (5.1), we partition the set of all buses $\mathscr{N}$ into S subsets with $\mathscr{S} = \{1, \ldots, S\}$ such that we have the set $\mathscr{N}_i = \{n_i^1, \ldots, n_i^{N_i}\} \subset \mathscr{N}$, where $\mathscr{N} = \cup_{i \in \mathscr{S}} = \mathscr{N}$ and $\mathscr{N}_i \cap \mathscr{N}_j = \emptyset$ for $i \neq j$. The last condition prohibits an overlap of the sub-networks. To nevertheless account for the connections between the sub-networks, and thus to obtain an equivalent representation of (5.16), we introduce the set $\mathscr{A}$ of auxiliary nodes in those lines which connect two neighboring sub-networks. Doing so, we obtain an enlarged set

$\mathscr{N}^A := \mathscr{N} \cup \mathscr{A}$, containing $A \in \mathbb{N}$ extra nodes. Following [55], recalling $x = (v, \theta)$ and $u = (p, q)$, we arrive at the separated, affinely coupled optimization problem

$$\min_{(x,u)} \sum_{i \in \mathscr{S}} k_i(u_i) \tag{5.17a}$$

subject to

$$F_i^{AC}(x_i, u_i, d_i) = 0 \qquad | \kappa_i \quad \forall i \in \mathscr{S}, \tag{5.17b}$$

$$\underline{x}_i \leq h(x_i, u_i, d_i) \leq \bar{x}_i \qquad | \xi_i \quad \forall i \in \mathscr{S}, \tag{5.17c}$$

$$\sum_{i \in \mathscr{S}} A_i(x_i, u_i, d_i)^\top = 0 \qquad | \lambda. \tag{5.17d}$$

In the above formulation, (5.17a) represent the costs of local power generation with $k_i(u_i) = \alpha_i p_i^2 + \beta_i p_i + \gamma_i$ for $i \in \mathscr{G}_i = \mathscr{N}_i \cap \mathscr{G}$ (set of local generators), (5.17b) are the power flow equations (5.11), constraints (5.17c) compactly encode (5.16b), and (5.17d) establish the consensus coupling constraints

$$\forall (k, l) \in \mathscr{A} \times \mathscr{A}, \ k \neq l : \ \theta_k = \theta_l, \quad v_k = v_l, \quad p_k = -p_l, \quad q_k = -q_l.$$

As discussed in [55], problem (5.17) is amenable to Algorithm 2, cf. [55, Algorithm 1]. It has been proven in [55, Theorem 1] that, under some (technical) assumptions, the application of Algorithm 2 to problem (5.17) is successful in the sense that it terminates after finite iterations for a user-defined accuracy tolerance. Moreover, under assumptions on the step-size update, the result [55, Theorem 2] ensures local quadratic convergence of the iterates to the optimal value.

5.3.2.5   Multi-stage AC-OPF

The power flow equations (5.11) are formulated at steady-state, at least from the viewpoint of control theory. Hence, (5.15), (5.16), (5.17) are static optimization problems. In this setting, however, it is not taken into account that a large-scale power plant cannot change its setpoint arbitrarily fast. To account for these limitations, in [10, Section 3.1] multi-stage and predictive OPF is under consideration. We denote with $u_l(t) = (p_l^g(t), q_l^g(t))$ the control at time instant $t \in \mathbb{N}$. Then, the dynamical limitations (typically only on the active power generation) can, e.g., be formulated as $p_l^g(t + 1) - p_l^g(t) \in [\Delta \underline{p}_l, \Delta \bar{p}_l]$ for $l \in \mathscr{G}$, leading to control constraints

$$u(t + 1) - u(t) = \delta u(t) \in \delta \mathscr{U} := (\times_{l \in \mathscr{G}} [\Delta \underline{p}_l, \Delta \bar{p}_l]) \times \mathbb{R}^{n_u^\sigma - |\mathscr{G}|} \subset \mathbb{R}^{n_u^\sigma}.$$

Exemplary, for a time horizon $0 < \hat{T} \in \mathbb{N}$ and $T = \{0, \dots, \hat{T}\}$, in virtue of (5.15) we may then state the multi-state AC-OPF

$$\min_{(x(\cdot), u(\cdot), \delta u(\cdot)) \in \mathbb{R}^{(n_x^\sigma + 2n_u^\sigma)\hat{T}}} \sum_{k \in T} J(u(k)) + \|\delta u(k)\|^2$$

subject to $\forall k \in T$

$$u(k + 1) = u(k) + \delta u(k),$$

$$\delta u(k) \in \delta \mathscr{U}$$

$$(x(k), u(k)) \in (\mathscr{X}^{AC} \times \mathscr{U}^{AC}) \cap \mathscr{F}^{AC}(d(k)),$$

$$x(k) \in \mathscr{C}^{AC},$$

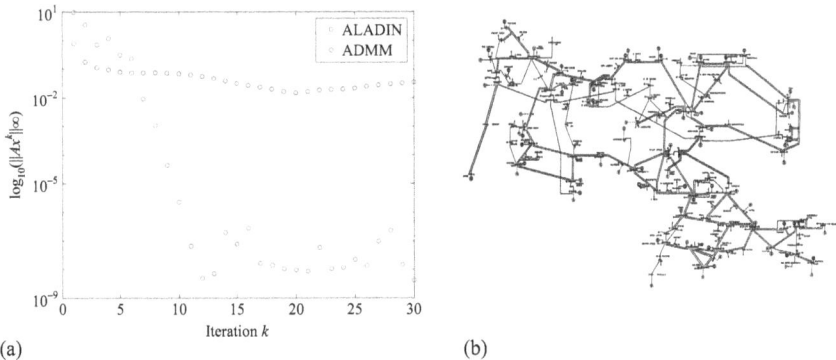

Figure 5.1   (a) Logarithmic constraint violation for ADMM and ALADIN.
(b) Example of the IEEE 118 bus test system.

where the second addend in the objective function penalizes frequency and amount of power generation. The multi-state AC-OPF already has a flavor of predictive control presented in Section 5.3.1, and its form is similar to that of an optimal control problem, cf. [86].

5.3.2.6   Numerical example

To showcase functioning of the proposed methods, we briefly present a numerical example of an AC-OPF problem (5.17). We simulate the IEEE 118-bus test case, considered in [55]. Figure 5.1b shows the map of the IEEE 118-bus test system. We compare application of Algorithm 1 (ADMM) with Algorithm 2 (ALADIN). Exemplary, in Figure 5.1a violation of the coupling constraints (5.17d) is depicted. It can be seen that Algorithm 2 is superior to Algorithm 1 in terms of constraint satisfaction. The simulation has been performed using the MATLAB® toolbox ALADIN$-\alpha$ [66]. We refer to [55,66], where more detailed and comprehensive numerical case studies are conducted. Therein, also different problems ranging from 5-bus to 300-bus test systems are considered and different variants of ALADIN are compared with respect to number of iterations against relative error, communication effort, and convergence rate. Notice that the ALADIN$-\alpha$ toolbox [66] comprising instructive examples for different distributed optimization problems is available open-source [87].

5.4   Outlook and open problems

ADMM and ALADIN are only two distributed algorithms that can be applied to power and energy systems. Indeed, more recent advances have considered distributing interior point methods and sequential quadratic programming, see [64] and [49] for details and for numerical results for AC-OPF problems (also in comparison to ADMM and ALADIN). A very much open problem in distributed optimization is the formal analysis of the interplay between the actually considered decomposition and the realized convergence speed of the algorithms. In general, ALADIN (Algorithm 2) exhibits convergence guarantees for non-convex optimization problems. However, the guarantees often rely on a globalization step, which as such is difficult to distribute between agents.

Some open-source codes and tools for distributed optimization are available. In case of ALADIN, [66] provides an open-source MATLAB implementation of different variants tailored for algorithmic prototyping, while [57] provides code tailored to power system applications and TSO-DSO coordination. Similarly, ADMM codes have been published, cf. [24,66] and references therein. However, in comparison to the quite large number of powerful open-source codes for solving NLPs and optimal control problems, there is still demand for further tool development.

For both—numerical optimization algorithms and for modeling and control of distributed energy systems—data-driven and learning approaches are of increasing interest [88]. With respect to optimization, see, e.g., [89], where recurrent neural networks are employed to reduce the number of iterations and [90] for learning of parameters for ADMM for recent case studies on power systems. Within this framework, recent extensions to stochastic systems are key to properly take uncertainties into account [91].

From an applications perspective, scalable algorithms for distributed stochastic OPF, distributed approaches to reactive power dispatch (which involves discrete decision variables [8]), and the solution of problems involving coupled power systems and gas networks are of interest. Moreover, efficient numerical implementations which can be applied on small-scale computational units (a.k.a. embedded systems) in distribution grids and multi-energy problems [92] are of growing relevance. Another topic of interest is to go from separate open-source grid models [93] and open-source datasets of load and generation of time series such as [94,95] to coupled grid models and time-series data, which appear to be not yet available. Likewise, open-source datasets and benchmarks for multi-energy systems such as [92] appear to be still rare.

Acknowledgments

L. Lanza gratefully acknowledges support by the Carl Zeiss Foundation (VerneDCt – Project-ID 2011640173).

References

[1]   Westermann D, Schlegel S, Sass F, *et al.* Curative actions in the power system operation to 2030. In: *International ETG-Congress 2019; ETG Symposium.* VDE; 2019. p. 1–6.

[2]   Boie I, Kost C, Bohn S, *et al.* Opportunities and challenges of high renewable energy deployment and electricity exchange for North Africa and Europe—scenarios for power sector and transmission infrastructure in 2030 and 2050. *Renewable Energy.* 2016;87:130–144.

[3]   Rasheduzzaman M, Mueller JA, and Kimball JW. Reduced-order small-signal model of microgrid systems. *IEEE Transactions on Sustainable Energy.* 2015;6(4):1292–1305.

[4] Gnärig L, Gensior A, Laza SB, Carrasco M, and Reincke-Collon C. Model reduction using singular perturbation methods for a microgrid application. In: *2022 24th European Conference on Power Electronics and Applications (EPE'22 ECCE Europe)*. Piscataway, NJ: IEEE; 2022. p. 1–10.

[5] Yazdanian M, and Mehrizi-Sani A. Distributed control techniques in microgrids. *IEEE Transactions on Smart Grid*. 2014;5(6):2901–2909.

[6] Zhao C, Mallada E, and Dörfler F. Distributed frequency control for stability and economic dispatch in power networks. In: *2015 American Control Conference (ACC)*. Piscataway, NJ: IEEE; 2015. p. 2359–2364.

[7] Ortega Á, and Milano F. Frequency control of distributed energy resources in distribution networks. *IFAC-PapersOnLine*. 2018;51(28):37–42.

[8] Murray A, Engelmann A, Hagenmeyer V, and Faulwasser T. Hierarchical distributed mixed-integer optimization for reactive power dispatch. *IFAC-PapersOnLine*. 2018;51(28):368–373.

[9] Sauerteig P, Baumann M, Dickert J, Grundel S, and Worthmann K. In: Göttlich S, Herty M, Milde A, editors. *Reducing Transmission Losses via Reactive Power Control*. Cham: Springer International Publishing; 2021. p. 219–232.

[10] Faulwasser T, Engelmann A, Mühlpfordt T, and Hagenmeyer V. Optimal power flow: an introduction to predictive, distributed and stochastic control challenges. *at-Automatisierungstechnik*. 2018;66(7):573–589.

[11] Parisio A, Rikos E, and Glielmo L. A model predictive control approach to microgrid operation optimization. *IEEE Transactions on Control Systems Technology*. 2014;22(5):1813–1827.

[12] Mühlpfordt T, Roald L, Hagenmeyer V, Faulwasser T, and Misra S. Chance-constrained AC optimal power flow: a polynomial chaos approach. *IEEE Transactions on Power Systems*. 2019;34(6):4806–4816.

[13] Braun P, Grüne L, Kellett CM, Weller SR, and Worthmann K. Towards price-based predictive control of a small-scale electricity network. *International Journal of Control*. 2020;93(1):40–61.

[14] Hu J, Shan Y, Yang Y, *et al.* Economic model predictive control for microgrid optimization: a review. *IEEE Transactions on Smart Grid*. 2024;15(1):472–484.

[15] Oldewurtel F, Parisio A, Jones CN, *et al.* Use of model predictive control and weather forecasts for energy efficient building climate control. *Energy and Buildings*. 2012;45:15–27.

[16] Schiffer J, Seel T, Raisch J, and Sezi T. Voltage stability and reactive power sharing in inverter-based microgrids with consensus-based distributed voltage control. *IEEE Transactions on Control Systems Technology*. 2015;24(1):96–109.

[17] Bai H, Zhang H, Cai H, and Schiffer J. Voltage regulation and current sharing for multi-bus DC microgrids: a compromised design approach. *Automatica*. 2022;142:110340.

[18] Dörfler F, Simpson-Porco JW, and Bullo F. Breaking the hierarchy: distributed control and economic optimality in microgrids. *IEEE Transactions on Control of Network Systems*. 2015;3(3):241–253.

[19]   Schiffer J, Zonetti D, Ortega R, Stanković AM, Sezi T, and Raisch J. A survey on modeling of microgrids—from fundamental physics to phasors and voltage sources. *Automatica*. 2016;74:135–150.

[20]   Molzahn DK, Dörfler F, Sandberg H, *et al.* A survey of distributed optimization and control algorithms for electric power systems. *IEEE Transactions on Smart Grid*. 2017;8(6):2941–2962.

[21]   Sauerteig P, and Worthmann K. Towards multiobjective optimization and control of smart grids. *Optimal Control Applications and Methods*. 2020;41(1):128–145.

[22]   Eichfelder G. Twenty years of continuous multiobjective optimization in the twenty-first century. *EURO Journal on Computational Optimization*. 2021;9:100014.

[23]   Halffmann P, Schäfer LE, Dächert K, Klamroth K, and Ruzika S. Exact algorithms for multiobjective linear optimization problems with integer variables: a state of the art survey. *Journal of Multi-Criteria Decision Analysis*. 2022;29(5–6):341–363.

[24]   Stomberg G, Ebel H, Faulwasser T, and Eberhard P. Cooperative distributed MPC via decentralized real-time optimization: Implementation results for robot formations. Control Engineering Practice. 2023;138:105579.

[25]   Worthmann K, Kellett CM, Grüne L, and Weller SR. Distributed control of residential energy systems using a market maker. *IFAC Proceedings Volumes*. 2014;47(3):11641–11646.

[26]   Yang T, Yi X, Wu J, Yuan Y, *et al.* A survey of distributed optimization. *Annual Reviews in Control*. 2019;47:278–305.

[27]   Rantzer A. Dynamic dual decomposition for distributed control. In: *2009 American Control Conference (ACC)*. Piscataway, NJ: IEEE; 2009. p. 884–888.

[28]   Chatzipanagiotis N, Dentcheva D, and Zavlanos MM. An augmented Lagrangian method for distributed optimization. *Mathematical Programming*. 2015;152:405–434.

[29]   Houska B, Frasch J, and Diehl M. An augmented Lagrangian based algorithm for distributed nonconvex optimization. *SIAM Journal on Optimization*. 2016;26(2):1101–1127.

[30]   Falsone A, Margellos K, Garatti S, and Prandini M. Dual decomposition for multi-agent distributed optimization with coupling constraints. *Automatica*. 2017;84:149–158.

[31]   Varagnolo D, Zanella F, Cenedese A, Pillonetto G, and Schenato L. Newton-Raphson consensus for distributed convex optimization. *IEEE Transactions on Automatic Control*. 2015;61(4):994–1009.

[32]   Wang Y, and Nedić A. Differentially-private distributed optimization with guaranteed optimality. In: *2023 62nd IEEE Conference on Decision and Control (CDC)*. Piscataway, NJ: IEEE; 2023. p. 4162–4169.

[33]   Darup MS, Alexandru AB, Quevedo DE, and Pappas GJ. Encrypted control for networked systems: an illustrative introduction and current challenges. *IEEE Control Systems Magazine*. 2021;41(3):58–78.

[34] Schlüter N, Binfet P, and Darup MS. A brief survey on encrypted control: From the first to the second generation and beyond. *Annual Reviews in Control*. 2023:100913.

[35] Mbuwir BV, Spiessens F, and Deconinck G. Distributed optimization for scheduling energy flows in community microgrids. *Electric Power Systems Research*. 2020;187:106479.

[36] Zeilinger MN, Pu Y, Riverso S, Ferrari-Trecate G, and Jones CN. Plug and play distributed model predictive control based on distributed invariance and optimization. In: *2013 IEEE 52nd Conference on Decision and Control (CDC)*. Piscataway, NJ: IEEE; 2013. p. 5770–5776.

[37] Dörfler F, Simpson-Porco JW, and Bullo F. Plug-and-play control and optimization in microgrids. In: *2014 IEEE 53rd Conference on Decision and Control (CDC)*. Piscataway, NJ: IEEE; 2014. p. 211–216.

[38] Crisostomi E, Liu M, Raugi M, and Shorten R. Plug-and-play distributed algorithms for optimized power generation in a microgrid. *IEEE Transactions on Smart Grid*. 2014;5(4):2145–2154.

[39] Braun P, Faulwasser T, Grüne L, Kellett CM, Weller SR, and Worthmann K. Hierarchical distributed ADMM for predictive control with applications in power networks. *IFAC Journal of Systems and Control*. 2018;3:10–22.

[40] Jiang Y, Sauerteig P, Houska B, and Worthmann K. Distributed optimization using ALADIN for MPC in smart grids. *IEEE Transactions on Control Systems Technology*. 2020;29(5):2142–2152.

[41] Jiang Y, Kouzoupis D, Yin H, Diehl M, and Houska B. Decentralized optimization over tree graphs. *Journal of Optimization Theory and Applications*. 2021;189:384–407.

[42] Engelmann A, Shin S, Pacaud F, and Zavala VM. Scalable Primal Decomposition Schemes for Large-Scale Infrastructure Networks. arXiv preprint arXiv:221211571. 2022.

[43] Engelmann A, Bandeira MB, and Faulwasser T. Approximate Dynamic Programming with Feasibility Guarantees. arXiv preprint arXiv:230606201. 2023.

[44] Everett III H. Generalized Lagrange multiplier method for solving problems of optimum allocation of resources. *Operations Research*. 1963;11(3):399–417.

[45] Giselsson P, Doan MD, Keviczky T, De Schutter B, and Rantzer A. Accelerated gradient methods and dual decomposition in distributed model predictive control. *Automatica*. 2013;49(3):829–833.

[46] Botkin-Levy M, Engelmann A, Mühlpfordt T, Faulwasser T, and Almassalkhi MR. Distributed control of charging for electric vehicle fleets under dynamic transformer ratings. *IEEE Transactions on Control Systems Technology*. 2021;30(4):1578–1594.

[47] Glowinski R, and Marroco A. Sur l'approximation, par éléments finis d'ordre un, et la résolution, par pénalisation-dualité d'une classe de problèmes de Dirichlet non linéaires. *Revue française d'automatique, informatique, recherche opérationnelle Analyse numérique*. 1975;9(R2):41–76.

[48]   Gabay D, and Mercier B. A dual algorithm for the solution of nonlinear varia-
       tional problems via finite element approximation. *Computers & Mathematics
       with Applications*. 1976;2(1):17–40.
[49]   Stomberg G, Engelmann A, and Faulwasser T. Decentralized non-convex
       optimization via bi-level SQP and ADMM. In: *2022 IEEE 61st Con-
       ference on Decision and Control (CDC)*. Piscataway, NJ: IEEE; 2022.
       p. 273–278.
[50]   Boyd S, Parikh N, Chu E, Peleato B, and Eckstein J. Distributed optimiza-
       tion and statistical learning via the alternating direction method of multipliers.
       Foundations and Trends® in Machine learning. 2011;3(1):1–122.
[51]   Erseghe T. Distributed optimal power flow using ADMM. *IEEE Transactions
       on Power Systems*. 2014;29(5):2370–2380.
[52]   Magnússon S, Weeraddana PC, and Fischione C. A distributed approach
       for the optimal power-flow problem based on ADMM and sequential
       convex approximations. *IEEE Transactions on Control of Network Systems*.
       2015;2(3):238–253.
[53]   Stomberg G, Engelmann A, and Faulwasser T. A compendium
       of optimization algorithms for distributed linear-quadratic MPC.
       at-Automatisierungstechnik. 2022;70(4):317–330.
[54]   Braun P, Grüne L, Kellett CM, Weller SR, and Worthmann K. A dis-
       tributed optimization algorithm for the predictive control of smart grids. *IEEE
       Transactions on Automatic Control*. 2016;61(12):3898–3911.
[55]   Engelmann A, Jiang Y, Mühlpfordt T, Houska B, and Faulwasser T. Toward
       distributed OPF using ALADIN. *IEEE Transactions on Power Systems*.
       2018;34(1):584–594.
[56]   Zhai J, Dai X, Jiang Y, Xue Y, *et al.* Distributed optimal power flow for VSC-
       MTDC meshed AC/DC grids using ALADIN. *IEEE Transactions on Power
       Systems*. 2022;37(6):4861–4873.
[57]   Mühlpfordt T, Dai X, Engelmann A, and Hagenmeyer V. Distributed
       power flow and distributed optimization—formulation, solution, and open
       source implementation. *Sustainable Energy, Grids and Networks*. 2021;26:
       100471.
[58]   Chanfreut P, Maestre J, Krishnamoorthy D, and Camacho E. ALADIN-based
       Distributed Model Predictive Control with dynamic partitioning: An appli-
       cation to Solar Parabolic Trough Plants. arXiv preprint arXiv:230502821.
       2023.
[59]   Dai X, Lian Y, Jiang Y, Jones CN, and Hagenmeyer V. Hypergraph-based fast
       distributed AC power flow optimization. In: *2023 62nd IEEE Conference on
       Decision and Control (CDC)*. Piscataway, NJ: IEEE; 2023. p. 4572–4579.
[60]   Zhai J, Dai X, Jiang Y, *et al.* Distributed optimal power flow for VSC-MTDC
       meshed AC/DC grids using ALADIN. *IEEE Transactions on Power Systems*.
       2022;37(6):4861–4873.
[61]   Jiang Y, Zanon M, Hult R, and Houska B. Distributed algorithm for
       optimal vehicle coordination at traffic intersections. *IFAC-PapersOnLine*.
       2017;50(1):11577–11582.

[62] Shi J, Zheng Y, Jiang Y, Zanon M, Hult R, and Houska B. Distributed control algorithm for vehicle coordination at traffic intersections. In: *2018 European Control Conference (ECC)*. Piscataway, NJ: IEEE; 2018. p. 1166–1171.

[63] Engelmann A, Jiang Y, Houska B, and Faulwasser T. Decomposition of non-convex optimization via bi-level distributed ALADIN. *IEEE Transactions on Control of Network Systems*. 2020;7(4):1848–1858.

[64] Engelmann A, Stomberg G, and Faulwasser T. An essentially decentralized interior point method for control. In: *2021 60th IEEE Conference on Decision and Control (CDC)*. Piscataway, NJ: IEEE; 2021. p. 2414–2420.

[65] Engelmann A. *Distributed Optimization with Application to Power Systems and Control*. KIT Scientific Publishing; 2022.

[66] Engelmann A, Jiang Y, Benner H, Ou R, Houska B, and Faulwasser T. ALADIN-α – An open-source MATLAB toolbox for distributed non-convex optimization. *Optimal Control Applications and Methods*. 2022;43(1): 4–22.

[67] Burk D, Völz A, and Graichen K. Distributed optimization with ALADIN for non-convex optimal control problems. In: *2020 59th IEEE Conference on Decision and Control (CDC)*. Piscataway, NJ: IEEE; 2020. p. 4387–4392.

[68] Houska B, and Jiang Y. Distributed optimization and control with ALADIN. *Recent Advances in Model Predictive Control: Theory, Algorithms, and Applications*. 2021;p. 135–163.

[69] Grüne L, and Pannek J. *Nonlinear Model Predictive Control*. Springer; 2017.

[70] Allgöwer F, Findeisen R, and Nagy ZK. Nonlinear model predictive control: From theory to application. *Journal-Chinese Institute of Chemical Engineers*. 2004;35(3):299–316.

[71] Vazquez S, Leon JI, Franquelo LG, *et al.* Model predictive control: a review of its applications in power electronics. *IEEE Industrial Electronics Magazine*. 2014;8(1):16–31.

[72] Sultana WR, Sahoo SK, Sukchai S, Yamuna S, and Venkatesh D. A review on state of art development of model predictive control for renewable energy applications. *Renewable and Sustainable Energy Reviews*. 2017;76:391–406.

[73] Hans CA, Braun P, Raisch J, Grüne L, and Reincke-Collon C. Hierarchical distributed model predictive control of interconnected microgrids. *IEEE Transactions on Sustainable Energy*. 2018;10(1):407–416.

[74] Schwenzer M, Ay M, Bergs T, and Abel D. Review on model predictive control: an engineering perspective. *The International Journal of Advanced Manufacturing Technology*. 2021;117(5–6):1327–1349.

[75] Sharadga H, Hajimirza S, and Balog RS. Time series forecasting of solar power generation for large-scale photovoltaic plants. *Renewable Energy*. 2020;150:797–807.

[76] Richter L, Bauer F, Klaiber S, and Bretschneider P. Day-ahead electricity load prediction based on calendar features and temporal convolutional networks. In: *International Conference on Time Series and Forecasting*. Springer; 2021. p. 243–253.

[77] Viehweg J, Worthmann K, and Mäder P. Parameterizing echo state networks for multi-step time series prediction. Neurocomputing. 2023;522:214–228.

[78]   Worthmann K, Kellett CM, Braun P, Grüne L, and Weller SR. Distributed and decentralized control of residential energy systems incorporating battery storage. *IEEE Transactions on Smart Grid*. 2015;6(4):1914–1923.

[79]   Braun P, Grüne L, Kellett CM, and Worthmann K. Model predictive control and distributed optimization in smart grid applications. In: *Handbook of Smart Energy Systems*. Springer; 2023. p. 1239–1263.

[80]   Frank S, and Rebennack S. An introduction to optimal power flow: theory, formulation, and examples. *IIE Transactions*. 2016;48(12):1172–1197.

[81]   Mühlpfordt T, Faulwasser T, Roald L, and Hagenmeyer V. Solving optimal power flow with non-Gaussian uncertainties via polynomial chaos expansion. In: *2017 IEEE 56th Annual Conference on Decision and Control (CDC)*. Piscataway, NJ: IEEE; 2017. p. 4490–4496.

[82]   Bienstock D, Escobar M, Gentile C, and Liberti L. Mathematical programming formulations for the alternating current optimal power flow problem. *Annals of Operations Research*. 2022;314(1):277–315.

[83]   Frank S, Steponavice I, and Rebennack S. Optimal power flow: a bibliographic survey I: formulations and deterministic methods. *Energy Systems*. 2012;3:221–258.

[84]   Frank S, Steponavice I, and Rebennack S. Optimal power flow: a bibliographic survey II: non-deterministic and hybrid methods. *Energy Systems*. 2012;3:259–289.

[85]   Capitanescu F. Critical review of recent advances and further developments needed in AC optimal power flow. *Electric Power Systems Research*. 2016;136:57–68.

[86]   Faulwasser T, and Engelmann A. Toward economic NMPC for multi-stage AC optimal power flow. *Optimal Control Applications and Methods*. 2020;41(1):107–127.

[87]   ALADIN-α MATLAB toolbox. Date of access 1. March 2024. Available from: https://alexe15.github.io/ALADIN.m/.

[88]   Bertozzi O, Chamorro HR, Gomez-Diaz EO, Chong MS, and Ahmed S. Application of data-driven methods in power systems analysis and control. *IET Energy Systems Integration*. 2023.

[89]   Baumann M, Grundel S, Sauerteig P, and Worthmann K. Surrogate models in bidirectional optimization of coupled microgrids. *at-Automatisierungstechnik*. 2019;67(12):1035–1046.

[90]   Zeng S, Kody A, Kim Y, Kim K, and Molzahn DK. A reinforcement learning approach to parameter selection for distributed optimal power flow. *Electric Power Systems Research*. 2022;212:108546.

[91]   Faulwasser T, Ou R, Pan G, Schmitz P, and Worthmann K. Behavioral theory for stochastic systems? A data-driven journey from Willems to Wiener and back again. *Annual Reviews in Control*. 2023;55:92–117.

[92]   Sass S, Faulwasser T, Hollermann DE, *et al.* Model compendium, data, and optimization benchmarks for sector-coupled energy systems. *Computers & Chemical Engineering*. 2020;135:106760.

[93] Meinecke S, Sarajlić D, Drauz SR, *et al.* Simbench—a benchmark dataset of electric power systems to compare innovative solutions based on power flow analysis. *Energies*. 2020;13(12):3290.

[94] Ratnam EL, Weller SR, Kellett CM, and Murray AT. Residential load and rooftop PV generation: an Australian distribution network dataset. *International Journal of Sustainable Energy*. 2017;36(8):787–806.

[95] Spalthoff C, Sarajlic D, Kittl C, *et al.* SimBench: Open source time series of power load, storage and generation for the simulation of electrical distribution grids. In: *International ETG-Congress 2019; ETG Symposium*. VDE; 2019. p. 1–6.

Chapter 6

Integrating distributed energy resources in real-world sector-coupled microgrids: challenges, strategies, and experimental insights

Friedrich Wiegel[1], Sophie An[1], Jan Wachter[1],
Sebastian Beichter[1], Anne-Christin Süß[1],
Ömer Ekin[1] and Veit Hagenmeyer[1]

This chapter provides a comprehensive overview of the state of the art of energy management systems (EMS) in the context of microgrids and focuses on practical applications and challenges in real-life scenarios. The contribution centers around implementation and evaluation of a rule-based EMS within the Smart Energy System Laboratory (SESCL) at Karlsruhe Institute of Technology (KIT). Utilizing a specially developed EMS testbench, the behavior of the system is investigated in scenarios that reflect non-ideal conditions of real components. The real-time measurement campaign includes a systematic evaluation under two different scenarios: high renewable energy generation with low local consumption and vice versa. In addition to demonstrating and quantifying the performance of different EMS operating strategies, the experiments reveal unpredictable behaviors and challenges when integrating simulation-tested systems into practice. These results highlight the complexity associated with the transition from simulation-tested models to real-world applications.

6.1 Introduction

Addressing the challenges posed by the restructuring of the energy system toward a sustainable, reliable but also economically feasible energy supply can be achieved by the deployment of distributed energy resources (DERs) [1,2]. DERs are power generation units, preferably based on renewable resources such as photovoltaic systems and wind turbines, but also conventional fuel-based generators powered by gas or diesel, as well as a range of energy storage systems (ESS) [3,4]. This development can also be seen on the low voltage (LV) level of the grid: The large-scale integration of new devices such as electric vehicle charging (EVC), heat pumps as well as local generation in the form of rooftop-photovoltaic (PV) systems leads to vastly changed

---

[1]Institute for Automation and Applied Informatics, Karlsruhe Institute of Technology Karlsruhe, Germany

consumption patterns. Such developments have the potential to cause congestion [5], voltage imbalances [6] or power quality issues [7,8]. Furthermore, advancements in information technology have enabled the monitoring and data collection from these grid participants, transforming the lower voltage level from a passive load to an active grid asset. The quickly rising number of such active energy resources vastly increases the control complexity which urgently needs to be addressed. For this, microgrids (MG), as power systems with defined boundaries, provide a viable solution to enable the integration of DERs while increasing the overall reliability of the power grid [9]. This requires to ensure power reliability for consumers, maintaining the power quality, achieving economic feasibility within the microgrid, and optimal utilization of the available resources to minimize the need for grid reinforcement. Effective control strategies are required, encompassing challenges across various timescales, from short-term power balance to long-term energy supply security. Part of this solution is the application of an energy management system (EMS) for the considered microgrid, which can be seen as the link between short-term control concerned with operational stability and long-term planning of resources.

The research community can contribute to the solution of this problems in several ways: by proposing conceptual solutions, by developing new technologies, and by demonstrating through real-world applications that existing solutions are viable and have reached a high Technology Readiness Level (TRL), thus promoting awareness and acceptance in society. To foresee arising problems during these fast-paced changes, it is essential to leverage advanced testing laboratory infrastructures to provide hands-on experience and enable realistic test scenarios of advanced control strategies for microgrid EMS. Real-world application of EMS faces multiple challenges, including the need for appropriate communication infrastructure and interfaces, unpredictable behavior of real hardware, and the requirement to provide flexible solutions tailored to the specific component mix of a microgrid while maintaining compatibility with new components.

Overall, the presented contributes to tackling the issues by (1) providing an in-depth overview of current EMS technologies, (2) detailing the implementation and validation of an EMS testbed, (3) evaluating rule-based EMS in various scenarios to understand the behavior of real-world components, and (4) highlighting unexpected challenges in integrating EMS into real-world applications.

The remainder of this chapter is structured as follows. Section 6.2 provides a brief overview of the relevant literature and related, hardware-based case studies on EMS systems. Next, Section 6.3 describes the laboratory environment with its communication and automation infrastructure, which forms the core of the EMS testbench environment. Section 6.4 then describes the microgrid under study and its individual components in detail. Section 6.5 first presents the experimental scenarios considered and then discusses the results obtained. Finally, Section 6.6 concludes the chapter.

6.2   State of the art EMS and relevant case studies

With the multitude of variable and active participants within a EMS take on the task of directing their operation toward a common goal for the participants under its control. The EMS must achieve the predetermined objective while still following the constraints of the given grid structure and reacting to the varying circumstances given by the intermittent generation of renewable devices.

In general, the overarching challenge for EMS systems is two-fold: (i) generating an optimal schedule for the available resources under uncertainty of the generation and load forecasts and (ii) reacting to the current state of the considered system, e.g., power balance or reaction to fault situations. These two essential functions operate on different timescales: whereas the schedule generation is typically in the range of hours-ahead to day-ahead, the reaction to the current state of the system must be in the seconds to minutes range. Modern EMS systems can therefore be seen as energy and power management systems. The latter function overlaps with the timescales of newly emerging grid ancillary services that are required to cope with the faster dynamics of inverter-dominated power grids. For a stable operation of a MG, the EMS needs to work on top of the faster control structures of both, each single device as well as any possible services, that the MG provides for the main grid.

For the first identified task of EMS, the optimal scheduling on a larger timescale, many publications have focused on possible implementations of EMS in MGs with different objectives and optimization methods in recent years. Extensive work has been undertaken on energy scheduling and forecast-based optimal management of a MG. Works such as Zia *et al.* [10], Meng *et al.* [11] and Ahmad *et al.* [12] give a more detailed overview of recent works based on their chosen solution algorithms for the optimal energy allocation, the control strategy and the uncertainty forecasts used in the decision making of the EMS. For further information on these aspects of EMS in MGs we refer to the studies cited therein.

The present work focuses on establishing a real experimental MG with an EMS and to showcase the communication structure and control needed to establish a coherent energy strategy for the MG as a whole. Special attention is put on the influence of the non-optimal behavior of realistic loads as well as the control and communication structures that are needed to provide a well-suited power management to react to this behavior within a short time frame. Research toward the real-world impact of advanced EMS systems as well as arising challenges of the implementation of such approaches is strongly linked to case studies as presented in this section. For this, we give a brief overview of selected case studies including real hardware in this field. Table 6.1 summarizes the case studies and gives information about the used algorithm, considered devices, the power-scale and the used communication protocols (if the respective information is available). As can be seen from the case studies in Table 6.1, the need for communication to orchestrate the different resources is a major challenge for implementation, as different protocols need to be integrated into the overall system. The provision of specialized research infrastructure solving this task enables researchers as well as industrial partners to test new EMS systems under realistic conditions without the need to deal with low-level communication. Especially the fast-timescale tasks need to be verified under realistic communication conditions, since the setpoint sending as well as obtaining measurement data is time-critical.

Table 6.1 Overview of related case studies and use cases

Use case	Description	References
Survey on smart energy management in cities	Broad survey of relevant technologies and case studies toward smart energy management and smart cities.	[13]
EMS for the microgrid of the marina of Ballen, Samsø, Denmark	Model Predictive (MPC) based scheduling of the generation devices, storages, and flexible loads under consideration of the non-flexible generation and demand. Various PV systems, battery energy storage system (BESS), uncontrollable loads, and several controllable loads (HVAC system, Sauna, waste-water pump, etc.). EMS system is studied based on real measurement data of the devices. The devices are full-scale in the kW range. Communication is not specified.	[14]
Laboratory microgrid and smart office buildings at the Sapienza University of Rome (Lambda Lab)	Centralized, rule-based EMS that dispatches the available resources in real-time. PV array, diesel generator, BESS as well as controllable and uncontrollable loads are considered, as well as sector-coupling elements, e.g., heat pump and fuel-cell. The real-time EMS is implemented in the lab using the real components to verify the approach. Simulation results are presented to conduct long-time economic studies. Combination of different communication protocols is used (KNX, Modbus on a TCP/IP network).	[15,16]
Laboratory microgrids at UCLA Smart Grid Energy Research Center (SMERC) and Korea Institute of Energy Research (KIER)	Centralized, optimization-based EMS system is implemented for different microgrid scenarios and testbeds. PV system, BESS, electric vehicle charging (EVC) as well as controllable loads (e.g., dimmable lights) are considered. Unavailable resources are emulated via power amplifiers and simulation models. Different communication protocols, e.g., IEC61850 and OPC/UA are used.	[17,18]
Small-scale laboratory microgrid	Centralized, optimization-based EMS is experimentally studied. Genetic algorithms are used to solve the optimization problem, which considers emissions as well as economic considerations. Different generation resources, e.g., wind turbine generators (WTG), PV, fuel cell, are integrated with local loads as well as energy storage systems. All devices are realized in a reduced-power scale. The communication is based on ZigBee.	[19]
Microgrid testbed at the Canadian Renewable Energy Laboratory (CANREL)	A rule-based and an optimization-based EMS system are compared for the same system. A mixture of emulation-based devices and real devices is used for the experiments, in which renewable (WTG, PV) and conventional generation (Diesel engine) are considered along with storage systems (BESS) and controllable loads. All devices were full-scale in the kW range. Different commercial communication protocols, e.g., Modbus, are used.	[20,21]
Smart poly-generation microgrid of the University of Genoa	Optimization-based EMS with a user-definable objective function that schedules the available resources. The approach offers modularity in the deployed models for the grid and devices. Considered devices include PV generation, microturbine, BESS as well as EVC and uncontrollable load. The used communication is not detailed.	[22]
Microgrid laboratory testbench	Two-layer economic MPC framework for optimization-based EMS. The microgrid consists of a PV system, a BESS, and a gasoline-fueled generator as well as non-controllable loads. All devices are full-scale and connected in a	[23]

The essential components besides the EMS and communication infrastructure, in the considered use cases can be summarized as follows:

- **Generation:** Renewable, such as PV or wind power, and conventional generation, e.g., fuel-based generators, must be treated differently by the EMS. Renewables do not have fuel costs, and therefore must not be curtailed if possible, the operational costs emerge from the wear of storage units in the case of overproduction. However, fuel-based generation should only be used for contingency operations or due to grid stability considerations.
- **Storage:** The importance of storage is directly related to the intermittency of renewable generation and due to the relevant investment costs of storage devices, the optimal use of the capacity by EMS systems is crucial.
- **Loads:** Loads can be broadly categorized as controllable and uncontrollable. Uncontrollable loads are devices that the EMS cannot influence. Controllable loads can be influenced by the EMS, however, the extent can be different, ranging from schedulable with power setpoints to mere interruptible loads that can only be turned on or off.

6.3 Applied research ecosystem

6.3.1 Smart Energy System Control Laboratory

The Smart Energy System Control Laboratory (SESCL), part of the Energy Lab* project [24], is dedicated to the development and testing of advanced power systems and control algorithms in a realistic environment [25]. Galvanically isolated from the public grid, it enables the investigation and evaluation of operating and control strategies in critical stability scenarios. The research facility features a fully automated and user-friendly infrastructure with a microgrid with adjustable topology. This microgrid includes various energy system components such as generators, distributed energy resources (DER), battery energy storage systems (BESS), and smart buildings. The components within this grid can be interconnected in various configurations, following the physical characteristics of transmission lines by using an automated busbar matrix. This design facilitates on-the-fly adjustments such as load shedding, integration of additional generators, consumers, or prosumers, and changing the grid topology on demand. These features make the laboratory predestined for testing the solutions in context of EMSs.

6.3.2 Power system-in-the-loop

The objective of this contribution is to analyze the impact of different energy management strategies on a microgrid with small control reserve. In this case, the entire microgrid and not a single device is the subject of investigation in a realistic environment. As proposed in [26], the term power system in-the-loop (PSIL) is introduced to distinguish the coupling of single hardware components from the coupling of multidevice power hardware, cf. Figure 6.1. The multiple power hardware-in-the-loop

*www.elab2.kit.edu/english/index.php

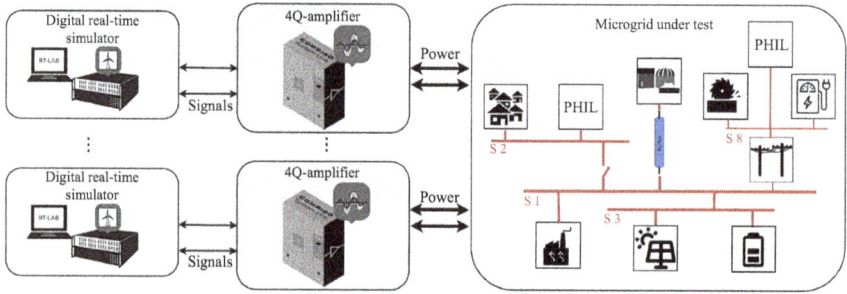

Figure 6.1   General power system-in-the-loop structure

(PHIL) setups that utilize power amplifiers coupled with a real-time simulation system are utilized to simulate controllable participants. The goal is to provide auxiliary services and assure the required system behavior, such as the droop behavior of the grid. To make the power amplifier behave like a real power system resource (e.g., substation with superimposed grid), a sufficiently accurate model of the desired resource is executed on the real-time simulation system. The calculated states are converted into the desired physical quantity (current and voltage) and provided by the power amplifier that interfaces the real microgrid.

6.3.3   Automation and communication architecture

Figure 6.2(a) outlines an automated software architecture of SESCL centered on a supervision system. This system oversees the grid by logically assigning participants to users with the correct permissions, conducting preliminary checks, managing grid topology, monitoring power, and ensuring operational safety. It also records events and communicates with director systems, power analyzers, data management systems, and intermediate services. The architecture also includes four director systems responsible for orchestrating experiments, aggregating data, and managing communications with the intelligent electronic devices (IEDs) [27] of grid participants. These systems use real-time scheduling and support manual or automated control commands. IEDs use a Generic Component Framework (GCFW) [25] to provide an abstraction layer that ensures compatibility with different hardware and enables remote procedure calls for smooth integration into the laboratory system. The data management framework includes several databases, in particular for administrative data, and time series databases for managing high-frequency raw data and other measurements, respectively. Intermediary services are used to optimize resource sharing and processing times by integrating open-source tools such as ELSA [25], Grafana,[†] and Tango Controls,[‡] and to ensure accurate time synchronization between devices. The Process Control Center (PCC) provides interfaces for administration, customer interaction, and monitoring, and provides real-time visualization and control of

---

[†] https://grafana.com/
[‡] https://www.tango-controls.org/

the system. It leverages web technologies for user-friendly access to experiment management and data visualization.

In SESCL, various automation and control approaches are developed, tested, and validated in the context of the LV power system. The lab's experimental range encompasses a wide array of applications, from studying heat dynamics to managing different levels of energy reserves: from tertiary and secondary down to primary and instantaneous reserves. To facilitate the collaboration required during the stages of development and testing—which includes real-time simulators, power hardware-in-the-loop setups, databases, and actual physical components—a specialized framework for data exchange is being established. This framework ensures the seamless real-time transfer of information and is designated as the Energy Lab Link Environment (ELLE), which is depicted in Figure 6.2(b).

ELLE is a state-of-the-art middleware platform that facilitates the swift integration and validation of automation, control, and energy management algorithms for microgrids and energy systems in the SESCL environment. This platform enables these algorithms to be tested in real-time with real power system components, which advances the rapid development of operating algorithms for power systems. The essential benefit is that researchers can work within a familiar ecosystem, leveraging well-known tools like MATLAB®/Simulink® and C/C++ for algorithm development, deployment, and validation on the real system. The process is even more accelerated by eliminating the tedious task of programming unique interfaces for each device.

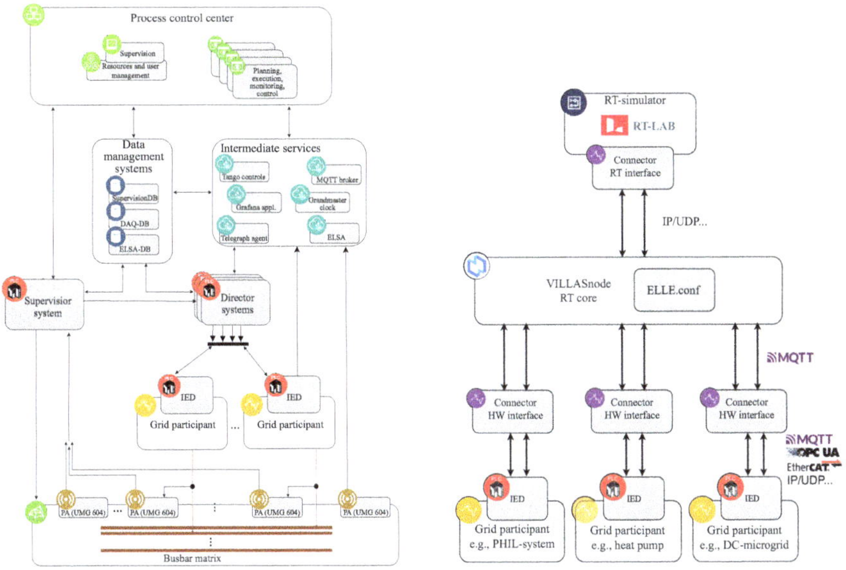

Figure 6.2   (a) Automation architecture [25] and (b) energy lab link environment structure

Its architecture supports the asynchronous control of grid participants, synchronous state and parameter acquisition of the same, and also the grid statuses. Moreover, it inherits all the safety features from the SESCL framework, which underscores its commitment to operational security and reliability. The core architecture of ELLE consists of several essential elements, (a) RT simulator, (b) VillasNode[§] [28] with real-time functionality, (c) connector library, and (d) grid participants with IED client equipped with GCFW. The key elements of ELLE are the connectors, consisting of a device driver, a data mapper and a frame builder, and an appropriate VILLASnode configuration defined in the *ELLE.conf* file. The VILLASnode acts as an agent that forwards the data from the RT connector to the HW connectors, which then communicate with the IEDs via various protocols such as EtherCAT, OPC UA, MQTT, etc. and vice versa, while ensuring the timing requirements and the correct sequence. Connectors provide an abstraction layer that ensures they are exchanged, interpreted, and assigned to the appropriate target. To do this, they use the functions available on each side (RT and HW).

6.4   Microgrid under study

To assess the effectiveness and feasibility of proposed energy management strategies, a microgrid with various DERs and dynamic loads is employed. The considered topology is inspired by the STAGE76[¶] student residence in Bruchsal, Germany. This modern, sustainable, and energy-efficient residential complex generates power using a rooftop solar panel system and Savonius wind turbines. It also incorporates battery storages, Electrical Vehicle Supply Equipment (EVSE), and reuses operational waste heat for heating purposes. In this setup, the sustainable design of the complex is adapted to a microgrid structure and enhanced with an efficient waste heat recovery heat pump and connection to a DC microgrid. The shift toward DC microgrids is driven by the prevalence of direct current usage in electricity storage systems, renewable energy sources, and most electricity consumers. These DC microgrids minimize the need for frequent AC to DC conversions, leading to more efficient energy utilization. The DC microgrid and the battery system are used as versatile ESS. The power consumption profile of a residential complex is represented by programmable loads and real smart buildings of the Living Lab Energy Campus project [29]. To accurately simulate the system's performance, factors like voltage drops and line losses in the connecting cables are considered by incorporating a line replica. All components are connected at a single point of common coupling (PCC) to a substation of the main power grid. The layout of this microgrid is illustrated in Figure 6.3 and individual components are described in more detail in the following sections.

[§]https://villas.fein-aachen.org/docs/node/
[¶]www.stage76.de

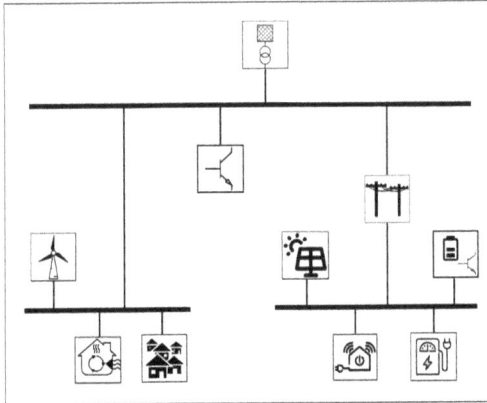

Figure 6.3   Microgrid topology used in the case study

Table 6.2   Substation model parameters

Parameter	Value	Description
$k_{Pf, sub}$	$0.1\ \frac{Hz}{kW}$	P/F droop coefficient
$k_{QV, sub}$	$0.1\ \frac{V}{kVAr}$	Q/V droop coefficient

6.4.1   Substation

The substation[||] acts as the connection point between the microgrid and a superimposed grid, operating with P/f and Q/V droop behavior in accordance with (6.1). Where P_n, Q_n denote the nominal values for active and reactive Power, the corresponding measured values are given by P_m, Q_m. The reference values f_D, V_D are used for the inverse Park transformation to transform a dq0 rotating reference frame to a three-phase signal. The parameters are summarized in Table 6.2, and the nominal values are chosen according to the nominal load of the devices in the microgrid.

$$f_D = k_{Pf} (P_n - P_m)$$
$$V_D = k_{QV} (Q_n - Q_m) \tag{6.1}$$

6.4.2   Wind turbine generator

In the utilized microgrid, the STAGE76 Savonius wind turbines are substituted by a three-bladed horizontal-axis wind turbine (HAWT) model, which offers a comparable performance. The wind turbine generator model is run on a real-time simulator

---

[||]To ensure consistent system behavior and the ability to run experiments regardless of weather or time of day, some microgrid participants are simulated using a real-time simulator that controls a power amplifier.

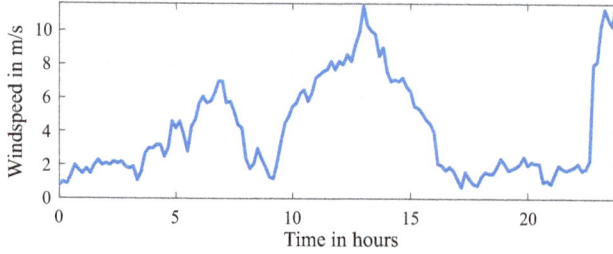

Figure 6.4   Exemplary windspeed profile for the geographic location of the considered microgrid

Table 6.3   WTG system parameters

Parameter	Value	Description
$S_{\text{wtg,n}}$	10 kVA	Nominal power
w_{nom}	$8\,\frac{\text{m}}{\text{s}}$	Nominal windspeed
w_{cutin}	$5\,\frac{\text{m}}{\text{s}}$	Cut-in windspeed
w_{cutout}	$15\,\frac{\text{m}}{\text{s}}$	Cut-out windspeed

to guarantee a reproducible behavior of the device. The deployed control strategy can be categorized as grid-supporting [30]. If no setpoints are commanded from the EMS, the device aims to achieve a maximum active power injection. However, if setpoints for active or reactive power are commanded, the devices follow the request if the constraints of the available power allow it. The system is modeled to mimic a typical behavior of WTG systems, with a wind speed w_s over available power $S_{\text{wtg,av}}$ profile as depicted in Figure 6.4 [31]. This yields the following equations for the available power

$$
S_{\text{wtg,av}} = \begin{cases} 0, & \text{for } w_s < w_{\text{cutin}} \\ S_{\text{wtg,n}} \frac{(w_s - w_{\text{cutin}})}{(w_{\text{nom}} - w_{\text{cutin}})}, & \text{for } w_{\text{cutin}} \le w_s < w_{\text{nom}} \\ S_{\text{wtg,n}}, & \text{for } w_{\text{nom}} \le w_s < w_{\text{cutout}} \\ 0, & \text{for } w_{\text{cutout}} \le w_s \end{cases} \tag{6.2}
$$

where $S_{\text{wtg,n}}$ is the nominal power of the system and w_{cutin}, w_{nom}, w_{cutout} the cut-in, nominal and cut-out wind speeds of the WTG, respectively. The parameters of the WTG system are summarized in Table 6.3.

6.4.3   Photovoltaic system

The PV system model is also operated on a real-time simulator, employing the identical control strategy as the WTG system: If no EMS command is given, the objective is to achieve maximum active power injection, otherwise the compliance with the

Table 6.4   PV system parameters

Parameter	Value	Description
$S_{pv,peak}$	10 kVA	Peak power
η_{nom}	1000 $\frac{W}{m^2}$	Nominal irradiance
T_{nom}	25 °C	Nominal system temperature
c_T	$-0.4\frac{\%}{°C}$	Temperature coefficient

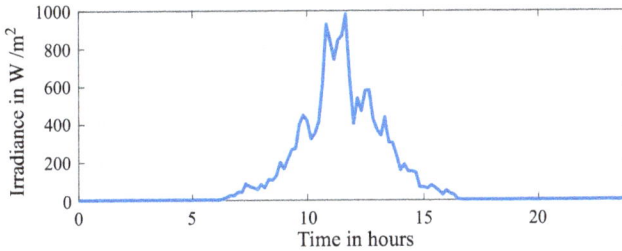

Figure 6.5   Exemplary 24 h profile of solar irradiation for the geographic location of the considered microgrid

commanded setpoints. The system is modeled to mimic a typical behavior of PV system, with an irradiation η_{irr} to available power $S_{pv,av}$ according to [32]

$$S_{pv,av} = S_{pv,peak}\frac{\eta_{irr}}{\eta_{nom}}(1 + (T_c - T_{nom})c_T), \qquad (6.3)$$

where $S_{pv,peak}$ is the power of the PV plant at nominal conditions defined by the irradiance η_{irr} and module temperature T_{nom} and c_T is the temperature degradation coefficient. Table 6.4 summarizes the PV system parameters.

6.4.4   Battery energy storage system

The BESS is developed using prototyping equipment for the power electronics that is supplied by a DC source. To ensure repeatable testing conditions, the battery behavior and state of charge (SoC) are modeled using the real-time controller of the inverter system. This is justified by the considered timescales of the experiments, the low-level behavior of the battery has only negligible influence on the power flow and the considered AC side dynamics. The relevant parameters of the BESS are summarized in Table 6.5. The BESS can be freely controlled by the EMS within the operational limits set by maximum active and reactive power, as well as the SoC.

Table 6.5   BESS system parameters

Parameter	Value	Description
$S_{bess, max}$	10 kVA	Maximal power
E_{bess}	15 kWh	BESS capacity
η_{bess}	50 W	BESS losses

Table 6.6   EVSE parameters

Parameter	Value	Description
$S_{wallbox,peak}$	22 kVA	Peak power
SOC	0–100 %	State of charge

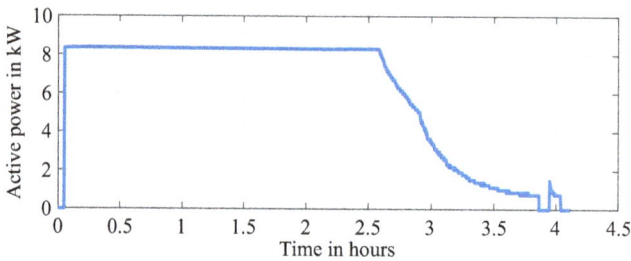

Figure 6.6   Exemplary charging curve of an EV (Nissan Leaf)

6.4.5   Electrical vehicle supply equipment

The EVSE is the connector between the EV and the electrical grid. It is responsible for the safe operation during charging and discharging of the EV. For this purpose, the charging station communicates the corresponding maximum active power and current values to the vehicle and provides further safety features. Due to the norm DIN EN IEC 61851-1 it is not possible to set the reactive power. Moreover, the limits of the current per phase to be set are either between 6 A and 32 A for a wallbox with a maximum power of 22 kW resulting in the output power as stated in

$$P_{wallbox} = \begin{cases} 0, & \text{for } I_{set} < 6A \\ P, & \text{for } I_{set} > 6A \end{cases} \tag{6.4}$$

The real charging power can subsequently vary greatly in terms of the number of phases used and the current set, depending on the vehicle to be charged and the state of charge of the vehicle, cf. Figure 6.6.

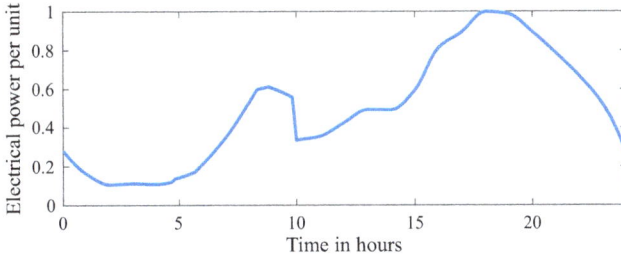

Figure 6.7   Electricity demand of residential load

6.4.6   Residential load

To represent the total electrical power consumption within a residential area, an adjustable passive RL-Load that follows an aggregated power profile, is used. The profile is generated using a planning tool for district projects.** The 50 occupied student residences are represented by the aggregated load profile shown in Figure 6.9 with a peak load of 15 kW at 7 pm, taking into account the fact that the students in the dormitory hardly have any household appliances with high power consumption and that we expect a high statistical smearing due to the low usage in the apartment. For the EMS in this study, the household load is considered uncontrollable, since all schedule-able loads that can be part of a household, e.g., heat pump, EV charging, or battery system are considered separately.

6.4.7   Heat pump

Heat pumps are vital in connecting the heating and power sectors, especially within microgrids. Their multifunctional role allows them to address heating requirements while aiding the local electricity grid. These devices convert electricity into heat, allowing electrical energy that can't be distributed to the local grid to be used for heating (Energy Conversion). In addition, heat pumps can be shut down to reduce peak electricity demand (Demand Response). By using thermal storage systems, they make an important contribution to peak load shifting in the residential sector.

For integration into an EMS, the Smart Grid Ready Label (SG-Ready) defines interfaces and operating modes for heat pumps and is available for a wide range of heat pumps on the German-speaking market. With the SG-Ready label, a heat pump covers the following four operation states [33]:

Blocking mode: The heat pump is switched off for a maximum of 2 h when the limits, e.g., temperature boundaries of the thermal energy storage, are respected.
Normal operation: The heat pump utilizes an internal control algorithm.
Boosted operation: The heat pump is recommended to be switched on. Following this recommendation will result in a higher electric power consumption compared to the normal operation mode.

**www.acad.npro.energy

Peak operation: The heat pump and available electric (backup) heaters are switched on to maximize the electric power consumption.

The used model of the heat pump is SG-Ready enabled and consists of three parts: The control unit (CU), the heat pump (HP), and the thermal energy storage (TES). The parameters of the heat pump model are summarized in Table 6.7.

The decision regarding whether the heat pump may follow the SG-Ready signals ($CU_{Command} = 2$) or its internal control algorithm ($CU_{Command} = 1$) is made in the control unit. The state of charge of the thermal energy storage (TS_{Status}) is taken into account with a specific focus on examining the following three states: $SOC_{TES} \geq 96\%$, $SOC_{TES} \leq 30\%$ and $30\% \leq SOC_{TES} \leq 96\%$

In the heat pump part, the conversion of the internal control algorithm and the operation modes of the SG-Ready signals into thermal power setpoints is defined based on the signal $CU_{Command}$. Figure 6.8 shows the processing of SG-Ready signals into thermal power setpoints. The necessary electrical power for providing the thermal power is determined through a calculation utilizing a constant COP.

The thermal energy storage part determines the current state of charge of the thermal energy storage (TS_{Status}) and sends this information to the control unit. The thermal load profile of a single-family house, which impacts the state of charge of the thermal energy storage (TES), is considered in this part. The single-family house is represented by the thermal load profile shown in Figure 6.9 with a peak load of 3250 W at 6h30 and 2580W at 21h30.

Table 6.7   HP system parameters

Parameter	Value	Description
P_{nom}	1.25 kW	Nominal active power
$\dot{Q}_{nom}$	5 kW	Nominal thermal power
COP	4	Coefficient of performance
C_{TES}	35 kWh	Heat capacity of the thermal energy storage

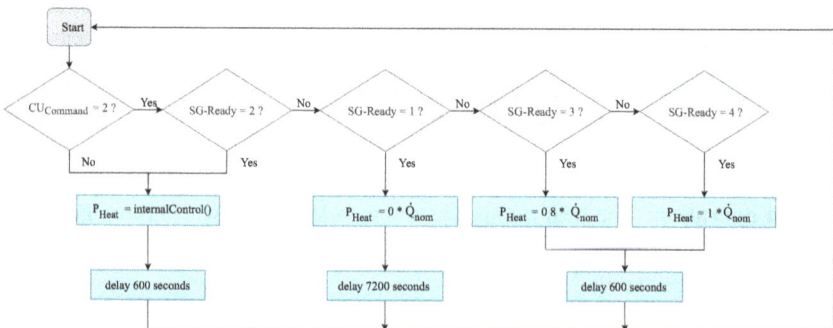

Figure 6.8   Flow chart of the heat pump operation states

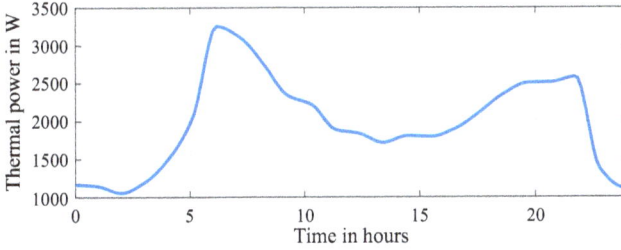

Figure 6.9   Heat demand of the building for 24 h

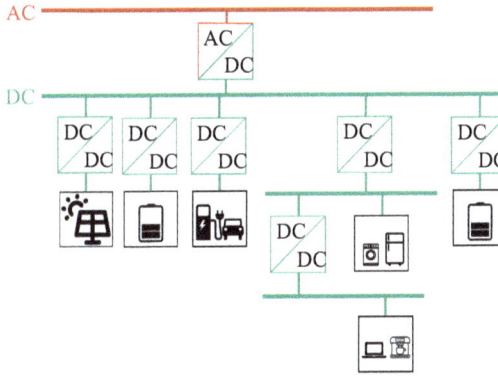

Figure 6.10   DC microgrid hardware setup

Table 6.8   DC microgrid parameters

Parameter	Min. value	Max. value	Description
P_{ref}	$-10\,kW$	$10\,kW$	Reference active power
$\dot{Q}_{ref}$	$-10\,kvar$	$10\,kvar$	Reference active power
SoC_{DCM}	0%	100%	Representative SoC of the DC microgrid

6.4.8   DC microgrid

The DC microgrid is engineered to be a flexible and supportive element in the broader microgrid infrastructure, contributing ancillary services to enhance the performance and resilience of the AC grid. Figure 6.10 illustrates the system setup and detailed information about the converters technical specifications and control technology is available in [34]. It consists of two batteries controlled to ensure a desired DC voltage and a PV system controlled by an MPPT tracker to provide as much power as possible. In terms of this study, this particular DC microgrid is designed to enhance the AC grid by providing virtual inertia, active and reactive power and facilitating load shedding. The interface implemented for this purpose corresponds to the parameters in Table 6.8.

6.4.9 Energy management systems

In the presented contribution, three versions of rule-based EMS are implemented. The first version acts as a standard household EMS, which manages the charging and discharging of a battery via a solar module and supplies a varying household load. The battery is managed based on a simple logic it charges, when the solar panels produce more power than consumed by the household until its maximum capacity is reached. Similarly, it discharges, when the household's need cannot be met by the solar power until the battery is emptied.

The second version of the EMS considers the Microgrid (MG) in its entirety, assigning power setpoints to each controllable unit within the system. The fundamental objective of this EMS's internal logic is to minimize the amount of power exchanged with the main electrical grid. For this reason, the active power measurements of all components are collected and communicated to a central EMS. Since the EMS is not supposed to replace any control actions, that are needed to maintain the voltage and frequency stability of the grid, the setpoints are given once every minute. This chosen time frame gives the opportunity to react to fluctuations in generation and load, while still being slow enough not to disturb the internal controls of the devices.

This EMS calculates appropriate setpoints for the components with a simple rule-based logic, which is shown in Figure 6.11. The system prioritizes the charging and discharging of the battery as a first action so that any other action disturbing the normal operation of the other grid participants is kept to a minimum. In the case of larger discrepancies between generation and load, that the battery cannot provide on its own or due to nearly full or empty capacity, other components are given suitable setpoints by the EMS. Possible actions that can be taken for

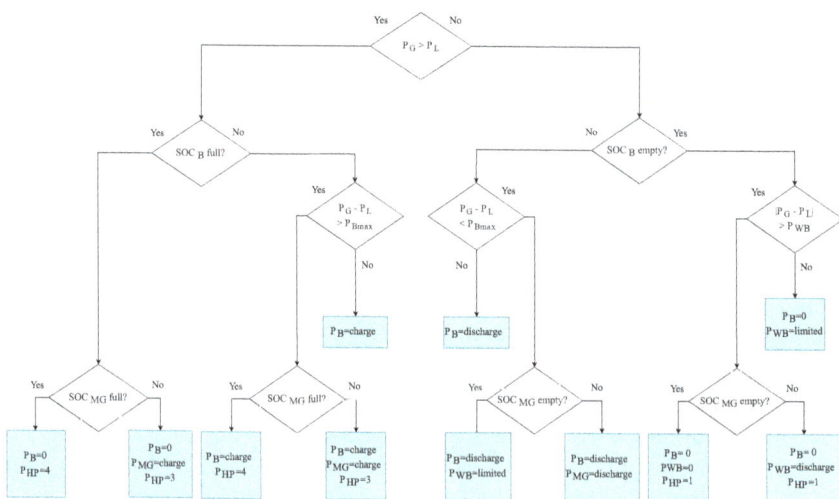

Figure 6.11   Decision rules for the active power management

overgeneration are increasing the consumption of the heat pump to fill its thermal storage or charging the DC-microgrid. In the opposite case, when there is a lack of renewable generation, the charging power of the electric vehicle can be limited, the heat pump can be completely shut down or the DC Microgrid can be discharged.

6.5 Experimental scenarios and results

This section presents the empirical data acquired during the experiments and explains the corresponding scenarios and associated meta-information. All the measurement data presented emanate from experiments executed in the real-time environment of Section 6.3. During the measurement campaign, for each delineated scenario, two distinct sets of measurements were systematically recorded under two varying contexts: (a) high generation from renewables with low local consumption, and (b) low generation from renewables with high local consumption. The duration of each measurement series is quantified as 2 h and 30 min. In each iteration, the electric vehicle is connected for charging at the 50-min mark and remains attached to the charging station until the end of the iteration.

6.5.1 Scenarios description

Reference Scenario: The initial point of all presented research is centered around the microgrid as illustrated in Figure 6.3, Section 6.4. Within this reference scenario measurement series, specific constraints are imposed on the grid: the BESS and the DC Microgrid are not integrated into the microgrid. There is no active EMS. Participants are able to feed into the grid or draw power from it in an uncoordinated manner.

Home EMS: The second scenario examined in these experiments focuses on the right part of the microgrid. In this context, all grid participants on the lower right bus behave like a modern household with a BESS integrated with the household load and solar cells, imitating the behavior of a modern prosumer. Within this framework, the EMS functions as a Home Energy Management System, responsible for managing the operations within the household, and does not interact with the rest of the microgrid. The charging and discharging of the battery depends on the PV generation, the load consumption, and the state of charge (SoC) of the BESS.

Microgrid EMS: In the final scenario, the EMS interacts with the entire microgrid, following the rule-based logic outlined in Section 6.4.9. It generates setpoints for all controllable units according to two different operational strategies. First, the "Load and Generation Scheduling" strategy is applied, where the EMS provides setpoints to grid participants every 15 min based on available information. The second mode of the EMS functions on a more rapid scale, aiming to achieve Load Balancing and minimize draw from the PCC. In this mode, the setpoints for the network participants are updated every 3 min. This operating strategy is called in the following as "Load Balancing" strategy.

6.5.2   Results and discussion

The measurement series shown in Figure 6.12 presents the power at the PCC of the microgrid (top), the generation profiles of renewables (middle), and the load profiles (bottom) under the reference scenario conditions. Thereby, positive values for the power denotes consumption for loads and production for the generators.

The scenario shown in Figure 6.12(a) shows a timeframe with high generation from the renewable sources. Since there is no energy storage included in this reference scenario, the generation from the wind turbine and the solar cell dominate the overall interaction with the grid and MG feeds power into the grid during the whole window of time, even after the electric vehicle is connected as a load at around 51 min into the experiment. In contrast, Figure 6.12(b) visualizes a period with low generation from RE. The power required to operate the loads is provided almost entirely by the main grid, at least from the start of the charging process.

These data sets serve as the foundation for the qualitative and quantitative assessment of the operational strategies of the EMS discussed in Section 6.5.1. In addition to the qualitative evaluation, evident from the reduction in power consumption or feed-in overtime at the PCC as shown in the graphs, the operational strategies are also quantitatively assessed using the root mean squared deviation.

As a first step, the effects of local home EMS on the PCC of the microgrid are examined. The corresponding measurement results are presented in Figure 6.13(a). Upon examining the PCC, it becomes apparent that there is no significant difference compared to the reference scenario (a). This assumption is supported by the qualitative evaluation of the data. In this case, the root mean squared deviation is even 17.33% or 1.024 kW worse than in the completely uncoordinated operation of the

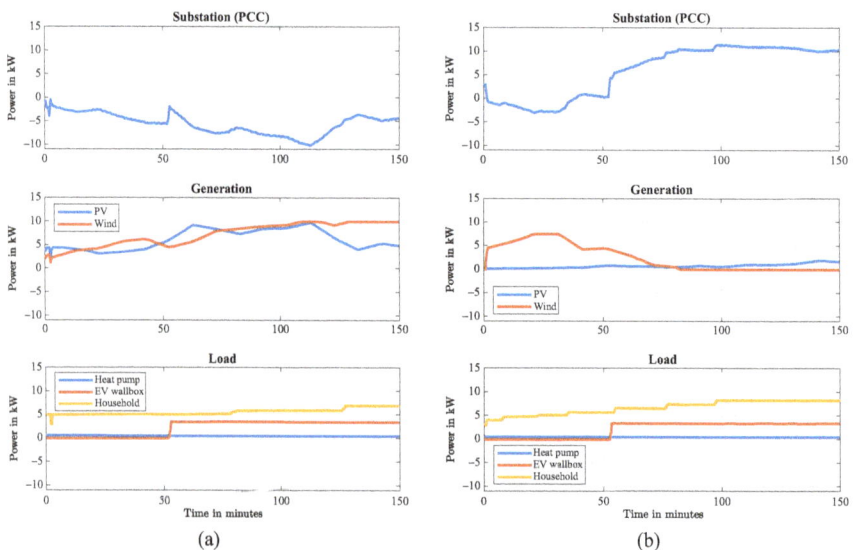

*Figure 6.12   Active power of all components during the reference scenarios:
(a) high RE generation and (b) low RE generation*

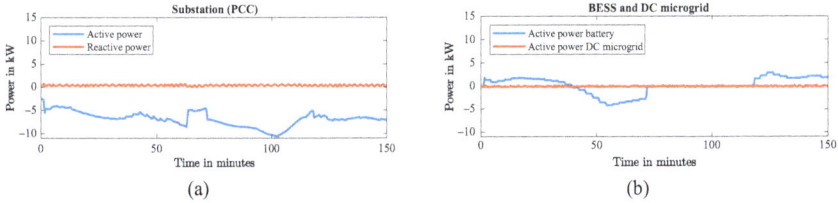

Figure 6.13   *Active power at the PCC (a) and of the battery and DC-MG*
(b) when using local HEMS in the scenario with low generation and
high load

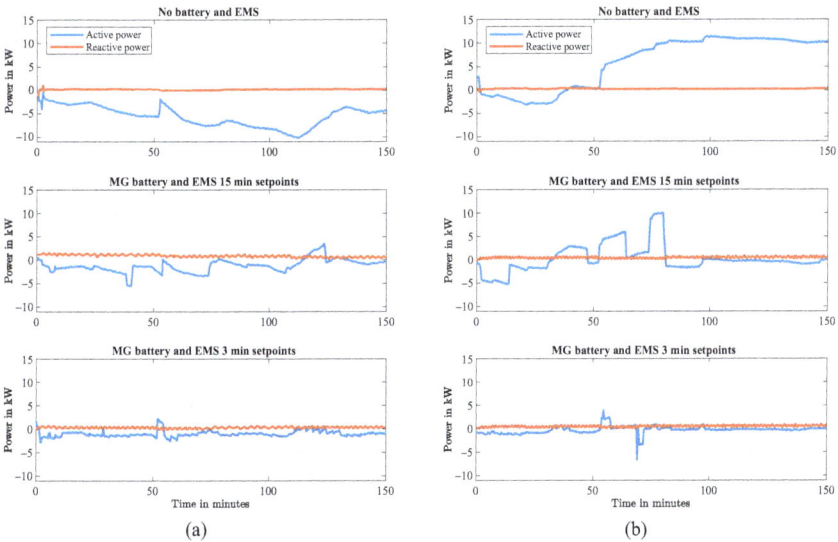

Figure 6.14   *Comparison of active and reactive power at the PCC for different*
EMS operation modes: (a) high RE generation and (b) low RE
generation

microgrid. Since the HEMS does not include the power exchange at the PCC into
its decision basis, the battery is charged and discharged as shown in Figure 6.13(b),
solely based on the discrepancies of the solar generation and the local load. This
leads to a net discharging of the battery of the grid over a timeframe, in which the
generation as a whole dominates.

Figure 6.14 gives an overview on how the different operation modes affected
the power exchange of the microgrid with the main grid. The resulting active and
reactive power at the substation are shown for the complete 150 min of experiment
time. The upper graphs show the active and reactive power at the substation of the

reference scenarios, which were already introduced in Figure 6.12, to enable a direct comparison.

The middle and bottom graphs illustrate the measurement results for the two Microgrid EMS operation modes, with the middle graph displaying the "Load and Generation Scheduling" operation, which generates setpoints every 15 min and the bottom the "Load Balancing" strategy, which works on a faster timescale of 3 min. In this comparison, it can be seen, that both central coordination strategies of the MG components manage to reduce the power exchange with the main grid during both time frames. Both the peak value and the root mean squared deviation of the power exchange is reduced in all four cases, the exact values are denoted in Table 6.9 and proves that a simple rule based EMS can effectively help to achieve a common goal of a MG. Visual examination of the results of the two operation strategies in direct contrast illustrates the differences that result from the two time scales. It can be seen that the fluctuation of the loads and the renewable generation undercuts the reaction time of 15 min. This time lag of the slower reaction causes larger discrepancies during the "locked" time between setpoints and leads to larger steps when a new setpoint is communicated. Especially in scenario (b), this effect is clearly visible. Referring back to Figure 6.12, this can be traced back to the larger changes in the load and almost nonexistent solar generation, which causes the changes in wind power to be much prevalent to the overall behavior of the system. In contrast, wind and solar power supplement each other to a certain degree in scenario (a), leading to a smoother generation curve in total.

Notably, in all four cases, the effect of the beginning of the charging process of the electric vehicle can be recognized as a load step at the PCC in minute 51 of the experiment. In both cases of the "Load Balancing" operation, this load step is smoothed out within a short time frame. Again, the two scenarios of the "Load and Generation Scheduling" operation show different effects. In scenario (a), the connection of the EV happen just shortly before a new setpoint is calculated, leading to a much faster reaction to the step and also a smaller peak. Whereas in scenario (b) the connection happens a few minutes after a new setpoint is calculated and simultaneous to an increase in the household load, thus leading to a much slower reaction and a larger peak. Another point of interest in scenario (b) happens, when the minimum SoC of the battery is reached and it stops providing power to the MG. This leads to the step in power at around minute 70, which is then mitigated in the with the next setpoint by limiting the wall box and dispatching the DC-MG. Figure 6.15 depicts the step caused by the charging of the EV as well as the drop out of the battery.

All these specific moments in addition to the general trends of the measurement results, showcase how the "Load Balancing" strategy takes advantage of the small numbers of the components within the grid and the simple decision method of the EMS to allow a fast decision time to calculate the setpoints. This makes it possible for the EMS to react beneficially to the fluctuating behavior of renewable power sources and human consumers without highly sophisticated forecasting methods.

Due to the fact that the DC microgrid operates on much smaller time scales in terms of dynamics, the interface for power exchange between the DC and AC

Table 6.9   Power exchanged at the substation

Operation strategy	Peak power	Absolute RMSE	Relative RMSE
Timeframe (a)			
Reference	−10.06 kW	5.91 kW	100%
"Load and Generation Scheduling"	−5.46 kW	1.76 kW	29.76%
"Load Balancing"	−4.38 kW	1.39 kW	23.44%
Timeframe (b)			
Reference (b)	11.19 kW	8.09 kW	100%
"Load and Generation Scheduling" (b)	10.07 kW	2.98 kW	36.79%
"Load Balancing" (b)	3.95 kW	0.83 kW	10.30%

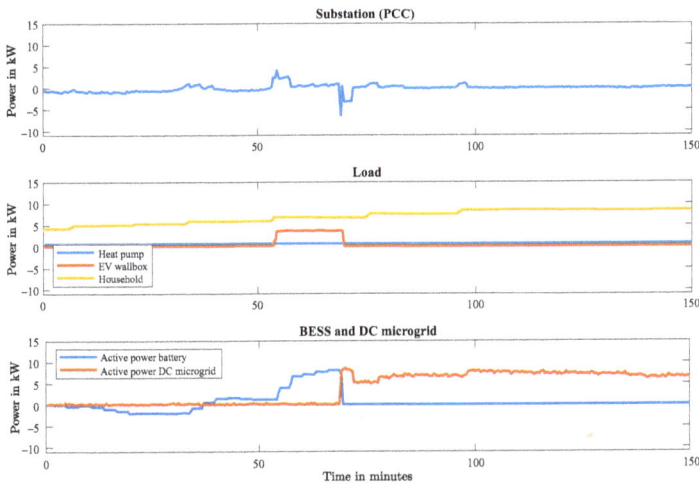

Figure 6.15   Active power at the PCC, of load, of BESS and of DC microgrid for "load balancing" operation with low generation and high load

microgrid requires separate consideration. Figure 6.16 shows the experimental result of the DC grid reacting to an active power reference of $P = -4$ kW between 0.1 sec and 0.2 sec and $P = 4$ kW between 0.1 sec and 0.2 sec. The DC Bus Voltages and Load Voltages deviation is below 2%. While supporting the AC grid with the reference active power, the DC Voltage is kept at its desired value.

Besides the results from the selected scenarios, the experiments also reveal the non-ideal, unpredictable behaviors of the components involved and their impact on the EMS's functionality. Two specific occurrences are highlighted to illustrate the challenges encountered when transitioning systems and technologies from

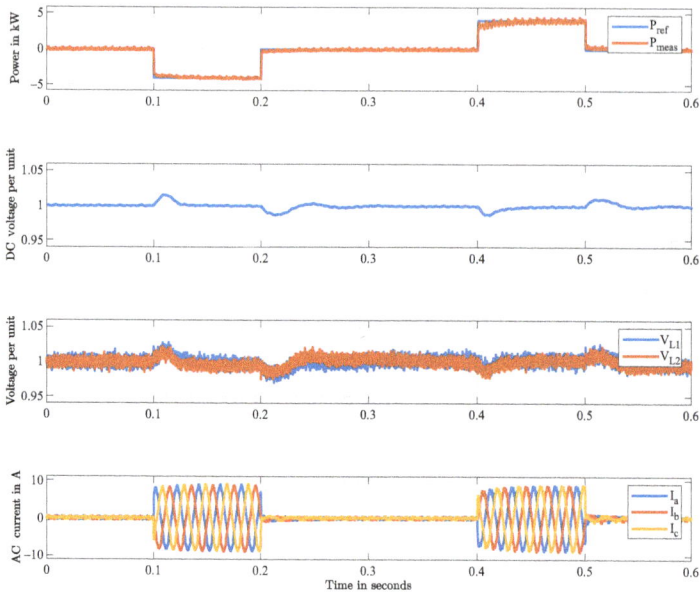

Figure 6.16   The DC microgrid supports the distribution AC grid while stabilizing the DC bus voltage

simulation-based testing to practical laboratory experiments, and of course also into the real-world application in the long run.

The first example focuses on the electric vehicle (EV) charging process. As already noted in the description of the EVSE 6.4.5, the permissible maximal power of the charging process and the range for the charging current is regulated by the wall box or the car itself. This leaves considerable discretion to the manufacturer in managing the EV's charging process. The vehicle shown in these experiments used a single phase for power consumption. This had a twofold impact: firstly, it influenced the logic of the EMS, when limiting the power of the EVSE as a load-reducing option. More critically, this single-phase load caused a significant imbalance in the three-phase voltages of the connected line. In this MG topology, the PV system inverter is connected close to the EVSE, as shown in Figure 6.3. When charging the car at its maximum rated power, the asymmetry of the voltages caused the solar module's converter control to disconnect from the grid and cease feeding in the generated power. Figure 6.17 shows the voltages measured at the point of connection of the solar module and the power of both the solar module and the EVSE around the time of connection.

In the final example, the sudden loss of communication with one of the participants is explained in more detail. The associated measurement series is depicted in Figure 6.18. This series reveals two crucial moments. At the 50th minute, the charging process begins. After a brief peak at the PCC, the sudden increase in

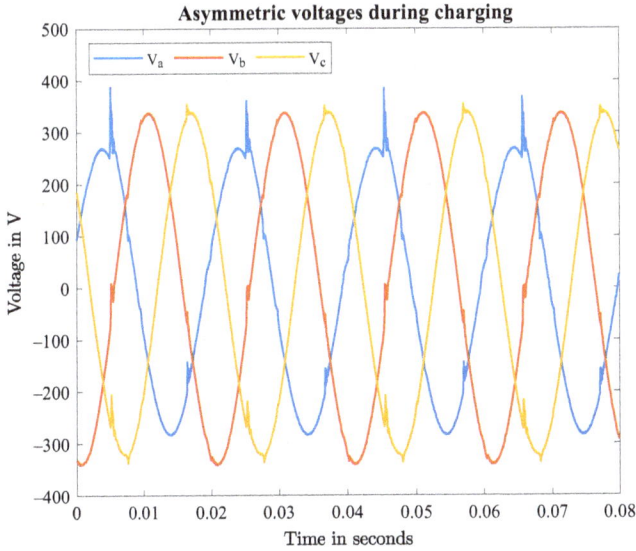

Figure 6.17   Voltages during an asymmetrical charging process at the EVSE

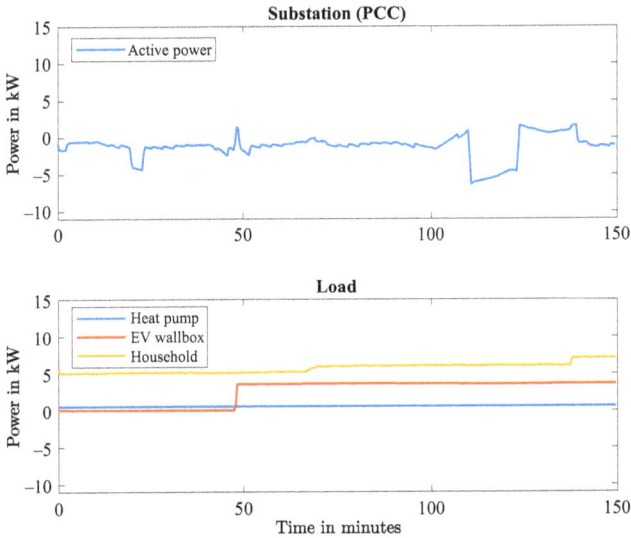

Figure 6.18   Active power at the PCC during the link loss

load is temporarily regulated. However, at the 110th minute, there is a loss of communication with the BESS and the DC Microgrid. As a result, these components enter a safety mode and become passive in their interactions with the Microgrid. Consequently, for about 10 min, the decrease in load at the PCC cannot be compensated for.

6.6   Conclusion

This chapter presents an in-depth review of the current state of EMS, focusing on their real-world application and performance under real-world conditions, assessed using a specialized EMS testbench, called ELLE. It starts with an overview of recent EMS trends, followed by an analysis of pertinent case studies that connect theoretical concepts with practical applications. The core of this study involves creating and utilizing a suitable test environment, along with the implementation and assessment of a rule-based EMS, to investigate the system's behavior in scenarios reflecting non-ideal conditions of real systems within the Smart Energy System Laboratory (SESCL) microgrid at the Karlsruhe Institute of Technology's Energy Lab. The measurement campaign, executed in real-time, plays a pivotal role in understanding EMS performance under varying conditions. This approach facilitates the inclusion of actual grid components like electric vehicle (EV) or renewable energy (RE) resources with their distinct inverter technologies. The specific design by different manufacturers is crucial as it profoundly influences the dynamic processes, impacting both the environmental conditions and the operational characteristics of the EMS. The experiments are conducted over 2-h 30-min intervals, systematically observing the impact of different energy scenarios: high renewable energy generation with low local consumption, and vice versa. These findings highlight the challenges in transitioning from simulation-tested systems to real-world applications, providing critical insights for future EMS deployments. The article underlines the need for further research, particularly in integrating commercially available components, to enhance the resilience and efficiency of EMS in facing real-world challenges. The aim is to pave the way for optimizing the performance of EMS in diverse and dynamic environments, thereby contributing significantly to the evolution of smart energy systems.

References

[1]   Alanne K, and Saari A. Distributed energy generation and sustainable development. *Renewable and Sustainable Energy Reviews*. 2006;10(6):539–558.

[2]   Kroposki B, Johnson B, Zhang Y, *et al.* Achieving a 100% renewable grid: operating electric power systems with extremely high levels of variable renewable energy. *IEEE Power and Energy Magazine*. 2017;15(2):61–73.

[3]   Child M, Kemfert C, Bogdanov D, *et al.* Flexible electricity generation, grid exchange and storage for the transition to a 100% renewable energy system in Europe. *Renewable Energy*. 2019;139:80–101.

[4]   Sikorski T, Jasinski M, Ropuszynska-Surma E, *et al.* A case study on distributed energy resources and energy-storage systems in a virtual power plant concept: technical aspects. *Energies*. 2020;13(12):3086.

[5]   Çakmak HK, and Hagenmeyer V. Using open data for modeling and simulation of the all electrical society in eASiMOV. In: *Proc. 2022 Open Source Modelling and Simulation of Energy Systems (OSMSES)*. Piscataway, NJ: IEEE; 2022. pp. 1–6.

[6]   Möller F, Meyer J, and Radauer M. Impact of a high penetration of electric vehicles and photovoltaic inverters on power quality in an urban residential grid. Part I – Unbalance. *Renewable Energy and Power Quality Journal.* 2016;14(6):817–822.

[7]   Müller S, Meyer J, Möller F, *et al.* Impact of a high penetration of electric vehicles and photovoltaic inverters on power quality in an urban residential grid. Part II – Harmonic distortion. *Renewable Energy and Power Quality Journal.* 2016;14(6):823–828.

[8]   Meyer J, Stiegler R, Schegner P, *et al.* Harmonic resonances in residential low-voltage networks caused by consumer electronics. *CIRED - Open Access Proceedings Journal.* 2017;2017(1):672–676.

[9]   Hatziargyriou N, Asano H, Iravani R, *et al.* Microgrids. *IEEE Power and Energy Magazine.* 2007;5(4):78–94.

[10]   Zia MF, Elbouchikhi E, and Benbouzid M. Microgrids energy management systems: a critical review on methods, solutions, and prospects. *Applied Energy.* 2018;222:1033–1055. Available from: https://linkinghub.elsevier.co m/retrieve/pii/S0306261918306676.

[11]   Meng L, Sanseverino ER, Luna A, *et al.* Microgrid supervisory controllers and energy management systems: a literature review. *Renewable and Sustainable Energy Reviews.* 2016;60:1263–1273. Available from: https://www.scie ncedirect.com/science/article/pii/S1364032116002380.

[12]   Ahmad S, Shafiullah M, Ahmed CB, *et al.* A review of microgrid energy management and control strategies. *IEEE Access.* 2023;11:21729–21757. Available from: https://ieeexplore.ieee.org/document/10050868/.

[13]   Pandiyan P, Saravanan S, Usha K, *et al.* Technological advancements toward smart energy management in smart cities. *Energy Reports.* 2023;10:648–677. Available from: https://www.sciencedirect.com/science/article/pii/S2352484 723010995.

[14]   Carli R, Dotoli M, Jantzen J, *et al.* Energy scheduling of a smart microgrid with shared photovoltaic panels and storage: the case of the Ballen marina in Samsø. *Energy.* 2020;198:117188. Available from: https://www.sciencedirec t.com/science/article/pii/S0360544220302954.

[15]   Kermani M, Adelmanesh B, Shirdare E, *et al.* Intelligent energy management based on SCADA system in a real Microgrid for smart building applications. *Renewable Energy.* 2021;171:1115–1127. Available from: https://www.scien cedirect.com/science/article/pii/S0960148121003566.

[16]   Kermani M, Shirdare E, Najafi A, *et al.* Optimal self-scheduling of a real energy hub considering local DG units and demand response under uncertainties. *IEEE Transactions on Industry Applications.* 2021;57(4):3396–3405.

[17]   Shi W, Lee EK, Yao D, *et al.* Evaluating microgrid management and control with an implementable energy management system. In: *2014 IEEE International Conference on Smart Grid Communications (SmartGridComm)*; 2014. p. 272.

[18]   Lee EK, Shi W, Gadh R, *et al.* Design and implementation of a microgrid energy management system. *Sustainability.* 2016;8(11):1143. Available from: https://www.mdpi.com/2071-1050/8/11/1143.

[19]   Elsied M, Oukaour A, Youssef T, *et al.* An advanced real time energy management system for microgrids. *Energy.* 2016;114:742–752. Available from: https://www.sciencedirect.com/science/article/pii/S0360544216311616.

[20]   Restrepo M, Cañizares CA, Simpson-Porco JW, *et al.* Optimization- and rule-based energy management systems at the Canadian renewable energy laboratory microgrid facility. *Applied Energy.* 2021;290:116760. Available from: https://linkinghub.elsevier.com/retrieve/pii/S0306261921002671.

[21]   Nasr-Azadani E, Su P, Zheng W, *et al.* The Canadian renewable energy laboratory: a testbed for microgrids. *IEEE Electrification Magazine.* 2020;8(1): 49–60.

[22]   Delfino F, Ferro G, Robba M, *et al.* An energy management platform for the optimal control of active and reactive powers in sustainable microgrids. *IEEE Transactions on Industry Applications.* 2019;55(6):7146–7156. Available from: https://ieeexplore.ieee.org/document/8700197/.

[23]   Clarke WC, Brear MJ, and Manzie C. Control of an isolated microgrid using hierarchical economic model predictive control. *Applied Energy.* 2020;280:115960. Available from: https://linkinghub.elsevier.com/retrieve/pii/S0306261920314148.

[24]   Hagenmeyer V, Kemal Çakmak H, Düpmeier C, *et al.* Information and communication technology in energy lab 2.0: smart energies system simulation and control center with an open-street-map-based power flow simulation example. *Energy Technology.* 2016;4(1):145–162. Available from: https://onlinelibrary.wiley.com/doi/abs/10.1002/ente.201500304.

[25]   Wiegel F, Wachter J, Kyesswa M, *et al.* Smart Energy System Control Laboratory – a fully-automated and user-oriented research infrastructure for controlling and operating smart energy systems. *at-Automatisierungstechnik.* 2022;70(12):1116–1133. Available from: https://doi.org/10.1515/auto-2022-0018 [cited 2023-11-15].

[26]   Nguyen VH, Tran QT, Guillo-Sansano E, *et al.* In: Strasser TI, de Jong ECW, Sosnina M, editors. *Hardware-in-the-Loop Assessment Methods.* Cham: Springer International Publishing; 2020. p. 51–66. Available from: https://doi.org/10.1007/978-3-030-42274-5_4.

[27]   McDonald J. Substation automation basics—the next generation. *Electric Energy T&D Magazine.* 2007:30–36.

[28]   Monti A, Stevic M, Vogel S, *et al.* A global real-time superlab: enabling high penetration of power electronics in the electric grid. *IEEE Power Electronics Magazine.* 2018;5(3):35–44.

[29]   Beichter S, Beichter M, Werling D, *et al.* Towards a real-world dispatchable feeder. In: *2023 8th IEEE Workshop on the Electronic Grid (eGRID)*; 2023. p. 1–6.

[30]   Rocabert J, Luna A, Blaabjerg F, *et al.* Control of power converters in AC microgrids. *IEEE Transactions on Power Electronics.* 2012;27(11):4734–4749.

[31]   Dupont E, Koppelaar R, and Jeanmart H. Global available wind energy with physical and energy return on investment constraints. *Applied Energy.*

2018;209:322–338. Available from: https://www.sciencedirect.com/science/article/pii/S0306261917313673.

[32]   Li B, Roche R, Paire D, *et al.* Sizing of a stand-alone microgrid considering electric power, cooling/heating, hydrogen loads and hydrogen storage degradation. *Applied Energy*. 2017;205:1244–1259. Available from: https://www.sciencedirect.com/science/article/pii/S0306261917311595.

[33]   Jacob A. Regularium Für Das Label "SG Ready" Für Elektrische Heizungs- Und Warmwasserwärmepumpen Und Kompatible Systemkomponenten; 2020.

[34]   Ekin O, Perez F, Damm G, *et al.* A real-time PHIL implementation of a novel nonlinear distributed control strategy for a multi-terminal DC microgrid. In: *2023 IEEE Belgrade PowerTech*; 2023. p. 1–6.

Part IV

Multi-Energy Management

Chapter 7

Risk-averse transactive energy management for a multi-energy microgrid

Yunyang Zou[1], Yan Xu[1] and Cuo Zhang[2]

This chapter proposes a novel energy management method for a multi-energy micro-grid (MEMG) which supplies electrical and thermal energy simultaneously. The proposed method leverages the concept of transactive energy (TE) and formulates the problem as a Stackelberg game-theoretic bi-level optimization model. At the upper level, the MEMG operator optimizes energy scheduling and pricing strategies, while, at the lower level, residential, commercial, and industrial agents optimize their energy trading strategies. The bi-level model is then reformulated as an equivalent single-level mixed-integer linear program (MILP) for computational tractability. To coordinate the strategies made in the day-ahead and intra-day energy markets, an adaptive stochastic optimization (SO) approach is adopted, by which a day-ahead stochastic MILP and an intra-day deterministic model are formed. Furthermore, the conditional value-at-risk (CVaR) measure is incorporated in the day-ahead stage to account for the MEMG operator's risk aversion towards uncertainties. To solve the models efficiently, this chapter develops an adaptive Progressive Hedging (PH) algorithm to decompose the day-ahead stochastic MILP into multiple scenario-based subproblems, which can then be solved in parallel. Moreover, an outer approximation (OA) algorithm is employed in the intra-day stage to linearize the bilinear objective function. The effectiveness of the proposed risk-averse transactive energy management method and the efficiency of the developed solution algorithms are validated through simulation results. These results demonstrate the ability of the method to co-optimize energy scheduling and trading in an MEMG context. The developed algorithms provide efficient solutions and contribute to the overall performance improvement of the MEMG energy management method.

[1]School of Electrical and Electronic Engineering, Nanyang Technological University, Singapore
[2]School of Electrical and Information Engineering, The University of Sydney, Australia

Nomenclature

Acronyms

BS(C/D)	Battery storage (charging/discharging)
DER	Distributed energy resource
ESS	Energy storage system
MEMG	Multi-energy microgrid
TS(C/D)	Thermal storage (charging/discharging)

Sets and indices

N_T/t	Set/Index of time slots
N_J/j	Set/Index of DGs
N_I/i	Set of exogenous networks ($i = 0$), residential ($i = 1$), commercial ($i = 2$), and industrial ($i = 3$) agents

Parameters

$\kappa^{(\cdot)}$	Unit operating cost
$\lambda_{0,t}^{elec}/\lambda_{0,t}^{thm}$	Prices for selling electrical/thermal energy to external networks in the day-ahead market
$\lambda_{exo,t}^{elec}/\lambda_{exo,t}^{thm}$	Prices for buying electrical/thermal energy from external networks in the day-ahead market
$\Lambda_{0,t}^{elec/thm}$	Prices for selling electrical/thermal energy to external networks in the intra-day market
$\Lambda_{exo,t}^{elec/thm}$	Prices for buying electrical/thermal energy from external networks in the intra-day market
P_t^{PV}/P_t^{WT}	Electrical power outputs of a PV/WT
$E_0^{BS/TS}$	Initial energy stored in a BS/TS
$E^{BS/TS,rate}$	Rated capacity of a BS/TS
η^{CCHP}	Thermal-to-electrical ratio of a CCHP unit
η^{BSC}/η^{BSD}	Charging/discharging efficiency of a BS
η^{TSC}/η^{TSD}	Charging/discharging efficiency of a TS
$\tau^{BS/TS}$	Decay rate of a BS/TS
$P_{i,t}^{exp}/\theta_{i,t}^{exp}$	Expected electrical energy demand and indoor temperature of EU agent i
$\alpha_{i,t}/\beta_{i,t}$	Unit discomfort cost for electrical/thermal energy demand deviation of EU agent i
γ_i	Minimum daily demand ratio of EU agent i
C_i^{air}	Heat capacity of the air
R_i^T	Thermal resistance of building shells
$\theta_{i,t}^{am}$	Ambient temperature
$\chi_{min/max}^{BS/TS}$	Minimal/maximal state of charge of a BS/TS
$(\cdot)^{min/max}$	Minimal and maximal limits

Variables

$\lambda_{i,t}^{elec}/\lambda_{i,t}^{thm}$	Electrical/thermal trading prices between the MEMG operator and EU agent i ($i \in N_I \backslash 0$) in the day-ahead market
$P_{i,t}^{elec}/P_{i,t}^{thm}$	Electrical/thermal trading quantities the MEMG operator sells to entity i in the day-ahead market
$\Lambda_{i,t}^{elec/thm}$	Electrical/thermal trading prices between the MEMG operator and EU agent i in the intra-day market
$\Delta P_{i,t}^{elec}/\Delta P_{i,t}^{thm}$	Electrical/thermal trading quantities the MEMG operator sells to (>0) or buys from (<0) entity i in the intra-day market
$P_{j,t}^{DG}$	Electrical power output of DG j
$P_t^{CCHP,el/th}$	Electrical/thermal power output of a CCHP unit
P_t^{BSC}/P_t^{BSD}	Charging/discharging power of a BS
P_t^{TSC}/P_t^{TSD}	Absorbing/releasing power of a TS
E_t^{BS}/E_t^{TS}	Stored energy in a BS/TS
$P_{i,t}^{el,e}/P_{i,t}^{th,e}$	Electrical/thermal trading quantities that agent i buys from external networks in the day-ahead market
$\Delta P_{i,t}^{el,e/th,e}$	Electrical/thermal trading quantities that agent i buys from (>0) or sells to (<0) external networks in the intra-day market
$P_{i,t}^{ed}/\theta_{i,t}^{td}$	Actual electrical energy demand and indoor temperature of EU agent i
$P_{i,t}^{td}$	Actual thermal energy demand of EU agent i

7.1 Introduction

A multi-energy microgrid (MEMG) is a system that combines various distributed generation units, such as photovoltaic (PV) systems, wind turbines (WT), and combined cooling heat and power (CCHP) units. This enables the MEMG to simultaneously provide multiple energy supplies to customers with enhanced energy efficiency [1]. The MEMG can be implemented at different scales (e.g., building, community, industrial park, and even distribution network) [2], together with flexible loads, such as thermostatically controlled loads (TCLs) and electric vehicles (EVs) participating in demand response programs.

Extensive research has been conducted on multi-energy management in the literature. However, most works have primarily focused on supply-side energy management while oversimplifying demand-side management. In [3], the MEMG operator treats power demands as stochastic parameters and manage thermal demands directly without compromising customers' comfort. In [4], the multi-energy building operator dispatches both electrical and thermal loads through direct-load-control (DLC). In [5], the MEMG operator treats the electrical power loads as stochastic parameters and proposes a risk-averse energy management method, but still manages thermal loads using DLC. Recently, the concept of transactive energy (TE) has emerged, which is defined as a set of economic and control mechanisms that enable the dynamic balance of supply and demand across the entire electrical infrastructure with price serving as a key operational parameter [6]. By using price as an indirect control signal, an appropriate pricing scheme can stimulate desired demand-side response, leading to the achievement of an expected energy supply-demand balance. Moreover, it can further reduce the MEMG's reliance on external networks. Therefore, the integration of a pricing scheme can effectively promote the demand-side management and enable an efficient transactive energy management.

There are three main methods commonly used for designing pricing schemes in transactive energy management: auction mechanism [7–9], dual price [10–12], and game theory [13–20]. In [7], a double-auction peer-to-peer energy trading scheme is proposed, specifically implemented within a distribution grid. In [10], the market clearing mechanism for a local energy network is designed using a dual decomposition approach. The authors in [13] optimize energy scheduling and trading decisions for multi-carrier energy hubs based on Nash bargaining theory. The operators in a multi-energy industrial park employ Stackelberg game theory to determine compensation prices for peak load shifting, as discussed in [14]. For a comprehensive literature review and detailed comparison of these three pricing methods, please refer to our review paper [21]. Among these methods, game theory is particularly noteworthy for its ability to effectively capture complicated strategic interactions among diverse stakeholders, making it a widely-embraced choice in the design of transactive energy management method.

Moreover, given that the microgrid operator typically holds a strategic advantage in local energy markets, where consumers can only make decisions based on the operator's actions, the application of Stackelberg game theory becomes particularly effective under such cases. For example, in [15], a Stackelberg game-theoretical bi-level model is developed to determine energy dispatch and pricing, aiming at maximizing the revenue of an owner who owns and shares solar PV and energy storage in an apartment building. In [16] an electrical-gas-thermal energy sharing mechanism based on Stackelberg game theory is proposed, aiming to maximize the profit of the energy hub operator. The authors in [17] formulate the transactive energy management of an MEMG, considering bidirectional electricity/heat prices, as a Stackelberg game model. In [18], a day-ahead energy pricing scheme, along with associated energy management, is expressed as a bi-level Stackelberg game model with the objective of maximizing the energy service provider's profit in a regional integrated energy system. Nonetheless, it is important to note that the aforementioned works do not account for the underlying impacts of uncertainties, such as stochastic renewable power generation.

To address the uncertainties involved in power system optimization, several methods have been proposed in the literature, including stochastic optimization (SO), robust optimization [22], interval optimization [23], and chance constrained programming [24]. Among these approaches, SO has gained significant attention in transactive energy management due to its ability to sufficiently utilize the probabilistic information of uncertainties while maintaining computational tractability. For instance, in [11], a transactive energy supported operation framework is proposed for interconnected MEMGs using SO approach. In [8], a combination of Vickrey auction theory and Stackelberg game is applied in the H_2 pricing mechanism to maximize the profit of a hybrid-renewable-to-H_2 provider. In [19], the energy management of a price-maker storage system is formulated as a stochastic bi-level model. However, it is worth noting that the day-ahead energy management decisions derived from a single-stage SO model can only guarantee optimality in terms of the expected value over all scenarios generated based on day-ahead forecasts. Consequently, as uncertainties are realized gradually, it also becomes necessary to provide intra-day energy management decisions to maintain optimal energy management.

In this regard, a two-stage energy management framework is necessary to coordinate the day-ahead and intra-day decisions. In [9], a two-stage SO energy management model for a microgrid is introduced, aiming to minimize the operation cost by coordinating energy trading and generation scheduling. The authors in [25] propose a two-stage hybrid stochastic and robust model that maximizes microgrid profits in the day-ahead market and minimizes imbalance costs in the real-time balancing market. However, this model lacks specific pricing schemes. To address the computational challenges posed by SO, scenario reduction techniques are commonly applied in SO. Nevertheless, these techniques may disregard low-probability-high-impact scenarios [5], leading to a significant deviation from expected profit or cost under some unfavorable scenarios. Microgrid operators typically exhibit risk aversion towards uncertainties, as they do not hope to suffer a large deviations from the expected value. Therefore, it is essential to consider and incorporate an appropriate risk measurement index into the day-ahead SO model for obtaining risk-averse day-ahead energy management decisions. It is also noteworthy that the aforementioned works formulate day-ahead problems as stochastic mixed-integer program (MIP) models, which are then directly solved by commercial solvers. However, such optimization models, especially those with a large number of integer variables, face significant computational difficulty, which is still intractable for the cutting-edge solvers. Consequently, the development of a decomposition algorithm is highly desirable to alleviate the computational burden in solving stochastic MIP problems.

To fill the above research gaps, this chapter proposes a novel risk-averse two-stage transactive energy management method for an MEMG, which enables the MEMG operator to coordinate day-ahead and intra-day actions and optimize energy scheduling and pricing strategies simultaneously at each stage. The proposed method and its technical features are summarized as follows.

- According to diverse user behaviors, consumers are categorized into residential, commercial, and industrial energy users (EUs). Unlike many existing works that predominantly concentrate on the supply-side energy management of MEMG while oversimplifying the demand-side management, this chapter formulates the transactive energy management of an MEMG as a Stackelberg game-based bi-level model, enabling the effective interactions between the MEMG operator and various EUs.

- A two-stage transactive energy management framework for the MEMG is designed, and an adaptive SO approach is proposed to coordinate the day-ahead and intra-day transactive energy management decisions considering the uncertainties from renewable generation and ambient temperature. Furthermore, a conditional value-at-risk (CVaR) measure is incorporated in the day-ahead stage for addressing the MEMG operator's risk aversion towards the uncertainties.

- To reduce the computation difficulty of the multi-scenario stochastic MIP problem, an adaptive Progressive Hedging (PH) algorithm is developed, which can decompose the day-ahead stochastic mixed-integer linear program (MILP) into multiple scenario-based subproblems, allowing them to be solved in parallel. Additionally, an outer approximation (OA) algorithm is employed to solve the intra-day bilinear problem without the need of introducing binary variables. Both algorithms contribute to a significant improvement in computation efficiency, enhancing the overall performance of the proposed transactive energy management method.

The remainder of the chapter is organized as follows. Section 7.2 details the proposed MEMG energy management framework. Section 7.3 presents the deterministic joint multi-energy scheduling and trading formulation for an MEMG based on a Stackelberg game. Following this, Section 7.4 introduces a risk-averse adaptive SO approach, which extends the deterministic model to a day-ahead risk-averse stochastic model and an intra-day deterministic model. To solve the optimization models, two iterative algorithms are presented in Section 7.5, one for the day-ahead optimization and another for the intra-day optimization. In Section 7.6, case studies are conducted to demonstrate the effectiveness and efficiency of the proposed method and algorithms. The conclusions are drawn in Section 7.7.

7.2   Proposed transactive energy management framework

The proposed transactive energy management framework for an MEMG is depicted in Figure 7.1. The local energy market within the MEMG involves four participants. The MEMG operator assumes the responsibility of scheduling all the physical DER units, including a CCHP unit, diesel generators (DGs), renewable energy generation units, and ESSs. Additionally, the MEMG operator also determines the energy trading prices to incentivize the desired demand-side response. Three agents are introduced to represent the aggregated residential, commercial, and industrial EUs within the MEMG, respectively. These EU agents engage in transactions with the MEMG operator through the local energy market to meet their electrical and thermal demands. It is important to note that the local energy market is a non-monopoly market, allowing all MEMG participants to directly trade energy with the external energy networks (i.e., utility grid and district thermal network). The selling/buying prices for trading energy with the external networks are given by a time-of-use (TOU) pricing scheme in this chapter and thus treated as known parameters. Other alternative pricing schemes for the external networks can also be applied within this framework.

The proposed framework encompasses both a local day-ahead and intra-day energy market. The local markets provide a platform for EU agents to engage in transactions with the MEMG operator, which can greatly facilitate the local power balance and reduce the MEMG's reliance on the external energy networks. In the day-ahead energy market, the MEMG operator utilizes the day-ahead probabilistic predictions of the uncertainties (i.e., renewable generation and ambient temperature) to determine 24-h energy scheduling and pricing strategies aimed at maximizing daily profit. Based on the prices released by the MEMG operator, the EU agents, acting as price takers, optimize their day-ahead energy procurement commitments considering both trading costs and comfort levels. Through the internal energy trading, the

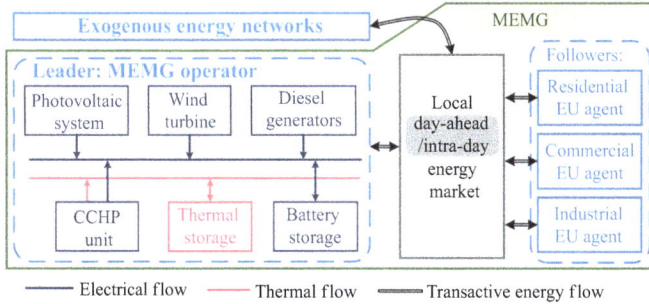

Figure 7.1   Structure of an MEMG with transactive energy

local energy generation can be consumed locally to the greatest extent possible. Note that in a non-monopoly market, reselling procured energy will not generate any profits. Hence, the MEMG operator, as an energy producer, only sells energy to others, while the EU agents, as energy consumers, just purchase energy from others in the day-ahead energy market.

The intra-day energy market operates on an hourly-ahead basis. In this market, the updated hourly-ahead point predictions of the uncertain parameters are utilized. To address the hourly-ahead power imbalance caused by uncertainties, the power output of the CCHP unit is adjusted for every hour. Considering the relatively slower response of DGs and the potential degradation of ESSs resulting from frequent changing/discharging, the day-ahead schedules on ESSs and DGs will remain unchanged during the intra-day stage. In addition to rescheduling the CCHP unit, the MEMG operator also releases newly hourly-ahead prices in the intra-day market. These prices guide the EU agents to adjust their day-ahead energy procurement commitments, thereby contributing to the intra-day power balance. Upon receiving the hourly-ahead price, the EU agents have the option to purchase additional energy to enhance their comfort level (if the hourly-ahead price gets lower compared with the day-ahead price) or to sell part of their day-ahead procurement commitments back to the MEMG operator for arbitrage (if the hourly-ahead price gets higher).

Note that the residential, commercial, and industrial agents reflect three representative load types in the real world, which have different preferences on energy demand and comfort requirement, and usually take different responsive strategies to market prices. For instance, the industrial agent typically prioritizes energy prices and cost-saving measures, with a lesser emphasis on maintaining a high comfort level. In contrast, the commercial agent places a greater emphasis on comfort level but may have a relatively weaker motivation to adjust their energy demand for bill saving. The residential agent tends to take into consideration both energy prices and comfort level when making energy trading related decisions.

7.3   Mathematical formulations

In the MEMG, the operator aims to maximize its operating profit, while the EU agents aim to minimize their individual energy consumption costs. Assuming that the

MEMG operator knows the EU agents' preference on energy demand and comfort requirement, it can take the lead in making strategies, hence is a strategy maker. The EU agents can only take strategies to optimize their objectives under the decisions from the MEMG operator, so are strategy takers. Hence, the trading interaction can be modeled by a Stackelberg game [26], in which the MEMG operator acts as a leader and EU agents as three followers. All participants in this game are assumed to be rational, and tend to adjust their strategies until the Stackelberg equilibrium is achieved. Note that the end EUs' preference information is collected and protected by their corresponding EU agents. The MEMG operator just knows the aggregated preference information provided by each EU agent. Only the EU agents know their respective end EUs' information based on contract agreement. Thus, the privacy of the end EUs can be effectively protected.

Generally, a Stackelberg game can be expressed as a bi-level optimization model [15]. For mathematical conciseness, this section firstly formulates the day-ahead joint energy scheduling and trading problem using a deterministic bi-level model. Then the model is extended into a two-stage stochastic model through an adaptive SO approach in Section 7.4.

7.3.1   Deterministic day-ahead bi-level model
7.3.1.1   Upper-level model for MEMG operator
The upper-level problem is to optimize the energy scheduling and pricing strategies for the MEMG operator, aiming to maximize its overall operating profit. The upper-level problem is mathematically expressed as follows.

$$\max \sum_{t \in N_T} \left[\sum_{i \in N_I} \left(\lambda_{i,t}^{elec} P_{i,t}^{elec} + \lambda_{i,t}^{thm} P_{i,t}^{thm} \right) - \sum_{j \in N_J} \kappa_j^{DG} P_{j,t}^{DG} \right. \tag{7.1}$$
$$\left. - \kappa^{BS} \left(P_t^{BSC} + P_t^{BSD} \right) - \kappa^{TS} \left(P_t^{TSC} + P_t^{TSD} \right) - \kappa^{CCHP} P_t^{CCHP,el} \right]$$

Subject to:

$$\sum_{j \in N_J} P_{j,t}^{DG} + P_t^{CCHP,el} + P_t^{PV} + P_t^{WT} - P_t^{BSC} + P_t^{BSD} = \sum_{i \in N_I} P_{i,t}^{elec}, \ \forall t \in N_T \tag{7.2}$$

$$P_t^{CCHP,th} - P_t^{TSC} + P_t^{TSD} = \sum_{i \in N_I} P_{i,t}^{thm}, \ \forall t \in N_T \tag{7.3}$$

$$P_{0,t}^{elec}, P_{0,t}^{thm} \geq 0, \ \forall t \in N_T \tag{7.4}$$

$$P_j^{DG,min} \leq P_{j,t}^{DG} \leq P_j^{DG,max}, \ \forall t \in N_T, \ \forall j \in N_J \tag{7.5}$$

$$R_j^{DG,min} \leq P_{j,t}^{DG} - P_{j,t-1}^{DG} \leq R_j^{DG,max}, \ \forall t \in N_T, \ \forall j \in N_J \tag{7.6}$$

$$P^{CCHP,min} \leq P_t^{CCHP,el} \leq P^{CCHP,max}, \ \forall t \in N_T \tag{7.7}$$

$$P_t^{CCHP,th} = \eta^{CCHP} P_t^{CCHP,el}, \ \forall t \in N_T \tag{7.8}$$

$$0 \leq P_t^{BSC} \leq P^{BSC,max}, \ \forall t \in N_T \tag{7.9}$$

$$0 \leq P_t^{BSD} \leq P^{BSD,max}, \ \forall t \in N_T \tag{7.10}$$

$$P_t^{BSC} P_t^{BSD} = 0, \ \forall t \in N_T \tag{7.11}$$

$$E_t^{BS} = E_{t-1}^{BS} \left(1 - \tau^{BS}\right) + \eta^{BSC} P_t^{BSC} - \frac{P_t^{BSD}}{\eta^{BSD}}, \ \forall t \in N_T \tag{7.12}$$

$$\chi_{min}^{BS} E^{BS,rate} \leq E_t^{BS} \leq \chi_{max}^{BS} E^{BS,rate}, \ \forall t \in N_T \tag{7.13}$$

$$E_0^{BS} = E_{24}^{BS} \tag{7.14}$$

$$0 \leq P_t^{TSC} \leq P^{TSC,max}, \ \forall t \in N_T \tag{7.15}$$

$$0 \leq P_t^{TSD} \leq P^{TSD,max}, \ \forall t \in N_T \tag{7.16}$$

$$P_t^{TSC} P_t^{TSD} = 0, \ \forall t \in N_T \tag{7.17}$$

$$E_t^{TS} = E_{t-1}^{TS} \left(1 - \tau^{TS}\right) + \eta^{TSC} P_t^{TSD} - \frac{P_t^{TSD}}{\eta^{TSD}}, \ \forall t \in N_T \tag{7.18}$$

$$\chi_{min}^{TS} E^{TS,rate} \leq E_t^{TS} \leq \chi_{max}^{TS} E^{TS,rate}, \ \forall t \in N_T \tag{7.19}$$

$$E_0^{TS} = E_{24}^{TS} \tag{7.20}$$

The objective function (7.1) is to maximize the overall operating profit of the MEMG operator by energy scheduling and pricing. Therein, the energy trading quantities with the EU agents, i.e., $P_{i,t}^{elec}$, $P_{i,t}^{thm}$ ($i \in N_I \backslash 0$), are optimized by the lower-lever EU agents, and thus treated as parameters here. Note that the operator does not buy energy from the external networks, since reselling procured energy cannot make profits in a non-monopoly market. The directions of energy flows trading with the external networks are stated in (7.4). Equations (7.2)–(7.3) are the power balance constraints. Equations (7.5)–(7.6) show the power output bounds and ramping limits for DGs. Equations (7.7)–(7.8) state the power output limits and energy conversion of the CCHP unit. Equations (7.9)–(7.14) are the operating constraints for BS and (7.15)–(7.20) are for TS.

7.3.1.2 Lower-level models for EU agents

Each lower-level problem aims to seek the optimal energy procurement strategy for each EU agent ($\forall i \in N_I \backslash 0$) based on the energy prices that are determined by the upper-level MEMG operator and treated as parameters at this level. The lower-level problems can be formulated as below, and the corresponding dual variables to each constraint are given after a colon.

$$\begin{aligned} \min \sum_{t \in N_T} &\left[\lambda_{i,t}^{elec} P_{i,t}^{elec} + \lambda_{i,t}^{thm} P_{i,t}^{thm} + \lambda_{exo,t}^{elec} P_{i,t}^{el,e} + \lambda_{exo,t}^{thm} P_{i,t}^{th,e} \right. \\ &\left. + \alpha_{i,t} \left(P_{i,t}^{ed} - P_{i,t}^{exp} \right)^2 + \beta_{i,t} \left(\theta_{i,t}^{td} - \theta_{i,t}^{exp} \right)^2 \right] \end{aligned} \tag{7.21}$$

Subject to:

$$P_{i,t}^{elec} + P_{i,t}^{el,e} = P_{i,t}^{ed} : \nu_{i,t}^{ed}; \ \forall t \in N_T \tag{7.22}$$

$$P_{i,t}^{thm} + P_{i,t}^{th,e} = P_{i,t}^{td} : \nu_{i,t}^{td}; \ \forall t \in N_T \tag{7.23}$$

$$P_{i,t}^{elec}, \ P_{i,t}^{thm} \geq 0 : \mu_{i,t}^{elec,m}, \ \mu_{i,t}^{thm,m}; \ \forall t \in N_T \tag{7.24}$$

$$P_{i,t}^{el,e}, \ P_{i,t}^{th,e} \geq 0 \ : \ \mu_{i,t}^{elec,e}, \ \mu_{i,t}^{thm,e}; \ \forall t \in N_T \tag{7.25}$$

$$P_{i,t}^{elec} \leq P_i^{elec,max} \ : \ \mu_{i,t}^{elec,max}; \ \forall t \in N_T \tag{7.26}$$

$$P_{i,t}^{thm} \leq P_i^{thm,max} \ : \ \mu_{i,t}^{thm,max}; \ \forall t \in N_T \tag{7.27}$$

$$P_{i,t}^{ed,min} \leq P_{i,t}^{ed} \leq P_{i,t}^{ed,max} \ : \ \mu_{i,t}^{ed,min}, \ \mu_{i,t}^{ed,max}; \ \forall t \in N_T \tag{7.28}$$

$$\sum_{t \in N_T} P_{i,t}^{ed} \geq \gamma_i \sum_{t \in N_T} P_{i,t}^{exp} \ : \ \mu_i^{ed} \tag{7.29}$$

$$P_{i,t}^{td} = C_i^{air} \left(\theta_{i,t-1}^{td} - \theta_{i,t}^{td} \right) + \frac{\left(\theta_{i,t}^{am} - \theta_{i,t}^{td} \right)}{R_i^T} \ : \ \nu_{i,t}^{in}; \ \forall t \in N_T \tag{7.30}$$

$$\theta_{i,t}^{td,min} \leq \theta_{i,t}^{td} \leq \theta_{i,t}^{td,max} \ : \ \mu_{i,t}^{td,min}, \ \mu_{i,t}^{td,max}; \ \forall t \in N_T \tag{7.31}$$

The objective function (7.21) is the utility function of EU agent i, which consists of the energy procurement cost (first four terms) and discomfort cost (last two terms) [4,27]. Equations (7.22)–(7.23) state that EU agent i can buy electrical/thermal energy from both the MEMG operator and the exogenous network. Equations (7.24)–(7.25) restrict the directions of energy trading, considering reselling energy cannot bring any profit for EU agents. Equations (7.26)–(7.27) are the distribution line/pipe capacity limits. Equations (7.28)–(7.29) describe the flexibility range of the electrical demand. Assuming the thermal energy is used for district cooling in summer, the temperature-dependent thermal demand is formulated in (7.30). The heating demand in winter can also be modeled similarly. Equation (7.31) means that the indoor temperature can be accepted within a certain range without much loss of comfort.

Notice that the different preferences of the EU agents are characterized by different parameter settings in model (7.21)–(7.31). The preference on energy demand is specified by $P_{i,t}^{ed,min}$, $P_{i,t}^{ed,max}$, $\theta_{i,t}^{td,min}$, $\theta_{i,t}^{td,max}$, γ_i, while the preference on comfort requirement is specified by $\alpha_{i,t}$, $\beta_{i,t}$, $P_{i,t}^{exp}$, $\theta_{i,t}^{exp}$.

7.3.2　Equivalent single-level MPEC model

The upper-level problem is optimized based on the results of the lower-level problems. Therefore, the interactive Stackelberg game problem can be rewritten as the bi-level optimization model as follows:

$$\max \ (7.1) \tag{7.32}$$

Subject to:

$$(7.2) - (7.20) \tag{7.33}$$

$$\boldsymbol{P}_i^{elec}, \ \boldsymbol{P}_i^{thm} \in \arg\min \ (7.21), \ \forall i \in N_I \backslash 0 \tag{7.34}$$

$$\text{Subject to: } (7.22) - (7.31), \ \forall i \in N_I \backslash 0 \tag{7.35}$$

where $\boldsymbol{P}_i^{elec} := \{P_{i,t}^{elec}\}_{t \in N_T}$ and $\boldsymbol{P}_i^{thm} := \{P_{i,t}^{thm}\}_{t \in N_T}$. The upper-level problem is to optimize the energy scheduling and pricing strategies for the MEMG operator based on the energy procurement strategy of each EU agent. The energy procurement strategies based on specific energy prices are optimized by the lower-level problems.

To solve the bi-level model (7.32)–(7.35), one needs to convert it into a single-level model. It is noticed that the lower-level models are convex optimization problems with differentiable objective and constraint functions, they thus can be equivalently replaced with their Karush-Kuhn-Tucker (KKT) conditions [28]. The resulting single-level model after KKT reformulation is known as a mathematical program with equilibrium constrains (MPEC). For more details on KKT reformulation, interested readers can refer to [29].

7.3.3 Model convexification and linearization

Three sources of nonlinearity are involved in the equivalent single-level MPEC model. They can be linearized following the convexification methods below.

Method #01: For bilinear constraints (7.11) and (7.17), they are loose enough and thus can be removed directly [22]. An Big-M approach can also be employed as an alternative way, which will be specified in following Method #02.

Method #02: For the complementarity conditions in the form of $0 \le g \perp \mu \ge 0$, a Big-M approach [19] can be adopted for linearization. The equivalent linearized constraints are $0 \le g \le \zeta M$ and $0 \le \mu \le (1 - \zeta) M$, where ζ is an auxiliary binary variable and M is set to a big positive number.

Method #03: For bilinear objective function (7.1), each bilinear term included can be linearized through discretization of energy trading prices. This method is inspired by the binary expansion method [30]. Taking bilinear term $\lambda_{i,t}^{elec} P_{i,t}^{elec}$ ($\forall t \in N_T$, $\forall i \in N_I \backslash 0$) as an instance, the detailed linearization procedure is summarized as below:

> Step 1: Discretize $\lambda_{i,t}^{elec}$ into a sequence of price parameters $\varpi_{i,t,n}^{elec}$, wherein n denotes the index of discretized price parameters. As pricing strategy of the MEMG operator in fact has to limited by existing buying and selling prices in the local energy market, we have $\varpi_{i,t,n}^{elec} \in [\lambda_{0,t}^{elec}, \lambda_{exo,t}^{elec}]$.

> Step 2: Replace $\lambda_{i,t}^{elec}$ with $\lambda_{i,t}^{elec} = \sum_{n \in N} \varphi_{i,t,n}^{elec} \varpi_{i,t,n}^{elec}$ and $\sum_{n \in N} \varphi_{i,t,n}^{elec} = 1$, where $\varphi_{i,t,n}^{elec}$ is introduced as an auxiliary binary variable.

> Step 3: Linearize the bilinear term $\lambda_{i,t}^{elec} P_{i,t}^{elec}$ in objective function (7.1) as $\sum_{n \in N} \varpi_{i,t,n}^{elec} P_{i,t,n}^{el}$, while introducing two linearized inequality constraints $0 \le P_{i,t,n}^{el} \le \varphi_{i,t,n}^{elec} M$ and $0 \le (P_{i,t}^{el} - P_{i,t,n}^{el}) \le (1 - \varphi_{i,t,n}^{elec}) M$. $P_{i,t,n}^{el}$ is an auxiliary continuous variable.

7.4 Risk-averse adaptive stochastic optimization modeling

To coordinate the day-ahead and intra-day decisions effectively, this section introduces an risk-averse adaptive SO approach [5]. Meanwhile, this approach addresses the issue of risk aversion faced by the MEMG operator in the presence of day-ahead uncertainties. Under the risk-averse adaptive SO, the deterministic model developed in Section 7.3 is extended into a day-ahead risk-averse stochastic optimization model and an intra-day deterministic optimization model.

7.4.1   Day-ahead risk-averse stochastic optimization model

In the day-ahead optimization, the ambient temperature, the power outputs of PV and WT are modeled as random variables. With their day-ahead probabilistic predictions, Latin hypercube sampling and simultaneous backward reduction techniques are utilized to generate a set of representative scenarios. For more details about the scenario generation and reduction procedures, interested readers can refer to [31].

Given the representative scenario set N_s, the day-ahead stochastic single-level MPEC model after linearization can be easily derived based on the deterministic one in Section 7.3. For the brevity on the page, only the compact form is presented here.

$$\max \ \mathbf{c}^T\mathbf{x} + \sum_{s \in N_s} \pi_s \mathcal{L}(\mathbf{x}, \, \mathbf{d}_s) \tag{7.36}$$

Subject to:

$$\mathbf{x} \in \chi \tag{7.37}$$

$$\mathcal{L}\,(\mathbf{x}, \, \mathbf{d}_s) = \max_{\mathbf{y}_s \in \Omega(\mathbf{x}, \, \mathbf{d}_s)} \ \mathbf{b}^T\mathbf{y}_s \tag{7.38}$$

$$\Omega\,(\mathbf{x}, \, \mathbf{d}_s) = \{\mathbf{y}_s | \mathbf{A}\mathbf{x} + \mathbf{B}\mathbf{y}_s \geq \mathbf{r}, \, \mathbf{E}\mathbf{x} + \mathbf{F}\mathbf{y}_s = \mathbf{d}_s\} \tag{7.39}$$

The adaptive SO approach divides all variables into two groups. $\mathbf{x}$ denotes the first-stage variables, including the scheduling decisions on DGs and ESSs as well as the day-ahead energy trading prices. $\mathbf{y}_s$ represents the second-stage variables covering all the remaining variables. The objective (7.36) aims to find a day-ahead solution $\mathbf{x}^*$ to maximize the expectation over all the representative scenarios. Therein, π_s is the probability of scenario s. In (7.37), χ defines the feasible region of the first-stage variables $\mathbf{x}$. Equations (7.38)–(7.39) indicate that $\mathbf{y}_s$ is optimized under the first-stage variables $\mathbf{x}$ and the corresponding uncertainty scenario $\mathbf{d}_s$.

The day-ahead optimization formulated in (7.36)–(7.39) can reach an optimal solution $\mathbf{x}^*$, which is solved by maximizing the expectation over all the representative scenarios. This means that $\mathbf{x}^*$ may be optimistic resulting in a disappointing profit under some unfavorable scenarios. In fact, a risk-averse MEMG operator does not hope to suffer a large deviation from the expected profit. To take into account the MEMG operator's risk aversion towards uncertainties, a CVaR measure [31] is adopted and incorporated in the day-ahead stochastic optimization model, which is expressed as (7.40). Therein, η is an auxiliary variable called value-at-risk (VaR). f_L^s represents the loss function. $[\,\cdot\,]^+$ is a projection operator, achieving $[\cdot]^+ = \max(\cdot, 0)$. $\sum_{s \in N_s} [f_L^s - \eta]^+ \pi_s$ calculates the expected risk above a specific VaR. ϑ defines the confidence level.

$$CVaR_\vartheta = \min_{\eta \in \mathbb{R}} \left\{ \eta + \frac{1}{1 - \vartheta} \sum_{s \in N_s} [f_L^s - \eta]^+ \pi_s \right\} \tag{7.40}$$

In this chapter, the uncertainty risk for the MEMG operator refers to the risk of a low profit in some unfavorable scenarios. Hence, $CVaR_\vartheta$ is rewritten in the form of "max" as (7.41). Therein, $f_P^s = c^T x + \mathcal{L}(x, d_s)$ denotes the profit under scenario s. $[\cdot]^-$ denotes the projector of $\min(\cdot, 0)$.

$$CVaR_\vartheta = \max_{\eta \in \mathbb{R}} \left\{ \eta + \frac{1}{1-\vartheta} \sum_{s \in N_s} [f_P^s - \eta]^- \pi_s \right\} \tag{7.41}$$

To consider the MEMG operator's risk aversion towards uncertainties, the objective function (7.36) is modified by incorporating the CVaR measure as (7.42). ρ is a user-defined weighting factor of CVaR.

$$\max_{x \in \mathcal{X}, y_s \in \Omega, \eta \in \mathbb{R}} c^T x + \sum_{s \in N_s} \pi_s \mathcal{L}(x, d_s) + \rho \cdot CVaR_\vartheta \tag{7.42}$$

In the modified objective function (7.42), the nonlinear term $[f_P^s - \eta]^-$ inside $CVaR_\vartheta$ can be replaced with $\Re_s$, while $\Re_s$ needs to satisfy the following two linear inequalities.

$$\Re_s \le 0, \Re_s \le f_P^s - \eta \tag{7.43}$$

where $\Re_s$ is an auxiliary variable.

7.4.2 Intra-day deterministic optimization model

In the intra-day optimization, the values of uncertain parameters are continuously updated using their hourly-ahead point predictions, but the day-ahead schedules on electrical/thermal storage and DGs remain unchanged during this stage. To address the intra-day (hourly-ahead) power imbalance, this stage is to reschedule the CCHP unit hourly, while determining the hourly-ahead trading prices in the intra-day market to guide the EU agents to adjust their day-ahead energy procurement commitments through intra-day energy trading. Notice that the intra-day energy trading decisions of the MEMG operator and EU agents should be made based on their day-ahead trading commitments. For instance, the quantities sold by the EU agents in the intra-day energy market cannot exceed their day-ahead energy procurement commitments.

The objective functions for the intra-day upper-level and lower-level problems at each time ($\forall t \in N_T$) are formulated in (7.44) and (7.45), respectively.

Intra-day objective function of upper-level MEMG operator:

$$\max \sum_{i \in N_I \setminus 0} \left(\Lambda_{i,t}^{elec} \Delta P_{i,t}^{elec} + \Lambda_{i,t}^{thm} \Delta P_{i,t}^{thm} \right) - \kappa^{CCHP} P_t^{CCHP,el}$$

$$+ \Lambda_{0,t}^{elec} [\Delta P_{0,t}^{elec}]^+ + \Lambda_{exo,t}^{elec} [\Delta P_{0,t}^{elec}]^- + \Lambda_{0,t}^{thm} [\Delta P_{0,t}^{thm}]^+ + \Lambda_{exo,t}^{thm} [\Delta P_{0,t}^{thm}]^- \tag{7.44}$$

Intra-day objective function of each lower-level EU agent ($\forall i \in N_I \setminus 0$):

$$\min \Lambda_{i,t}^{elec} \Delta P_{i,t}^{elec} + \Lambda_{i,t}^{thm} \Delta P_{i,t}^{thm} + + \alpha_{i,t} \left(P_{i,t}^{ed} - P_{i,t}^{exp} \right)^2 + \beta_{i,t} \left(\theta_{i,t}^{td} - \theta_{i,t}^{exp} \right)^2$$

$$+ \Lambda_{exo,t}^{elec} [\Delta P_{i,t}^{el,e}]^+ + \Lambda_{0,t}^{elec} [\Delta P_{i,t}^{el,e}]^- + \Lambda_{exo,t}^{thm} [\Delta P_{i,t}^{th,e}]^+ + \Lambda_{0,t}^{thm} [\Delta P_{i,t}^{th,e}]^- \tag{7.45}$$

Considering the intra-day bi-level model between the MEMG operator and EU agents has a similar formulation to the day-ahead one in Section 7.3, its constraints can be derived simply by the following two steps: (1) For constraints in the bi-level model developed in Section 7.3.1, replace $P_{i,t}^{elec/thm}$ with $P_{i,t}^{elec/thm*} + \Delta P_{i,t}^{elec/thm}$ and replace $P_{i,t}^{el,e/th,e}$ with $P_{i,t}^{el,e/th,e*} + \Delta P_{i,t}^{el,e/th,e}$, where $P_{i,t}^{elec/thm*}$, $P_{i,t}^{el,e/th,e*}$ denotes the energy trading commitments of the MEMG operator and EU agents made in the day-ahead energy market. (2) Decouple (7.29) into a single-period constraint as $P_{i,t}^{ed} \geq \Gamma_i P_{i,t}^{ed,fir*}$. Therein, $P_{i,t}^{ed,fir*}$ is the day-ahead electrical procurement of EU agent i and $\Gamma_i = \gamma_i \sum_{t \in N_T} P_{i,t}^{exp} / \sum_{t \in N_T} P_{i,t}^{ed,fir*}$.

Note that $\Lambda_{exo,t}^{elec/thm}$ are generally higher than $\Lambda_{0,t}^{elec/thm}$. Therefore, the operators $[\cdot]^{+/-}$ in the intra-day objective functions (7.44)–(7.45) can be easily addressed through introducing two non-negative auxiliary variables. For instance, we can let $\Delta P_{i,t}^{el,e/th,e} = B_{i,t}^{el,e/th,e} - S_{i,t}^{el,e/th,e}$, while $B_{i,t}^{el,e/th,e}$, $S_{i,t}^{el,e/th,e} \geq 0$. Then $[\Delta P_{i,t}^{el,e/th,e}]^+$ and $[\Delta P_{i,t}^{el,e/th,e}]^-$ can be replaced directly by $B_{i,t}^{el,e/th,e}$ and $-S_{i,t}^{el,e/th,e}$.

The intra-day equivalent single-level MPEC model can be derived from the intra-day bi-level model following the same way as described in Section 7.3.2, which is left to the readers.

7.4.3 Implementation of proposed risk-averse transactive energy management method

The flowchart in Figure 7.2 illustrates the implementation of the proposed risk-averse transactive energy management method. In the day-ahead energy market, the first step involves formulating the day-ahead upper-level model for the MEMG operator and the day-ahead lower-level models for the individual EU agents, as introduced in Section 7.3.1. As a price maker, the MEMG operator aims to set the day-ahead prices for each EU agent. Subsequently, the EU agents, acting as price takers, respond to the prices they receive by determining their energy trading quantities. To achieve the

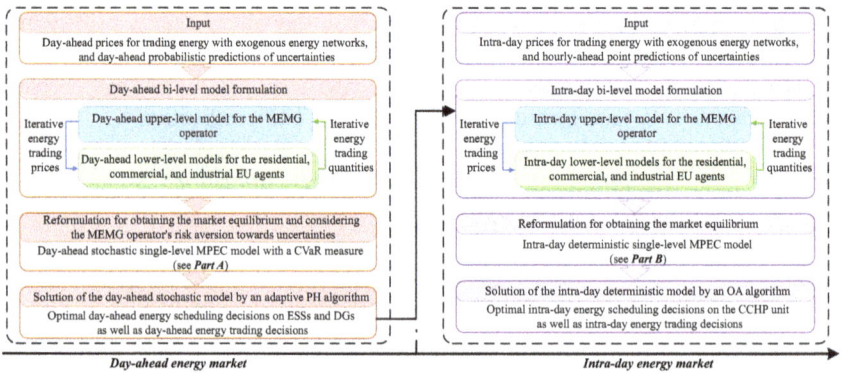

Figure 7.2   Implementation of the proposed risk-averse transactive energy management method

market equilibrium while considering the underlying impacts of uncertainties as well as the MEMG operator's risk aversion towards uncertainties, a day-ahead stochastic single-level MPEC model with a CVaR measure is then formulated using the day-ahead probabilistic predictions of uncertainties as inputs. The day-ahead risk-averse stochastic optimization model is a stochastic MILP. In following Section 7.5.1, an adaptive PH algorithm is developed to solve it.

In the intra-day energy market, the strategies are made based on the day-ahead energy scheduling decisions on ESSs and DGs, along with the day-ahead energy trading commitments. Similarly, the intra-day upper- and lower-level models are firstly formulated for the MEMG operator and EU agents respectively, which are interconnected by intra-day energy trading prices and quantities. Then in order to seek the market equilibrium in the intra-day energy market, the intra-day bi-level model is reformulated as an intra-day deterministic single-level MPEC model, which is later solved by an OA algorithm as introduced in following Section 7.5.2.

7.5   Solution methods

For achieving high-performance solutions, this section proposes two iterative algorithms for solving the day-ahead and intra-day optimization problems, respectively. An adaptive Progressive Hedging (PH) algorithm is developed to seek the day-ahead near-optimal solution in an effective and timely manner, which decomposes the day-ahead risk-averse stochastic MILP problem into scenario-based subproblems and solves them simultaneously in a parallel manner. In the intra-day stage, an Outer Approximation (OA) algorithm is employed to solve the intra-day bilinear problem without the need for introducing binary variables.

7.5.1   Adaptive PH algorithm for day-ahead risk-averse stochastic MILP

In existing works, most rely on commercial solvers, such as GUROBI, to directly solve stochastic problems. However, solving a stochastic MIP model may be computationally difficult for existing solvers, particularly when the MIP model involves a large number of integer variables. Therefore, many decomposition algorithms have been proposed to enhance computational efficiency. Two dominant classes of decomposition algorithms are stage-based algorithms (e.g., L-shaped method [32] and Benders decomposition [33]) and scenario-based algorithms (e.g., dual decomposition [34] and Progressive Hedging algorithm [35]). Compared with the stage-based decomposition, the scenario-based decomposition offers the advantage of distributing the computation load more evenly across subproblems, making it well-suited for parallel computing. Besides, stage-based algorithms like L-shaped and Benders methods encounter limitations when dealing with non-convex second-stage problems with integer variables [36]. In contrast, the PH algorithm imposes fewer restrictions on stochastic MIP and generally achieves faster convergence compared to the dual decomposition method [37]. The PH algorithm decomposes a stochastic MIP into multiple scenario-based subproblems that can be solved in parallel, resulting in a convergent near-optimal solution. However, the originla PH algorithm in [35] cannot balance the solution accuracy and convergence speed. Thus, in order for a better computation performance, an adaptive PH algorithm is further developed by adopting an adjustable penalty factor to achieve a high computation efficiency while guaranteeing an acceptable solution quality.

To demonstrate the adaptive PH algorithm, the day-ahead risk-averse stochastic MILP is expressed in a compact form as (7.46)–(7.47).

$$
\begin{aligned}
\max \boldsymbol{c}^T \boldsymbol{x} + \sum_{s \in N_s} \pi_s \boldsymbol{b}^T \boldsymbol{y}_s + \rho \cdot \left(\eta + \frac{1}{1-\vartheta} \sum_{s \in N_s} \pi_s \Re_s \right) \\
= \boldsymbol{c}^T \boldsymbol{x} + \rho \boldsymbol{y}_s + \sum_{s \in N_s} \pi_s \left(\boldsymbol{b}^T \boldsymbol{y}_s + \frac{\Re_s}{1-\vartheta} \right) \\
= \mathbb{C}^T \boldsymbol{x} + \sum_{s \in N_s} \pi_s \mathbb{B}^T \boldsymbol{y}_s
\end{aligned}
\tag{7.46}
$$

Subject to:

$$
(\boldsymbol{x}, \, \boldsymbol{y}_s) \in \mathcal{Q}_s, \ \forall s \in N_s
\tag{7.47}
$$

where $\boldsymbol{x} := \{\boldsymbol{x}, \eta\}$ represents the first-stage variables and $\boldsymbol{y}_s := \{\boldsymbol{y}_s, \Re_s\}$ denotes the second-stage variables. $\mathbb{C}$ and $\mathbb{B}$ are the augmented coefficient vectors associated with $\boldsymbol{x}$ and $\boldsymbol{y}_s$, respectively. Equation (7.47) defines the constraints under each scenario.

The proposed adaptive PH algorithm for solving the day-ahead risk-averse stochastic MILP model (7.46)–(7.47) is summarized in Algorithm 1. It starts from solving the subproblems under each scenario and calculating the expected value $\bar{\boldsymbol{x}}$ as well as the multiplier Ξ_s with an initial penalty factor $\mathcal{P}$. Step 7 adjusts the discrete distribution of energy trading prices to reduce the number of integer variables. After the initialization phase (Steps 1–7), the augmented subproblems are solved in Step 11 and $\bar{\boldsymbol{x}}$, Ξ_s are updated in Steps 15 and 16. The iterative phase (Steps 8–17) terminates until the stopping criteria is satisfied in Step 18, i.e., all first-stage decisions $\boldsymbol{x}_s$ converge to $\bar{\boldsymbol{x}}$. Note that when $\tau \geq \tau_1$, a bigger penalty factor $\mathcal{P}_1$ can be used to accelerate the convergence process, but the initial penalty factor $\mathcal{P}$ should still be set to a small value for ensuring high solution accuracy.

7.5.2   OA algorithm for intra-day bilinear problem

A linearization method for tackling the bilinear terms in the objective function has been presented in Section 7.3.3 (i.e., Method #03), which we refer to as the price discretizing method in this chapter for the sake of discussion. This method can effectively eliminate bilinear terms and transform the objective function into a linear form. However, it should be noted that the price discretizing method introduces a substantial number of 0–1 binary variables, which significantly increases the computational burden. Hence, for solving the intra-day bilinear optimization problem, an OA algorithm is employed. The OA algorithm is an iterative algorithm, which first linearizes the bilinear terms in the objective function around intermediate solution points and

---

Algorithm 1: The adaptive PH algorithm

---

1 Initialize the iteration index $\tau = 0$. Set an initial penalty factor $\mathcal{P}$ and a
   tolerance level ε.

2 **for** $s = 1 : N_s$ **do**

3 $\quad\quad$ $\alpha_s^{(\tau)} := \text{argmax}_\alpha \{\mathbb{C}^T\alpha + \mathbb{B}^T y_s : (\alpha, y_s) \in \mathcal{Q}_s\}$

4 $\quad\quad$ $\overline{\alpha}^{(\tau)} := \sum_{s \in N_s} \pi_s \alpha_s^{(\tau)}$

5 $\quad\quad$ $\Xi_s^{(\tau)} := \mathcal{P}(\alpha_s^{(\tau)} - \overline{\alpha}^{(\tau)})$

6 **end**

7 Adjust the discretized values of trading prices according to the results under
   each scenario in the initialization phase.

8 **repeat**

9 $\quad$ $\tau \leftarrow \tau + 1$

10 $\quad$ **for** $s = 1 : N_s$ **do**

11 $\quad\quad$ $\alpha_s^{(\tau)} := \text{argmax}_\alpha \{\mathbb{C}^T\alpha + \mathbb{B}^T y_s - \Xi_s^{(\tau-1)}\alpha - \frac{\mathcal{P}}{2}\|\alpha - \overline{\alpha}^{(\tau-1)}\|^2 :$
   $\quad\quad\quad$ $(\alpha, y_s) \in \mathcal{Q}_s\}$

12 $\quad\quad$ **if** $\tau \geq \tau_1$ **then**

13 $\quad\quad\quad$ $|$ Adjust the penalty factor, i.e., let $\mathcal{P} = \mathcal{P}_1$

14 $\quad\quad$ **end**

15 $\quad\quad$ $\overline{\alpha}^{(\tau)} := \sum_{s \in N_s} \pi_s \alpha_s^{(\tau)}$

16 $\quad\quad$ $\Xi_s^{(\tau)} := \Xi_s^{(\tau-1)} + \mathcal{P}(\alpha_s^{(\tau)} - \overline{\alpha}^{(\tau)})$

17 $\quad$ **end**

18 **until** $\sum_{s \in N_s} \pi_s \|\alpha_s^{(\tau)} - \overline{\alpha}^{(\tau)}\| \leq \varepsilon$;

---

then add the derived linear approximations into the OA formulation [38]. Unlike the price discretizing method, the OA algorithm not only improves the computational efficiency by avoiding the need for introducing binary variables, but also exports continuous price signals, enabling more accurate demand response control.

To facilitate the explanation of the OA algorithm, we present the intra-day bilinear model in a compact form as (7.48)–(7.49). The objective function consists of a linear subobjective $f(\Psi)$ and bilinear terms $\lambda^T P$. The constraints are all linear, which are achieved by using linearization methods #01 and #02 in Section 7.3.3.

$$R = \max\ f(\Psi) + \lambda^T P \tag{7.48}$$

Subject to:

$$H(\Psi, \lambda, P) = 0,\ G(\Psi, \lambda, P) \geq 0 \tag{7.49}$$

The detailed procedure of OA algorithm for solving the intra-day bilinear optimization problem (7.48)–(7.49) is summarized as below.

Step 1: Give an initial λ (i.e., λ_1). Set the lower bound $LB = -\infty$ and upper bound $UB = +\infty$. Initialize the index of iteration $m = 1$ and define a convergence tolerance δ.

Step 2: Given $\lambda = \lambda_m$, solve the problem $R(\lambda_m)$ and derive an optimal solution (Ψ_m, P_m). Then, update $LB = R(\lambda_m)$ and define the linearization of the bilinear terms $\lambda^T P$ at point (λ_m, P_m) as

$$L_m(\lambda, \ P) = \lambda_m^T P_m + (\lambda - \lambda_m)^T P_m + (P - P_m)^T \lambda_m \qquad (7.50)$$

Step 3: Check if the convergence tolerance is satisfied. If $UB - LB \leq \delta$, then stop the iteration and output the current optimal solution (Ψ_m, λ_m, P_m). Otherwise, proceed to Step 4.

Step 4: Solve the linearized form of the original bilinear problem, which is defined as below:

$$RL(\lambda_m, P_m) = \max f(\Psi) + \Theta \qquad (7.51)$$

Subject to:

$$\Theta \leq L_k(\lambda, \ P), \ \forall k \in \{1, \ldots, m\} \qquad (7.52)$$

$$H(\Psi, \lambda, P) = 0, \ G(\Psi, \lambda, P) \geq 0 \qquad (7.53)$$

Then update $UB = RL(\lambda_m, P_m)$ and let $m = m + 1$. Return the optimal solution λ_m to Step 2.

7.6   Case study

7.6.1   Testing system

The day-ahead optimization for the MEMG operates over a 24-h time horizon with 1-h intervals. The stochastic scenarios are generated according to different prediction accuracy, taking into account maximum prediction errors of 30% for renewable generation and 5% for the ambient temperature, as illustrated in Figure 7.3. The intraday optimization is implemented on an hourly basis, utilizing hourly-ahead point predictions. The technical parameters for the generators and energy storage [22] are provided in Tables 7.1 and 7.2, respectively. Table 7.3 presents the parameters associated with residential, commercial, and industrial EUs. Selling and buying prices for energy trading with the external energy networks can be found in Figure 7.4(c), while the expected energy demand of each EU agent is shown in Figure 7.5. Regarding the risk aversion modeling, a confidence level ϑ of 0.9 is set, along with a CVaR weight factor ρ of 0.1 [5]. In the adaptive PH algorithm, the tolerance ε and the value of τ_1 are set to 10^{-4} and 15, respectively. The tolerance δ in the OA algorithm is set to 10^{-5}.

All tests in this section are implemented on an Intel(R) Xeon(R) E5-1630 3.70 GHz 64-bit PC with 16GB RAM and solved by GUROBI on MATLAB®.

7.6.2   Day-ahead optimization results

With the day-ahead probabilistic predictions, an initial set of 2000 scenarios are initially generated using Latin hypercube sampling method and then reduced to five representative scenarios through simultaneous backward reduction. The day-ahead

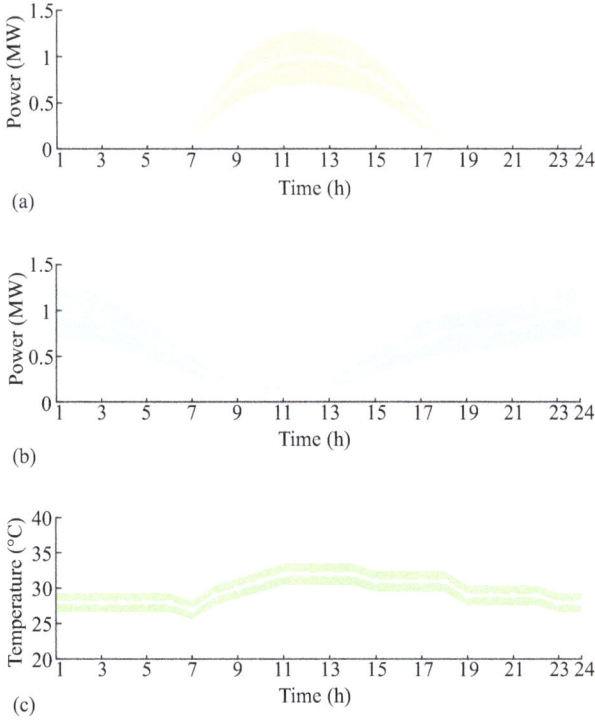

Figure 7.3 Day-ahead predictions for renewable generation and ambient temperature. (a) PV outputs. (b) Wind turbine outputs. (c) Ambient temperature.

Table 7.1 Technical parameters for generators within MEMG

Generator	P^{min} (MW)	P^{max} (MW)	$\kappa^{DG/CCHP}$ ($/MWh)
DG 1-3	(0.45, 0.3, 0.3)	(1.5, 1.0, 1.0)	(122, 134, 136)
CCHP	0	1.5	159

results are presented from the perspectives of the MEMG operator, EU agents, and computational performance.

7.6.2.1 Results for MEMG operator

Figure 7.4 provides insights into the day-ahead energy scheduling and pricing strategies of the MEMG operator. Analyzing the energy scheduling decisions (Figure 7.4(a) and (b)), it is evident that the MEMG operator primarily transacts with the EU agents, with only a small portion of energy being exported to the external networks (depicted by the red and blue lines). Notably, during the off-peak period of

Table 7.2   Technical parameters for energy storage within MEMG

Energy storage	$\chi_{min}^{BS(TS)}/$ $\chi_{max}^{BS(TS)}$	$\eta^{BSC(TSC)}/$ $\eta^{BSD(TSD)}$	$\tau^{BS(TS)}$	$E^{BS(TS),rate}$ (MW)	$\kappa^{BS(TS)}$ ($/MWh)
BS	0.2/0.9	0.98/0.98	0.01	2.0	30
TS	0.1/0.9	0.95/0.95	0.04	2.5	0.2

Table 7.3   Parameters for EUs within MEMG

Type of EUs	Residential EUs	Commercial EUs	Industrial EUs
α ($/MW^2)	826	900	600
β ($/°C^2)	0.366	0.6 (7:00-20:00) 0 (other periods)	0
γ	0.7	0	1
θ^{exp} (°C)	20	20	N/A
$\theta^{td,min/max}$ (°C)	17/25	18/23	N/A
C^{air} (MW/°C)	0.106	0.037	N/A
R^T (°C/MW)	14	6	N/A

electrical demand (1:00–6:00), the CCHP unit does not generate any thermal energy as shown in Figure 7.4(b). This is because both the MEMG and utility grid are in an over-generation state, such that the surplus electricity has to be accommodated at a negative feed-in tariff. Consequently, the MEMG operator minimizes the power output of CCHP unit during this off-peak period. Turning to the energy pricing decisions (Figure 7.4(c)), it can be observed that the electrical and thermal prices for transacting with the commercial EU agent (represented by the reddish-brown and purple lines) are equal to the prices for buying energy from the external energy networks. This parity arises from the fact that the commercial EU agent prioritizes comfort and is less sensitive to the prices released by the MEMG operator. Conversely, the electrical prices for the residential and industrial EU agents (shown by the red and yellow lines) are lower than the selling price of the utility grid during the off-peak period (1:00–6:00). This strategy aims to stimulate price-sensitive EUs to accommodate the surplus electrical generation during the off-peak period, otherwise the MEMG operator can only choose to export the surplus electricity into the grid at negative feed-in tariffs. Regarding the thermal prices for the residential EU agent (illustrated by the light blue line), they are slightly lower than the selling prices of the external thermal network only at certain time slots. It can be concluded that the day-ahead energy management adopted by the MEMG operator mainly relies on energy scheduling, while pricing strategy acts as a secondary means.

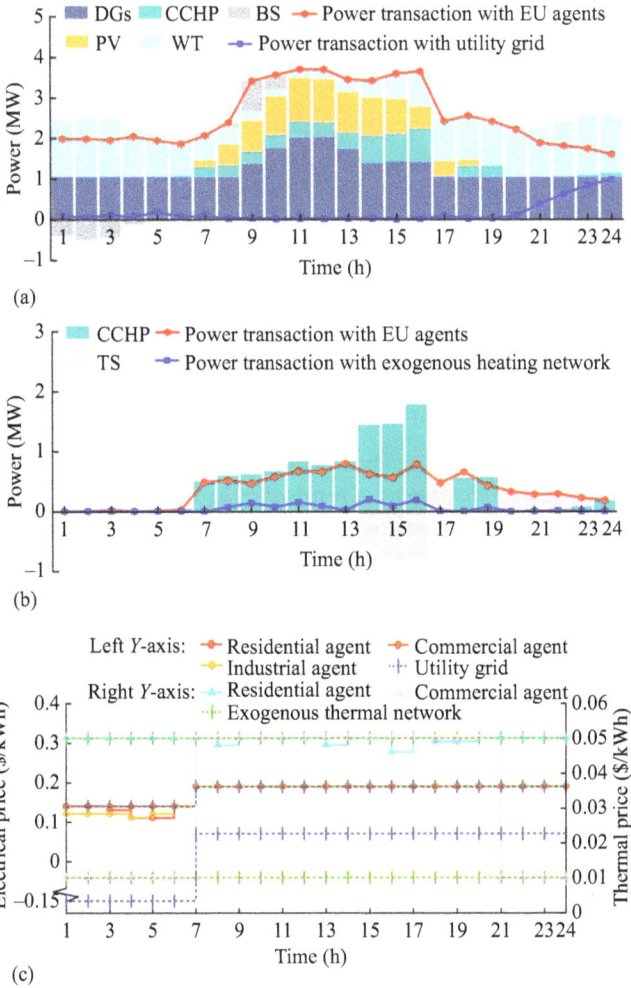

Figure 7.4   Day-ahead energy scheduling and pricing decisions adopted by MEMG operator. (a) Electrical power balance condition. (b) Thermal power balance condition. (c) Energy trading process with EU agents.

7.6.2.2   Results for EU agents

Figure 7.5 illustrates the day-ahead (DA) energy procurement commitments of different EU agents, shedding light on their specific preferences and behaviors. Figure 7.5(a) and (b) reveals that both residential and commercial EU agents procure less electricity than their expected demands in the day-ahead energy market, indicating a tendency towards power-shiftable electrical demands. Notably, the day-ahead electrical energy procurement commitment of the commercial EU agent closely aligns with its expected demand, which suggests that the commercial EUs are less sensitive to prices. In contrast, Figure 7.5(c) demonstrates that the price-sensitive industrial

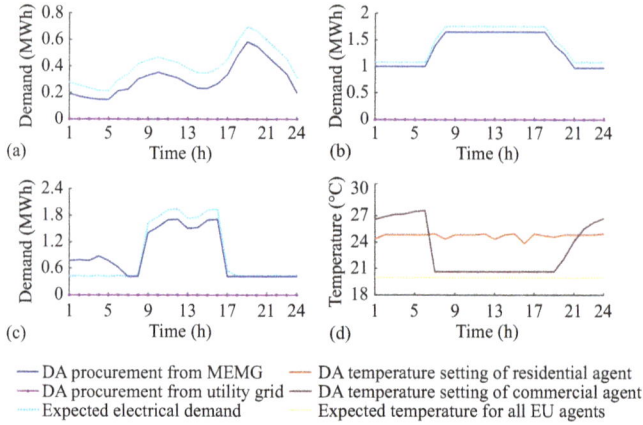

*Figure 7.5   Day-ahead energy procurement commitments made by EU agents. (a)
Electrical commitment of residential agent. (b) Pre-set demands of EU
agents. (c) Electrical commitment of residential agent. (d) Pre-set
temperatures of EU agents.*

EU agent shifts a portion of its electrical energy demand from the peak period (9:00–
17:00) to the off-peak period (1:00–6:00) with nearly unchanged daily total demand.
This suggests the electrical demand of the price-sensitive industrial EU agent tends
to be time-shiftable. Furthermore, Figure 7.5(d) compares the comfort requirements
of the residential and commercial EU agents regarding indoor temperature during
the working period (7:00–20:00). The figure illustrates that when the MEMG operator reduces the thermal price (Figure 7.4(c)), the relatively sensitive residential EU
agent tends to increase its purchase of cooling energy. This behavior aims to bring
the indoor temperature closer to the expected level, as indicated by the orange line in
Figure 7.5(d). In contrast, the commercial EU agent exhibits a more consistent and
higher comfort requirement on indoor temperature during the working period, with
less sensitivity to thermal price fluctuations.

7.6.2.3   Computational performance

Table 7.4 presents a comparison of computational performances using three different
solution methods: the direct use of off-the-shelf solver GUROBI, the PH algorithm,
and the proposed adaptive PH algorithm. The general method is to utilize GUROBI
to directly solve the day-ahead stochastic MILP problem, which ensures a 100%
optimal solution. However, this method is time-consuming due to the large number
of variables, particularly binary variables. In contrast, the PH algorithm effectively
addresses this issue by reducing the scale of the stochastic MILP problem through
scenario decomposition. It allows parallel solution of scenario-based subproblems,
leading to improved computational efficiency. PH algorithm 1 sets the penalty factor $\mathcal{P} = \mathcal{P}_1 = 100$ to ensure the accuracy of the near-optimal solution. However, it
fails to converge within 1000 iterations, equivalent to 3.21 h of solution time. Conversely, PH algorithm 2 sets $\mathcal{P} = \mathcal{P}_1 = 1000$. This adjustment enables obtaining a

Table 7.4   *Computational performance for solving the day-ahead risk-averse stochastic MILP*

Solution method	No. of variables	Solution time	Objective value ($)	Optimality accuracy
Direct use of GUROBI	14,862 continuous 4641 binary for all scenarios	79.70 h	7501.01	100%
PH algorithm 1	3938 continuous 2781 binary for each scenario	≥ 3.21 h	N/A	N/A
PH algorithm 2		0.56 h	7478.15	99.7%
Adaptive PH algorithm		0.71 h	7492.69	99.9%

near-optimal solution within 0.56 h, but the accuracy is only 99.7%. To achieve a faster convergence speed while guaranteeing an acceptable solution quality, the proposed adaptive PH algorithm introduces a time-varying penalty factor. Specifically, it sets $\mathcal{P} = 100$ and $\mathcal{P}_1 = 1000$. This modification yields a near-optimal solution with 99.9% accuracy within 0.71 h, which is fully compatible with the day-ahead computation requirement. Note that the solution time associated with the (adaptive) PH algorithms reflect the parallel computation time.

7.6.3   Intra-day optimization results

The energy trading decisions made by the MEMG operator and EU agents in the intra-day (ID) energy market are depicted in Figure 7.6. For illustrative purposes, the day-ahead (DA) energy trading prices are included as well, which are represented by the blue lines in Figure 7.6(a)–(c) and the orange lines in Figure 7.6(d)–(e).

The ID electrical energy trading results (Figure 7.6(a)–(c)) reveal that when the ID electrical price (represented by the gray line) falls below the DA price, the EU agents tend to purchase additional electricity in the intra-day energy market to raise their comfort level. Conversely, when the ID price rises, the EU agents tend to sell a portion of their DA procurement back to the MEMG operator for arbitrage. However, this observation is not completely valid for Figure 7.6(c), as the industrial EU agent cannot decrease its DA procurement during the intra-day stage due to the need to meet its daily demand, even when the ID price increases. The ID thermal energy trading results (Figure 7.6(d) and (e)) present that the EU agents still sell a portion of their DA thermal procurement at certain time slots when the ID price (represented by the dark brown line) falls below the DA price. This behavior is a consequence of the hourly-ahead ambient temperature reduction compared to the DA prediction, which reduces the ID thermal demands of the EU agents. In Figure 7.6(f), it can be observed

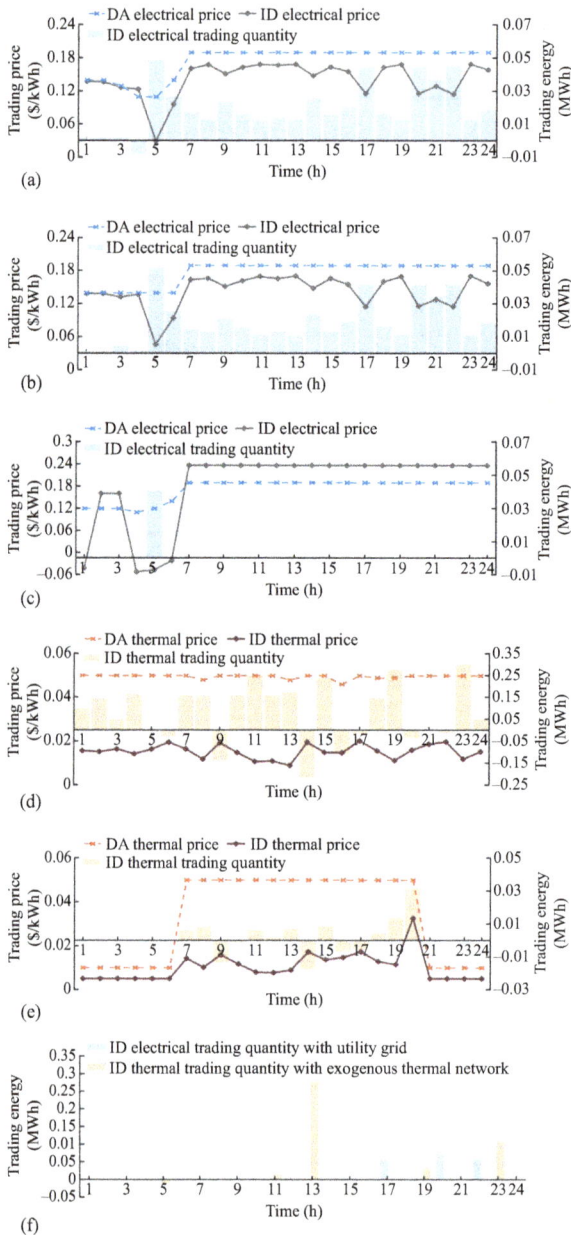

*Figure 7.6   Intra-day energy trading decisions of MEMG operator and EU agents.
(a) Intra-day electrical trading decisions of MEMG operator and
residential agent. (b) Intra-day electrical trading decisions of MEMG
operator and commercial agent. (c) Intra-day electrical trading
decisions of MEMG operator and industrial agent.*

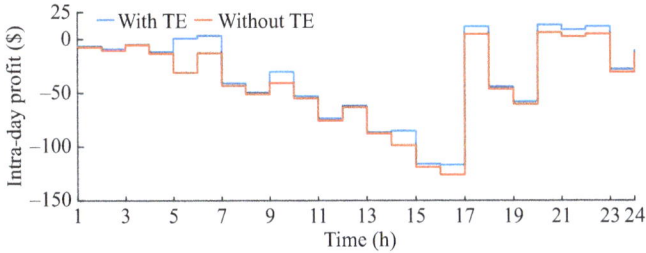

Figure 7.7   Intra-day operating profit of MEMG operator

Table 7.5   Computational performance for solving the intra-day bilinear problem

Bilinear problem solution method	Step size ($/kWh)	Convergence tolerance	Average computation time (s)
Price discretizing method	0.001	N/A	191.95
OA algorithm	$\to 0$	10^{-5}	13.51

that the MEMG operator engages in energy exchanges with the external energy networks in the ID energy market, but only at a few time slots. This implies that, the ID local power balance can be achieved by the pricing strategy of the MEMG operator at most of time. Hence, it can be concluded that the intra-day transactive energy management effectively reduces the MEMG's reliance on the external energy networks.

Figure 7.7 displays the intra-day operating profits of the MEMG operator with and without considering transactive energy (TE). Note that these profits exclude the day-ahead determined profit (which is calculated by subtracting the operating costs of DGs and ESSs from the sales revenue in the day-ahead energy market). The lower intra-day operating profit indicates that the MEMG operator incurs a higher cost to address the intra-day power imbalance resulting from uncertain renewable generation. Figure 7.7 demonstrates that without the intra-MEMG TE, the MEMG operator has to cost more to balance the intra-day power deviation, since it can only adjust the power output of the CCHP unit and engage in transactions with the external energy networks at unfavorable prices. In contrast, with the implementation of TE, the MEMG operator can guide the EU agents to adjust their day-ahead energy procurement commitments and participate in demand-side management, through releasing appropriate prices that benefit both parties. Therefore, with TE, the MEMG operator can not only reduce the costs spent on addressing intra-day power imbalance (Figure 7.7), but also reduce the MEMG's reliance on the external energy networks (Figure 7.6(f)).

Table 7.5 compares the computation performances of the price discretizing method and OA algorithm for solving the intra-day bilinear problem. It can be observed that OA algorithm offers the advantage of deriving continuous trading prices (i.e., step size $\to 0$) with a solution speed that is significantly faster compared to the price discretizing method. In other words, the OA algorithm can achieve a superior solution speed without sacrificing the accuracy of the solution, which thus enables more precise and flexible decision-making.

7.7 Conclusion

This chapter proposes a risk-averse transactive energy management method for an MEMG. The problem is formulated as a Stackelberg game-based bi-level model that can effectively capture the strategic interactions between the MEMG operator and residential, commercial, and industrial EU agents. To tackle the challenges posed by uncertainties and coordinate the strategies in the day-ahead and intra-day energy markets, a risk-averse adaptive SO approach is developed. This approach incorporates a CVaR measure, enabling the consideration of the MEMG operator's risk aversion towards uncertainties in the day-ahead stage. To ensure high computational efficiency, an adaptive PH algorithm is developed and an OA algorithm is adopted. The adaptive PH algorithm decomposes the day-ahead stochastic MILP problem into scenario-based subproblems, which can be solved in parallel. The OA algorithm solves the intra-day bilinear problem without the need for introducing binary variables, thereby achieving a superior solution speed without sacrificing the accuracy of the solution.

The simulation results demonstrate the effectiveness of the proposed risk-averse transactive energy management method as well as the efficiency of the adaptive PH and OA algorithms: (1) The proposed transactive energy management method can simultaneously provide optimal energy scheduling and pricing strategies for the MEMG operator, by which the demand-side management can be effectively utilized; (2) The transactive energy management method can guide the EU agents to adjust their day-ahead commitments by offering proper prices that are beneficial to both parties. The mutually beneficial pricing strategy can reduce the MEMG's reliance on the external networks; (3) The adaptive PH algorithm and OA algorithm effectively address the challenges associated with the multi-scenario stochastic MILP and bilinear problems, respectively. They both achieve a significant improvement in computation efficiency.

References

[1] Guelpa E, Bischi A, Verda V, *et al.* Towards future infrastructures for sustainable multi-energy systems: a review. *Energy.* 2019;184:2–21.

[2] Mancarella P. MES (multi-energy systems): an overview of concepts and evaluation models. *Energy.* 2014;65:1–17.

[3] Chen Y, Feng X, Li Z, *et al.* Multi-stage coordinated operation of a multi-energy microgrid with residential demand response under diverse uncertainties. *Energy Conversion and Economics.* 2020;1(1):20–33.

[4] Sharma S, Verma A, Xu Y, *et al.* Robustly coordinated bi-level energy management of a multi-energy building under multiple uncertainties. *IEEE Transactions on Sustainable Energy.* 2019;12(1):3–13.

[5]   Li Z, Wu L, and Xu Y. Risk-averse coordinated operation of a multi-energy microgrid considering voltage/var control and thermal flow: an adaptive stochastic approach. *IEEE Transactions on Smart Grid*. 2021;12(5):3914–3927.

[6]   Melton RB. Gridwise transactive energy framework. Pacific Northwest National Lab. (PNNL), Richland, WA (United States); 2013. Tech. Rep. PNNL-22946.

[7]   Sampath LPMI, Paudel A, Nguyen HD, *et al.* Peer-to-peer energy trading enabled optimal decentralized operation of smart distribution grids. *IEEE Transactions on Smart Grid*. 2021;13(1):654–666.

[8]   Zhang K, Zhou B, Chung CY, *et al.* A coordinated multi-energy trading framework for strategic hydrogen provider in electricity and hydrogen markets. *IEEE Transactions on Smart Grid*. 2022;14(2):1403–1417.

[9]   Lezama F, Soares J, Hernandez-Leal P, *et al.* Local energy markets: paving the path toward fully transactive energy systems. *IEEE Transactions on Power Systems*. 2018;34(5):4081–4088.

[10] Jiang X, Sun C, Cao L, *et al.* Peer-to-peer energy trading with energy path conflict management in energy local area network. *IEEE Transactions on Smart Grid*. 2022;13(3):2269–2278.

[11] Yang Z, Hu J, Ai X, *et al.* Transactive energy supported economic operation for multi-energy complementary microgrids. *IEEE Transactions on Smart Grid*. 2020;12(1):4–17.

[12] Li J, Khodayar ME, Wang J, *et al.* Data-driven distributionally robust co-optimization of P2P energy trading and network operation for interconnected microgrids. *IEEE Transactions on Smart Grid*. 2021;12(6):5172–5184.

[13] Wang Y, Huang Z, Li Z, *et al.* Transactive energy trading in reconfigurable multi-carrier energy systems. *Journal of Modern Power Systems and Clean Energy*. 2019;8(1):67–76.

[14] Liu N, Zhou L, Wang C, *et al.* Heat-electricity coupled peak load shifting for multi-energy industrial parks: a Stackelberg game approach. *IEEE Transactions on Sustainable Energy*. 2019;11(3):1858–1869.

[15] Fleischhacker A, Auer H, Lettner G, *et al.* Sharing solar PV and energy storage in apartment buildings: resource allocation and pricing. *IEEE Transactions on Smart Grid*. 2018;10(4):3963–3973.

[16] Peng Q, Wang X, Kuang Y, *et al.* Hybrid energy sharing mechanism for integrated energy systems based on the Stackelberg game. *CSEE Journal of Power and Energy Systems*. 2021;7(5):911–921.

[17] Ali A, Liu N, and He L. Multi-party energy management and economics of integrated energy microgrid with PV/T and combined heat and power system. *IET Renewable Power Generation*. 2019;13(3):451–461.

[18] Zhu X, Sun Y, Yang J, *et al.* Day-ahead energy pricing and management method for regional integrated energy systems considering multi-energy demand responses. *Energy*. 2022;251:123914.

[19] Nasrolahpour E, Kazempour J, Zareipour H, *et al.* A bilevel model for participation of a storage system in energy and reserve markets. *IEEE Transactions on Sustainable Energy*. 2017;9(2):582–598.

[20] Zou Y, Xu Y, Feng X, *et al.* Peer-to-peer transactive energy trading of a reconfigurable multi-energy network. *IEEE Transactions on Smart Grid.* 2022;14(3):2236–2249.

[21] Zou Y, Xu Y, Feng X, *et al.* Transactive energy systems in active distribution networks: a comprehensive review. *CSEE Journal of Power and Energy Systems.* 2022;8(5):1302–1317.

[22] Li Z, Xu Y, Fang S, *et al.* Robust coordination of a hybrid AC/DC multi-energy ship microgrid with flexible voyage and thermal loads. *IEEE Transactions on Smart Grid.* 2020;11(4):2782–2793.

[23] Jiang Y, Wan C, Chen C, *et al.* A hybrid stochastic-interval operation strategy for multi-energy microgrids. *IEEE Transactions on Smart Grid.* 2019;11(1):440–456.

[24] Tan J, Wu Q, Zhang M, *et al.* Chance-constrained energy and multi-type reserves scheduling exploiting flexibility from combined power and heat units and heat pumps. *Energy.* 2021;233:121176.

[25] Daneshvar M, Mohammadi-Ivatloo B, Zare K, *et al.* Two-stage robust stochastic model scheduling for transactive energy based renewable microgrids. *IEEE Transactions on Industrial Informatics.* 2020;16(11):6857–6867.

[26] Von Stackelberg H. *Market structure and equilibrium.* Springer Science & Business Media; 2010.

[27] Li J, Zhang C, Xu Z, *et al.* Distributed transactive energy trading framework in distribution networks. *IEEE Transactions on Power Systems.* 2018;33(6):7215–7227.

[28] Boyd SP, and Vandenberghe L. *Convex optimization.* Cambridge University Press; 2004.

[29] Zou Y, Xu Y, and Zhang C. A risk-averse adaptive stochastic optimization method for transactive energy management of a multi-energy microgrid. *IEEE Transactions on Sustainable Energy.* 2023;14(3):1599–1611.

[30] Ruiz C, Conejo AJ, and Gabriel SA. Pricing non-convexities in an electricity pool. *IEEE Transactions on Power Systems.* 2012;27(3):1334–1342.

[31] Xu Y, Dong ZY, Zhang R, *et al.* Multi-timescale coordinated voltage/var control of high renewable-penetrated distribution systems. *IEEE Transactions on Power Systems.* 2017;32(6):4398–4408.

[32] Grass E, Fischer K, and Rams A. An accelerated L-shaped method for solving two-stage stochastic programs in disaster management. *Annals of Operations Research.* 2020;284:557–582.

[33] Xiong P, and Jirutitijaroen P. Stochastic unit commitment using multi-cut decomposition algorithm with partial aggregation. In: *2011 IEEE Power and Energy Society General Meeting.* IEEE; 2011. p. 1–8.

[34] Carøe CC, and Schultz R. Dual decomposition in stochastic integer programming. *Operations Research Letters.* 1999;24(1–2):37–45.

[35] Arif A, Wang Z, Chen C, *et al.* A stochastic multi-commodity logistic model for disaster preparation in distribution systems. *IEEE Transactions on Smart Grid.* 2019;11(1):565–576.

[36] Kaisermayer V, Muschick D, Horn M, *et al.* Progressive hedging for stochastic energy management systems: the mixed-integer linear case. *Energy Systems*. 2021;12(1):1–29.

[37] Boyd S, Parikh N, Chu E, *et al.* Distributed optimization and statistical learning via the alternating direction method of multipliers. *Foundations and Trends® in Machine learning*. 2011;3(1):1–122.

[38] Bertsimas D, Litvinov E, Sun XA, *et al.* Adaptive robust optimization for the security constrained unit commitment problem. *IEEE Transactions on Power Systems*. 2012;28(1):52–63.

Chapter 8

Operation of multi-energy microgrids with laboratory validation

Vedran S. Perić[1], Ruihao Song[2], Verena Kleinschmidt[2] and Thomas Hamacher[2]

This chapter explores the advantages of integrated operation of district heating and cooling (DHC) and electric distribution systems in the context of multi-energy micro-grids. First, the chapter describes the potential of multi-energy operations from the perspective of flexibility. Second, it delves into the analysis of coordinated steady-state operation, showcasing the benefits of energy integration. Last, the chapter introduces CoSES laboratory infrastructure, which allows developing and validat-ing coordinating energy management schemes, ensuring real-world applicability and effectiveness.

8.1   Introduction

The aim of reducing greenhouse gas emissions drives the popularity of microgrid systems that efficiently manage renewable energy fluctuations. Conventional micro-grids are characterized by large amounts of renewable distributed energy resources (DER), like photovoltaic (PV) and wind energy, together with energy storage sys-tems. Further advancement involves microgrids with multiple energy vectors, such as heat, natural gas, or hydrogen [1], forming multi-energy microgrids (MEMGs) to boost efficiency [2].

Typical MEMGs consist of energy sources (PV and wind), energy storage sys-tems (batteries, super-capacitors, cooling/heat storage, or hydrogen tanks), energy conversion systems (electrolyzers, fuel cells, heat pumps, or natural gas power plants), transmission systems (electrical, thermal, and gaseous networks), and final consumption units (electrical, thermal, or hydrogen loads) [3]. MEMGs exploit the capabilities of power-to-X (heat, gas, etc.) units to use excess renewable electric energy to generate other energy vectors that provide greater flexibility and more accessible storage options. These versatile systems have been primarily developed

[1]Chair for Intelligent Energy Management, University of Bayreuth, Germany
[2]Munich Institute of Integrated Materials, Energy and Process Engineering (MEP), Technical University of Munich, Germany

in small applications like buildings, communities, and industries involving clean and renewable energy sources. Interconnected MEMGs enhance system stability, efficiency, and flexibility, addressing fluctuations in generation and consumption [2].

Several test projects are raised to demonstrate the availability of MEMG operations. In the United States, the University of California San Diego has a MEMG that includes a co-generation plant, solar PV systems, battery energy storage systems (BESS), and an extensive energy management system (EMS). The MEMG generates about 92% of the electricity used on campus, greatly reducing their reliance on the local utility. In China, a MEMG called Eco-City is tested in Tian'Jin. It includes a variety of energy sources, such as wind, solar, geothermal, and BESSs, with advanced grid control technology that allows for optimal dispatch. In Europe, Feldheim is a village in Germany that is entirely energy self-sufficient and uses a MEMG. It combines wind, solar, and biogas energy production with a local heating network and heat storage.

This chapter aims to explore the advantages of integrated DHC and electric systems. First, we will analyze the potential of multi-energy operations from the perspective of flexibility. Second, we will delve into the analysis of coordinated steady-state operation, showcasing the benefits of energy integration. Last, we will introduce the CoSES laboratory infrastructure, where we can develop and validate coordinating energy management schemes, ensuring real-world applicability and effectiveness.

8.2   Flexibility options in MEMGs

The main benefit of the integration of different energy sectors lies in increased flexibility, which enables the integration of larger renewable source capacities in the system [4]. Compared to decoupled energy systems, MEMGs offer conversion capabilities through the interconnection of different energy carrier networks. In addition, MEMGs are a promising option to reduce the uncertainty of renewable generation in energy systems. The benefits of MEMGs have been proved in several works like [5–7]. A general overview of MEMGs is given in [3].

The flexibility management in MEMGs has been extensively investigated in the literature from the point of integrated electricity and gas systems while less research has been done on systems comprising electricity, gas, and heat/cooling. Researchers have developed various MEMG concepts that range from small residential hubs to large-scale systems compared to distribution systems.

The literature shows different approaches to treating end-user energy prices within a MEMG. Price management is very important as pricing is seen as a main instrument for flexibility valuation in the system. While most of the prices are considered universal for a system and calculated based on given inputs, some researchers use locational marginal or nodal pricing. This offers the advantage that prices reflect network congestion effects and thus indicate the impact of individual resources or flexibility programs on an entire system.

The flexibility management in MEMGs in this chapter is categorized by different flexibility programs that can be divided into three groups—conversion, storage,

Table 8.1   Selection of flexibility resources by groups [4]

	Electricity	**Gas**	**Heat and Cooling**

Storage	Mechanical
	– Pumped heat electricity
	  storage (PHES)
	– Compressed air energy
	  storage (CAES)
	– Flywheel
	Electrical
	– Supraconducting magnetic
	  energy storage (SMES)
	– (Super-)Capacitor
	Electrochemical
	– Secondary battery
	– Flow battery
	– Plug-in-electric-vehicles (PEV)
		– Hydrogen storage
		– LNG storage
			Solar hydrogen
			Heat storage

Conversion	– Power-to-X – Power-to-Gas	– Power-to-Heat – Vehicle-to-Grid	
DSM	– Demand response	– Energy efficiency	

and demand-side management (DSM). Each group comprises various flexibility resources. Table 8.1 provides an overview of some of the most commonly used flexibility resources.

The flexibility provided by the supply side via up/downward regulation of power generation units is not included in the scope of this chapter.

8.2.1   Conversion units

The first group of flexibility is defined as energy conversion or the shifting of demand from one energy vector to another. Conversion units—bundled in a hub or as individual units—include combined heat and power (CHP), electric heaters or coolers, gas furnaces, heat pumps (HP) as well as techniques like power-to-gas or power-to-heat. These conversion devices or techniques can avoid network congestion or peak loadings and thus provide flexibility in a well-coordinated MEMG. In addition, co-generation in CHP units offers the potential to increase the efficiency of a power system [8].

In system operation with price-sensitive demand, the rate of conversion depends on the price spread of the energy carriers. A typical example of this happens during high electricity demand and, consequently, high electricity prices. Consumers would then prefer to buy gas and convert it to electricity via, e.g., a gas turbine. This effect appears for example in the MEMG in [5] where the authors apply a mixed integer linear programming (MILP) to minimize total costs for electricity and gas consumption. The prices for gas are assumed as real market prices, while electricity can be purchased via a time-of-use (TOU) tariff. The conversion capabilities of the MEMG allow a higher share of renewable generation in the form of PV and wind generation within the system.

In [9], implementing a proposed integrated demand response (DR) program in an electricity and gas network leads consumers to convert gas to electricity during peak hours. The system is modeled as an ordinal potential game and a distributed algorithm is used to determine the unique Nash equilibrium. Another example of the application of game theory with unique Nash equilibrium is presented in [10], where EMS coordinates the electricity and gas consumption in a smart energy hub framework modeled as a non-cooperative game, resulting in gas consumption increase during electricity peak demand.

The authors in [11] present conversion to electricity supply with a probabilistic scheduling model for the operation of energy hubs. They argue that if a thermal DR program is applied to a hub, the amount of gas input increases while the electricity input decreases. The purchased gas is then used to produce heat and electricity in a CHP unit.

In addition to electricity demand, the heat demand also can be covered through conversion. This is demonstrated in [12], where interconnected energy hubs with varying wind generation and a DR program in place are approximated with a stochastic linearization model. The heat demand is covered by CHP during the day and via an electric boiler during the night with low electricity prices. Another example of thermal demand covered by conversion is the regional integrated energy system in [13]. The authors implement a day-ahead and intraday optimization schedule that is divided into several layers. The operation optimized in this way includes the conversion of electricity into thermal energy (heat and cooling) as well as into gas.

In [14], the authors analyze that in a system with high wind capacity installed, CHP can increase wind curtailment needs. They propose to solve this issue with a power-to-hydrogen system. Excess power from wind energy is transferred to hydrogen via an electrolyzer and further converted into heat. Thus, the CHP scheduling is less dependent on a fixed electricity outcome and can operate at a more feasible operation point. The electricity is thereby used for heat via power to hydrogen or electric boiler instead of being curtailed.

One of the main advantages of conversion as a tool for flexibility is, that it does not incorporate a change in consumers' comfort. Furthermore, the authors in [9] argue that conversion could decrease the amount of required DSM and thus reduce the dependency on customers' motivation to participate in DSM or DR programs.

However, conversion benefits can only be fully realized in an environment where prices are dynamic. Changes in market prices are the only feasible signal for economic shortages of supply or grid congestion. The resulting price spread mirrors the current availability of different energy carriers and local transmission and conversion capacities.

8.2.2 Storage

A second type of flexibility in MEMGs is storage devices. These include gas storage and thermal storage as well as electricity storage or electrochemical storage. All these technologies allow storing supply surplus for later demand and thus provide flexibility by temporal unbundling of supply and demand.

In [6], the authors show that integrated energy companies (IECs) can profit from multi-energy markets with simultaneous trade of different energy sources. One factor that enables higher profits in the constructed market is battery storage. The IEC gains additional flexibility by charging batteries at peak load and discharging at valley loads flattening the overall load curves. In [15], for a MEMG within a distributed energy network, the optimal scheduling of heat and cooling storage is determined in partial and global optimization scenarios using the NAGA-II algorithm. The performed case study shows that thermal storage leads to a reduction in electric peak load.

Another option for storage is electric vehicles through vehicle-to-grid (V2G) technologies. In [16], the authors argue that electric vehicles, among others, allow a more efficient operation of combined cooling and heat power (CCHP) systems resulting in a significant benefit for operational cost and consumer energy costs. The plug-in hybrid electric vehicle (PHEV) is therefore employed as a storage capacity within a residential energy hub besides CCHP, renewable resources, batteries, and thermal storage. However, the authors argue that many consumers hesitate to integrate their PHEV in such a system because of battery degradation concerns.

A commonly mentioned form of thermal storage is based on the thermal inertia of buildings. The possibility of using buildings and their electro-thermal heating systems as response units for flexibility is being discussed in [17]. Furthermore, the authors present an approach toward the physical modeling of such systems, including demand for space heating, hot water, electricity, and thermal inertia of different components.

Same as for conversion, one advantage of storage in MEMGs is the unchanged comfort for consumers due to unchanged end-user consumption. An exception to this could be the use of thermal inertia as a storage unit depending on the resulting deviation of temperatures from "comfort regions." Other often mentioned advantages of storage technology in MEMGs lie in cost reductions and improved efficiency due to the temporary unbundling of demand and supply. The authors in [12] show that the introduction of natural gas and heat storage in an energy system can significantly reduce costs. From an investment point of view, adding heat storage to a hub significantly increases the profitability of the associated investment [18]. Furthermore, the risk profile of the investment is reduced by the additional flexibility.

Limitations of the storage technology as a flexibility tool are mainly of an economic nature. This is, for example, shown in [19], where scheduling follows an economic approach using an energy storage pricing policy model. Optimization shows that the ESS should only release energy during peak periods with high energy prices. During other times coal- and gas-powered generators are more cost-efficient options. Another limitation observed in [20] is that, unlike DSM, the storage did not reduce pollution in their analyzed system setup.

8.2.3   Demand side management

The third group of flexibility programs is defined as DSM. This includes all actions where a system's end consumers change their consumption profile, with or without a change in their overall energy consumption. Part of DSM is demand response (DR), which defines programs aimed at curtailing or shifting loads either incentive-based (like emergency or ancillary DR) or price-based (like time of use, real-time pricing). The DR programs allow access to the latent flexibility of the demand [21] and can be used to control electricity, gas, and heat/cooling demand within a MEMG and thus constitute the focus of DSM in this chapter.

According to [11] most research attention lies on DR for electricity loads. These flexible loads can either be managed separately for each consumer or collected through a so-called aggregator like in [22]. However, the assumption that aggregators would share the information on their energy consumption models in this work is rather unrealistic.

In contrast to papers analyzing the influence of a DSM program on a MEMG, the authors in [23] determine the required market signals to trigger a desired amount of demand adaptation. In their work, they present a DR simulation tool based on PSCAD, non-linear programming, and a price elasticity approach for consumers' load reduction to maximize the retailer's profit.

The above discussion focused on a supply-side view, where DR is used to improve the overall efficiency of the system. However, DR schemes can also be analyzed from the point of view of consumers, as shown in [24]. Here, the authors develop a strategy for consumers to exploit an existing DR program to minimize their energy purchasing costs. Therefore, the scheduling of household appliances and electric vehicles is determined depending on expected hourly electricity prices. In a conducted case study, consumers could reduce their electricity bill by 8% with optimal load profiles.

The feasibility of DR programs from a system point of view depends on the system and market environment. DR programs are only profitable in a defined range of costs, expected benefits, and reduced load. In [25], these areas are identified via so-called DR profitability maps.

Another important factor in the design of DR programs is their timing regarding how fast certain resources can be employed. The authors in [26] analyze the impact of a natural gas transportation network on the dispatch of several gas-fired units. They find that an hourly DR can contribute to optimal unit dispatch while the authors in [27] show that the order of different optimizations within a model also matters for the DR scheduling. Accordingly, co-optimization of DR and unit commitment leads to higher usage of DR capacities compared to DR application after the unit commitment decision of an MEMG.

Although less frequent, there are also works analyzing DSM on thermal loads. For example, in the aforementioned probabilistic scheduling model, the authors found that thermal DR programs increase gas consumption and reduce electricity consumption [11].

Among the frequently mentioned advantages of DR in MEMGs are reduced peak load [9,12,16,28], lower system operation cost [7,12,28,29], and higher efficiency of medium and large scale hubs [11,13]. The authors in [22,30] found that larger participation of consumers in DR results in decreased price volatility. Another effect on prices was found in [28,30], where DR reduced energy generation costs and locational marginal electricity and gas prices. Consequently, utility companies could earn higher profits as argued in [10]. Works like [29] indicate that through DR programs in MEMGs pollution levels could be decreased. Last, the authors in [18] found that similar to heat storage, adding a DSM program to a hub significantly increases the profitability of an investment and reduces the risk through additional flexibility.

Despite the numerous advantages stated in different research works, there are also some limitations to the usage of DR in MEMGs. One limiting factor is the complex market design that has to orchestrate the management of the available demand flexibilities at various times and the compensation schemes for conducted load reductions. Many of these options are vulnerable to deception by consumers who do not accurately report their desired level of load before modification for the DR program. Another limitation is the fact that the benefits of DR are highly dependent on consumer participation. Assumptions of participation rates in the analyzed papers range from 10% in [26] to 20% in [28,30]. In [31], the authors review several studies on DR program participation, response, and persistence. Despite high variations in results, they find that participation in opt-in programs is rather low with more than half of the analyzed programs showing rates below 10%. In addition, the desired load response is barely realized, and there are even undesired effects in consumer behavior. Regardless, the authors in [32] argue that even a small reduction of peak loads can result in a significant long-term benefit. They estimate that the avoided cost of capacity and energy through a 5% decrease in peak loads could add up to $3 billion per year in the United States.

A more specific negative effect of DR is being observed in [27]. The authors argue that DR may cause abrupt and unwarranted price increases due to displacing less flexible generating units and triggering additional transmission congestion.

8.3   Steady-state operation optimization

There have been numerous attempts to formulate a practical optimization model of an MEMG that is suitable for real-time operation [33,34]. These models typically result in MILP problems like in [17], where the model consists of generation, storage, buildings, power constraints, and reserve allocation with a focus on the physical modeling approach. Other MILP formulations model hubs with (co-)generation, storage units, and the objective to minimize (environmental) cost [16,35]. While the MEMG formulation in [7] is also based on MILP, the uncertain parameters of the hub operation are addressed in a stochastic model. The focus of the analysis lies in the impact of additional sources or storage units on optimal hub scheduling. Another approach for a multi-objective model is the Naga II algorithm in [15], where heat and cold storage is modeled in separate hubs with a focus on load forecasting. Another option to display the scheduling of MEMG components can be obtained from [6], where a day ahead multi-energy market is modeled. The formulation includes heat and power supply, storage, and various flexible loads that can either be shifted, reduced, or converted. While some of the approaches to model MEMGs have similarities, they are all based on specific stand-alone models.

8.3.1   Software tools for MEMG optimization

Among the available open-source frameworks, the authors in [36] identified Calliope, OEMOF, and URBS as suitable tools for the grid-based modeling of MEMGs. Calliope, a multi-scale energy system modeling framework, offers dispatch and investment optimizations [37]. The focus lies on simulating a high share of renewable generation while offering a high spatial and temporal resolution. The OEMOF modeling framework discussed in [36] and presented in [38] refers to the open energy modeling framework. It consists of a collection of Python packages covering the power, heat, and mobility sectors. However, the district heating optimization is in the early stage of development and underlies frequent changes [39]. The linear optimization model for distributed energy system URBS [40] was developed for the optimization of urban energy systems. Compared to the aforementioned tools, URBS focuses on DER and is especially valuable in determining the optimal size of storage components. Further alternatives for modeling MEMGs are the open-source tools OpenDSS and MESMO (previously named FLEDGE). OpenDSS [41] is a distribution system simulator offering DER integration. The Flexible Distribution Grid Demonstrator MESMO [42] offers optimization of power, heat systems, and gas systems considering various DERs such as renewable generation or flexible loads. Comparing OpenDSS and MESMO, relevant differences are the availability of a thermal system and the possibility to calculate distribution locational marginal prices (DLMP) to analyze network status offered by MESMO.

Focusing on research of flexibility provided by MEMGs, relevant criteria for a tool include the availability of multiple energy carrier networks and models for different load, generation, storage, and conversion units. All components should allow the integration of DR schemes and generation units should reflect the intermittency of renewables. The extension of the tool with additional models for distributed resources and the development of its own test cases and scenarios should be possible. The results provided by the tool should allow evaluation of the system behavior, the flexibility provided by the system design, and the influence of individual components. Last, an active community of users and developers ensures continuous enhancement of the tool.

With its modular structure, MESMO allows users to add specific DER components. Test cases can be built based on example scripts that are openly available. Another advantage of MESMO is the possibility to derive DLMP and their decomposition for both the electric and thermal grid. This is especially useful when analyzing the inherent flexibility in different system designs and the ability of a system to prevent DLMP increases.

8.3.2   MESMO model: electric grid

The formulation of the electric grid model in MESMO is explained in detail in [43,44] and is expressed as follows:

$$u_t = u^{ref} + M^{u,p}\Delta p_t + M^{u,q}\Delta q_t \tag{8.1}$$

$$\left|s_t^f\right|^2 = \left|s^{f,ref}\right|^2 + M^{s^f,p}\Delta p_t + M^{s^f,q}\Delta q_t \tag{8.2}$$

$$\left|s_t^t\right|^2 = \left|s^{t,ref}\right|^2 + M^{s^t,p}\Delta p_t + M^{s^t,q}\Delta q_t \tag{8.3}$$

$$p_t^{ls} = p_t^{ls,ref} + M^{p^{ls},p}\Delta p_t + M^{p^{ls},q}\Delta q_t \tag{8.4}$$

$$q_t^{ls} = q_t^{ls,ref} + M^{q^{ls},p}\Delta p_t + M^{q^{ls},q}\Delta q_t \tag{8.5}$$

The matrices M^x represent the sensitivity matrices of the respective parameters to changes in the active and reactive power injections Δp_t and Δq_t at time step t. In (8.1), the linearization of voltage magnitude u_t is expressed. The vector u^{ref} denotes the operating conditions at the linearization point. The matrices $M^{u,p}$ and $M^{u,q}$ express the sensitivity for changes in nodal voltages for a given change in active and reactive power injections. Equations (8.2) and (8.3) express the linearization of the squared branch flow from and to the respective nodes $\left|s_t^f\right|^2$, $\left|s_t^t\right|^2$. Here, $M^{s^f,p}$, $M^{s^f,q}$, $M^{s^t,p}$, and $M^{s^t,q}$ are the explicit sensitivity matrices with regard to line flows in power injections. The system losses p_t^{ls}, q_t^{ls} of active and reactive power are similarly linearized in (8.4) and (8.5) with matrices $M^{p^{ls},p}$, $M^{p^{ls},q}$, $M^{q^{ls},p}$, and $M^{q^{ls},q}$ containing the system loss sensitivity to changed power injections.

8.3.3   Thermal grid model

The thermal grid formulation in MESMO can be obtained from [44] and is expressed as follows:

$$h_t = h^{ref} + M^{h,p^{th}}\Delta p_t^{th} \tag{8.6}$$

$$v_t = v^{ref} + M^{v,p^{th}}\Delta p_t^{th} \tag{8.7}$$

$$p_t^{pm} = p^{pm,ref} + M^{p^{pm},p^{th}}\Delta p_t^{th} \tag{8.8}$$

Corresponding to the electric grid formulation, the thermal flow model is a linear approximation. The pressure head h_t at time step t is linearized in (8.6) with a reference point of linear operation h^{ref} and the sensitivity matrix $M^{h,p^{th}}$ providing the pressure head values for changes in the thermal power at the flexible loads Δp_t^{th}. The reference properties are obtained by solving a reference thermal power flow problem formulated in [44]. In (8.7) and (8.8) the branch flow volume v_t and the electric distribution pumping power demand p_t^{pm} are approximated similar with reference points v^{ref}, $p^{pm,ref}$ and sensitivity matrices $M^{v,p^{th}}$, $M^{p^{pm},p^{th}}$. The formulation assumes a radially connected thermal grid with distributed pumping and constant temperatures.

8.3.4   Gas grid model

The developed gas grid model is subject to several assumptions, including the assumption that there is no elevation difference of the pipes and no compression or pressure reduction in the stations. Furthermore, the gas temperature is assumed to be uniform, and thus there is no heat transfer between the transported gas and the surrounding soil. The pipeline flow dynamics are based on [38–40].

The gas grid formulation in MESMO is expressed as follows:

$$h_t^g = h^{g,ref} + M^{v^g,c^g} \Delta c_t^g \tag{8.9}$$

$$v_t^g = v^{g,ref} + M^{v^g,c^g} \Delta c_t^g \tag{8.10}$$

Here h_t^g and v_t^g denote the branch volume flow at gas grid branches b^g for time step t and the pressure at gas grid nodes n^g for time step t. The reference or nominal operation point of the gas grid is denoted by $h^{g,ref}$ and $v^{g,ref}$. The matrix M^{v^g,c^g} denotes the sensitivity matrix for the change of the respective properties to a change in gas consumption, which is denoted by Δc_t^g.

The reference flow problem of the gas grid is expressed as follows:

$$v^{g,ref} = (A^{n^g,b^g})^{-1} A^{n^g,der} p^{g,ref} \tag{8.11}$$

$$h^{g,ref} = [(A^{n^g,b^g})^T]^{-1} (f^{fr,ref} \odot f^0) \tag{8.12}$$

$$f_{b^g}^{fr,ref} = \frac{0.25}{\left[\ln \left(\frac{\varepsilon_{b^g}}{3.7 d_{b^g}} + \frac{5.74}{(Re_{b^g}^{ref})^{0.9}} \right) \right]^2} \tag{8.13}$$

Here A^{n^g,b^g} is the gas branch to node incidence matrix with entry 1 if branch b^g starts at node n^g, entry 1 if branch b^g ends at node n^g and entry 0 if none of those applies. The DER to gas node incidence matrix is denoted by matrix $A^{n^g,der}$ with entry 1 if the DER is connected at node ng. The vector $f^{fr,ref}$ denotes the Darcy–Weisbach friction factor with entry $f_{b^g}^{fr,ref}$ for each branch b^g, and $p^{g,ref}$ denotes the nominal gas load. The friction factor is calculated using the Swamee–Jain formula [40] approximating the correlation of the Colebrook–White equation. Thus, the calculation only holds for a Reynold coefficient and branch parameters within the following ranges:

$$4 \times 10^3 \le Re_{b^g}^{ref} \le 10^8 \tag{8.14}$$

$$10^{-6} \le \frac{\varepsilon_{b^g}}{d_{b^g}} \le 10^{-2} \tag{8.15}$$

The vector f^{f0} describes the form factor with entry $f_{b^g}^{f0}$ for each branch b^g. The nodal flow balance for gas grid nodes and the branch pressure are based on the Darcyß–Weisbach equation:

$$\sum_{der \in DER^{n^g}} c_{der}^{g^{ref}} = \sum_{b^g \in B^{n^g,2}} v_{b^g}^{g,ref} - \sum_{b^g \in B^{n^g,1}} v_{b^g}^{g,ref} \tag{8.16}$$

$$h_{n^g,b^g,2}^{g,ref} - h_{n^g,b^g,1}^{g,ref} = f_{b^g}^{f0} f_{b^g}^{fr,ref} \tag{8.17}$$

The presented gas grid model is used to simulate a natural gas distribution grid. However, the model can also be used for other gases like hydrogen in the future through the adaptation of several parameters.

8.3.5 Flexible DER model

The flexible DER model implemented in MESMO is expressed in state space form as follows:

$$x_{f,t+1} \leq A_f x_{f,t} + B_f^u u_{f,t} + B_f^d d_{f,t}$$
$$y_{f,t} \leq C_f x_{f,t} + D_f^u u_{f,t} + D_f^d d_{f,t} \tag{8.18}$$

Here $x_{f,t}$, $u_{f,t}$, $d_{f,t}$, and $y_{f,t}$ are the state, input, disturbance, and output vectors of the flexible load f at time step t. A_f and C_f are the state and output matrix of flexible load f, and B_f^u, D_f^u, B_f^d, and D_f^d are the input and feedthrough matrices on the input and disturbance vectors of flexible load f. A detailed description of a state space model for a flexible building can be obtained from [45]. This model is implemented for the electric grid in MESMO. For the MEMG in this chapter, the model has been implemented for the thermal grid simultaneously. Based on that, the hereafter formulated models for HP and CHP as additional DER are implemented.

8.3.6 HP model

The integration of HPs in MEMGs offers additional flexibility through conversion between the electric and thermal grid. The HP used in this chapter additionally participates in the implemented DR program through an attached storage unit. The thermal output vector of the state space form has the following additional restrictions:

$$y_{HP,t}^{th} = u_{HP,t}^{el} * \eta_{HP}$$
$$y_{HP,t}^{th-} \leq y_{HP,t}^{th} \leq y_{HP,t}^{th+}, \tag{8.19}$$

where $y_{HP,t}^{th}$ is the thermal output vector of the HP and $u_{HP,t}^{el}$ is the electric input vector of the HP. The thermal output is restricted by $y_{HP,t}^{th+}$ and $y_{HP,t}^{th-}$ which are specified as the nominal thermal power and the minimum thermal power of the flexible loads. The relationship of electric input and thermal output is defined by the HP efficiency η_{HP} expressed as follows:

$$\eta_{HP} = \frac{q}{P_C + P_P + P_F}, \tag{8.20}$$

where q is the resulting heat and P_C, P_P, and P_F are the input power for the compressor, pump, and fans (in evaporator and condenser).

8.3.7 CHP model

The integration of CHP offers highly efficient co-generation of electricity and heat. The modeled CHP links the natural gas network through its input and could prospectively also be fueled with green hydrogen. The defined CHP in this chapter participates in the implemented DR program through an attached storage unit. Additional constraints for the output vectors in state space form are added as follows:

$$y_{CHP,t}^{th} = y_{CHP,t}^{el} * \frac{\eta_{CHP}^{th}}{\eta_{CHP}^{el}}$$

$$y_{CHP,t}^{th-} \leq y_{CHP,t}^{th} \leq y_{CHP,t}^{th+}$$

$$y_{CHP,t}^{el-} \leq y_{CHP,t}^{el} \leq y_{CHP,t}^{el+}, \qquad (8.21)$$

where $y_{CHP,t}^{th}$ and $y_{CHP,t}^{el}$ are the thermal and electric output vectors of the CHP at time step t. They are limited by $y_{CHP,t}^{th+}$ and $y_{CHP,t}^{el+}$ (specified as the nominal thermal and active power of the CHP in the model configuration) and the minimum thermal and active power $y_{CHP,t}^{th-}$, $y_{CHP,t}^{el-}$. The relationship of the CHP outputs is defined by the ratio of the thermal and electric efficiency η_{CHP}^{th} and η_{CHP}^{el}, which are defined as follows:

$$\eta_{CHP}^{th} = \frac{q}{q_{el}}$$

$$\eta_{CHP}^{el} = \frac{p}{Fuel}, \qquad (8.22)$$

where q and q_{el} are the utilizable system output heat and the output heat of the electricity generating unit. The utilizable output power is denoted as p and the fuel input for the electricity generating unit as *Fuel*.

8.3.8   MEMG optimization: case study

In this section, a case study of a MEMG operation using MESMO is demonstrated in a 24-h scheduling problem. To show the behavior of the system in different circumstances, the test case comprises several scenarios. The test system layout is an extension of the system used in [15]. All three networks have the same layouts with nodes 0–19 as shown in Figure 8.1.

The electric grid supplies 19 households as flexible loads on each node 1–19. Intermittent electric generation is included in the form of PVs at nodes 2, 5, 13, and 16. The thermal grid supplies 19 households with flexible thermal loads at nodes 1–19. Solar-thermal generators at nodes 1, 8, 11, and 17 represent households with intermittent thermal generation. Equally, the gas grid supplies 19 households as flexible gas loads at nodes 1–19. Further DER models are included to couple the three networks. The HPs at nodes 7 and 15 allow conversion from electricity to thermal energy. The CHPs at nodes 6 and 18 allow the conversion of gas to electricity and thermal energy. The household load profiles are based on standard load profiles from [43]. German wholesale data is used to determine the source node's electricity price.

For the optimization of the described test system, the MESMO considers the operational constraints of the electric, thermal, and gas grid as well as the DER model in order to derive the economic dispatch of the DER and is expressed as in (8.23):

$$\min_{p_t, q_t, p_t^{th}} \sum_{t \in T} c_t^{ref} (1^T p_t + \frac{1}{\eta^{ch}} 1^T p_t^{th}) \qquad (8.23)$$

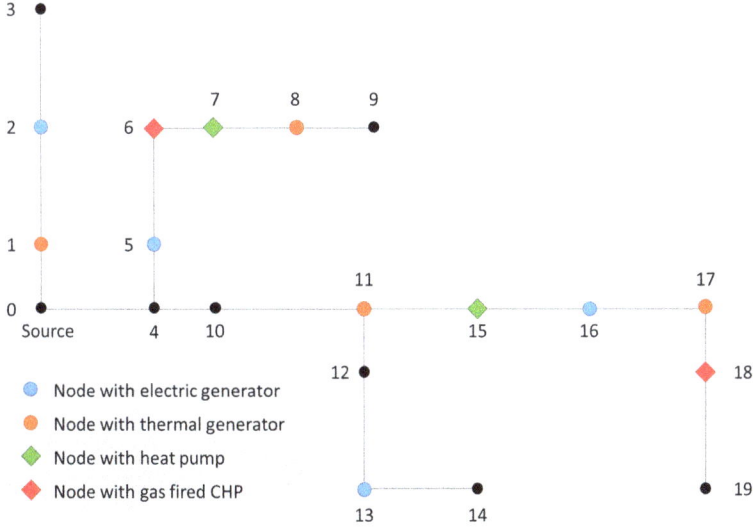

Figure 8.1   Test case layout of the electric and thermal grid [34]

s.t. $(\forall t \in T)(1)–(22)$

$$\boldsymbol{h}^{th-} \leq \boldsymbol{h}_t^{th} \; : \boldsymbol{\mu}_t^h$$

$$\boldsymbol{h}^{g-} \leq \boldsymbol{h}_t^{g} \; : \boldsymbol{\mu}_t^h$$

$$\boldsymbol{v}_t^{g} \leq \boldsymbol{v}^{g+} \; : \boldsymbol{\mu}_t^{v+}$$

$$\boldsymbol{v}_t^{th} \leq \boldsymbol{v}^{th+} \; : \boldsymbol{\mu}_t^{v+}$$

$$\boldsymbol{p}_t^{th,src} - \frac{1}{\eta^{ch}}\mathbf{1}^T\boldsymbol{p}_t^{th} = p_t^{pm} \; : \lambda_t^{p^{pm}}$$

$$\boldsymbol{u}^- \leq \boldsymbol{u}_t \leq \boldsymbol{u}^+$$

$$\left|\boldsymbol{s}_t^{f}\right|^2 \leq \left|\boldsymbol{s}_t^{f,+}\right|^2$$

$$\left|\boldsymbol{s}_t^{t}\right|^2 \leq \left|\boldsymbol{s}_t^{t,+}\right|^2$$

$$p_t^{src} - \mathbf{1}^T\boldsymbol{p}_t = p_t^{ls}$$

$$q_t^{src} - \mathbf{1}^T\boldsymbol{q}_t = q_t^{ls}$$

$$\boldsymbol{y}_{f,t}^- \leq \boldsymbol{y}_{f,t} \leq \boldsymbol{y}_{f,t}^+$$

$$\boldsymbol{y}_{CHP,t}^{th} = \boldsymbol{u}_{CHP,t}^{g} * \eta_{CHP}^{th}$$

$$\boldsymbol{y}_{CHP,t}^{el} = \boldsymbol{u}_{CHP,t}^{g} * \eta_{CHP}^{el},$$

where c_t^{ref} denotes the electric energy price at the reference node and $\boldsymbol{y}_{CHP,t}^{th}$, $\boldsymbol{y}_{CHP,t}^{el}$ are the thermal and electric output vectors of the CHP at time step t. They are limited

Table 8.2   Optimization scenario overview

Model components	Scenarios		Scenarios	
	1	2	3	4
Electric grid including intermittent generation	×	×	×	×
Thermal grid including intermittent generation	×	×	×	×
Gas grid			×	×
Coupling of electric and thermal grid	×	×	×	×
Coupling of electric, thermal, and gas grid			×	×
Constrained operation		×		×

by $y_{CHP,t}^{th+}$ and $y_{CHP,t}^{el+}$ (specified as the nominal thermal and active power of the CHP in the model configuration) and the minimum thermal and active power $y_{CHP,t}^{th-}$, $y_{CHP,t}^{el-}$.

The combined optimal operation problem is solved in four different scenarios summarized in Table 8.2. Scenarios 1 and 2 show the thermal-electric MEMG (MEMG 1) with intermittent generation and sector coupling through conversion in the form of HPs. In Scenarios 3 and 4, MEMG 2 consists of an electric, thermal, and gas system with intermittent generation and an additional sector coupling in the form of CHPs. In Scenarios 2 and 4, both MEMGs are facing an operational constraint of network congestions which is implemented by a branch flow constraint between nodes 10 and 11 of the electric and thermal grid. The optimization in these four scenarios allows to compare the reaction of MEMG 1 and MEMG 2 to the network congestions and implicitly shows which MEMG has the higher inherent flexibility. The chosen modification of the branch flow constraint should simulate different incidents in a network like lower renewable generation than expected (e.g., through deviations from projected wind occurrence or unexpected maintenance on generating units) or higher demand in parts of the network than expected. This would lead to less electricity reaching the constrained part of the network.

MESMO provides a wide range of results, including the branch flows and power magnitude vectors of the grids. The behavior of the DER can be analyzed using their active or reactive power, thermal power, or gas consumption. Another interesting indicator for the performance of different system architectures is the electric and thermal DLMP. In this chapter, a selection of the available results is presented to demonstrate how MESMO can be used to compare different MEMG setups.

Figure 8.2 shows the branch flow in the gas system. It shows peaks in the morning, around noon, and in the evening due to the standard load profile used for the flexible gas loads representing the households. In Scenarios 2 and 4, distribution network congestion in the electric and thermal grid is simulated through a branch flow constraint at branch 10. To analyze the effect of the coupling of the electric and gas grid through the CHP, the electric DLMP spread can be used. Figure 8.3 shows the average spread of the electric DLMP at nodes 11–19, comparing the constrained and unconstrained operation of each MEMG. The spread in MEMG 1 is higher during peak hours at noon and in the evening. The spread of MEMG 2 is slightly lower

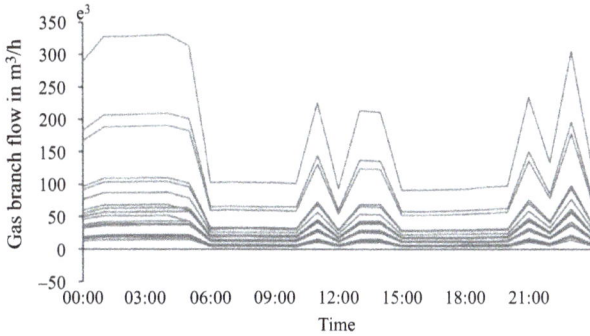

Figure 8.2   Branch flow of the gas system at branches 0–18 [34]

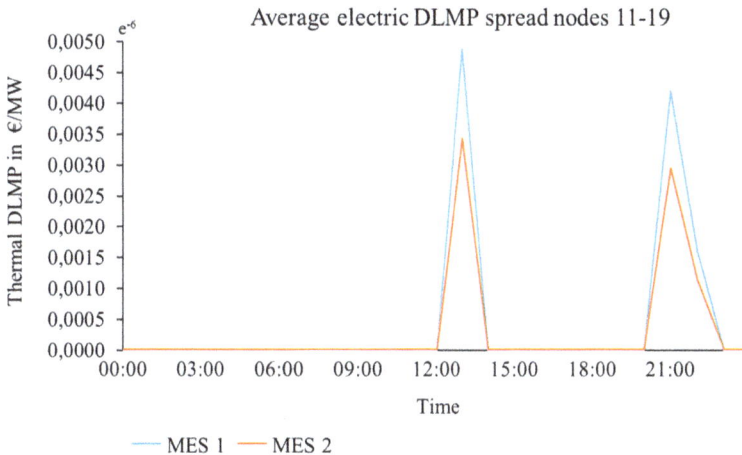

Figure 8.3   Electric DLMP spread for MEMG 1 and MEMG 2 [34]

which means, that the nodal price increase in congested operations is lower. This indicates that the coupling with the gas grid in MEMG 2 is beneficial for the electric grid when balancing congestion in the network.

Another interesting analysis is the behavior of the conversion DER in the congested network (although this cannot be used to directly compare the two MEMG against each other). The HP at node 15 in MEMG 1 is located in the congested part of the network. As shown in Figure 8.4 it is only activated in constrained operation to provide additional heat during the night and at noon. In MEMG 2, there is an additional conversion unit in the form of a gas-fired CHP. The CHP located in the congested network part is similarly only activated in constrained operation. Figure 8.5 shows that the CHP provides additional power and heat during the night and at noon.

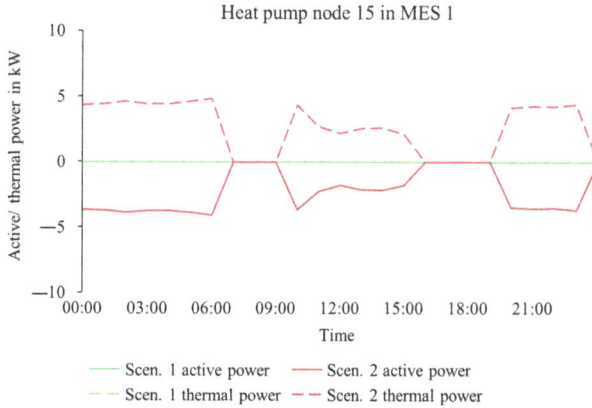

Figure 8.4   Scheduling of HP at node 15 in MEMG 1 in constrained (Scenario 2) and unconstrained (Scenario 1) operation [34]

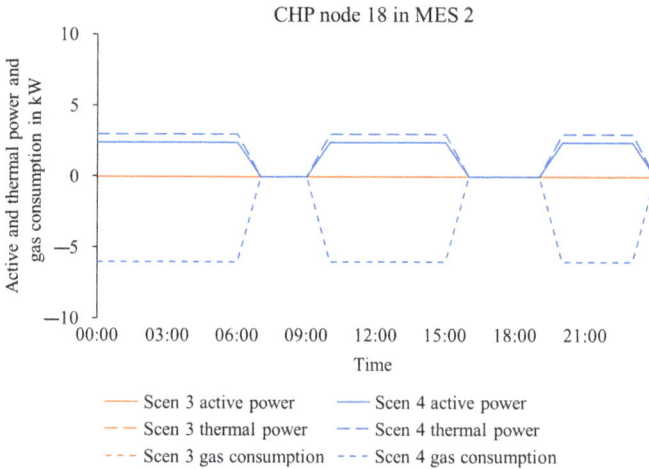

Figure 8.5   Scheduling of CHP at node 18 in MEMG 2 in constrained (scen. 4) and unconstrained (scen. 3 operation) [34]

8.4   MEMG in a laboratory environment

Although the MEMG operation and benefits have been shown in numerous works, experimental validation and technology demonstration are rare. The constructed optimization problem is often based on significant simplifications of the energy grid, generation equipment, and load profiles. It is crucial to apply the developed operational strategies in a laboratory environment to test their feasibility and performance, and hence reveal the challenges during the implementation process.

The Combined Smart Energy System (CoSES) laboratory [46] at the Technical University of Munich operates a MEMG that consists of four single-family houses

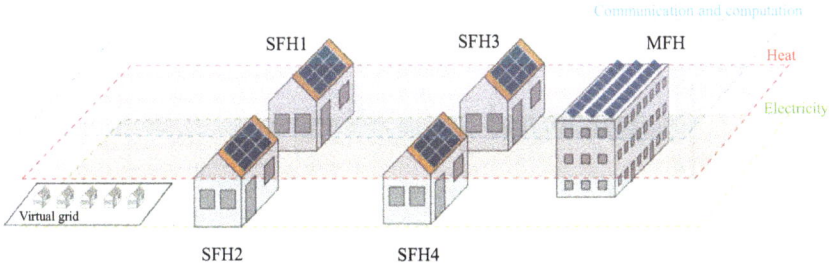

Figure 8.6   CoSES laboratory overview [46]

(SFH1-SFH1) and one multi-family house (MFH) with integrated heat, electric, and communication layers (Figure 8.6). This chapter presents the infrastructure of the laboratory as an example of how current operational strategies can be validated and demonstrated in a more realistic environment. Each house in the laboratory represents a prosumer in terms of both electricity and heat. The buildings are connected with distribution electric and heat grids. In addition, the electric and heat grids are coupled through the CHPs and HPs, which essentially integrate these two energy systems into one entity. Two electric vehicle charging stations include a transportation system in the test MEMG as well.

8.4.1   Electric grid

A single-line diagram of the default configuration of the experimental MEMG is shown in Figure 8.7. A medium voltage (20 kV) feeder supplies the experimental MEMG system. Two OLTC transformers (250 KVA each) are supplying the laboratory experimental grid that consists of a maximum of 10 low-voltage (LV) buses. The topology of the LV grid is formed by connecting these 10 LV buses with 12 available cables that are laid in the foundations of the building. The total length of all cables is around 1.8 km, with individual lengths between 100 m and 250 m and conductor (copper) diameters between 70 and 150 mm^2. The cables' endings are accessible at the switchboard, where they can be connected to any of the 10 LV buses, which makes the realization of an arbitrary MEMG topology possible. In addition to the cables that form the topology of the experiment grid, several electrical components (Figure 8.8) are available in the laboratory, which can be connected to any of the 10 LV buses:

1.   Electric batteries (2 × 13 kWh).
2.   Egston inverter systems for prosumer emulating.
3.   Photovoltaic panels (18 kW).
4.   Synchronous and induction motor/generator emulator (30 kW).
5.   2 × EV charging stations.

The Egston load emulator (COMPISO system units—CSU) consists of a transformer (galvanic isolation), rectifier, DC bus, and seven three-phase bipolar inverters connected to the common DC bus. A bipolar inverter represents a fully controllable

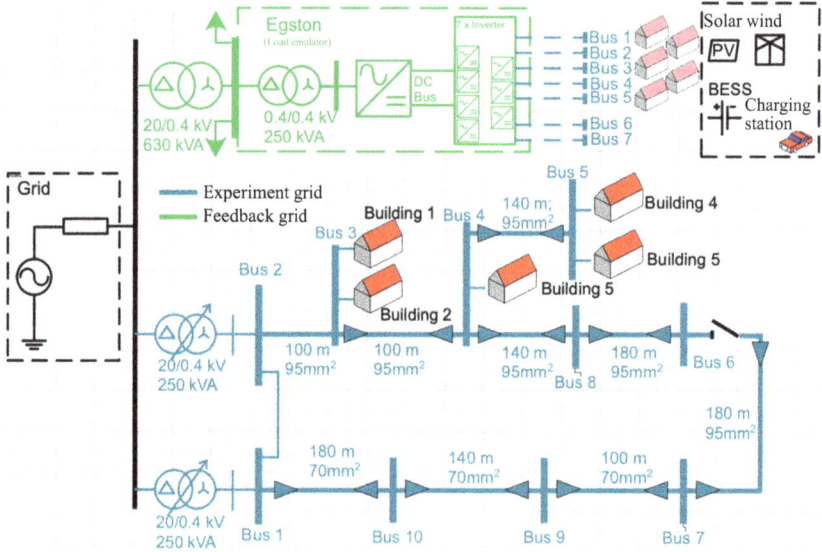

Figure 8.7   Electric grid single line diagram in CoSES MEMG [46]

Figure 8.8   Available electrical components: Egston CSU units (left), photovoltaic panels (middle), and EV charging stations (right)

load/generation with a maximum current of 126 A and voltage of 433 V. Five out of seven inverters are used as five building prosumers. The other two inverters are used as additional load/generation that can be connected to arbitrary LV bus, normally used to emulate additional renewable resources located outside of the buildings. The concept of circulating power is accomplished through a feedback grid that is supplied through the 630 kVA transformer. The feedback grid supplies the Egston load emulator and auxiliary equipment that is not part of the experimental grid.

8.4.2   Heat grid

A two- and three-temperature level bidirectional heat grid connects five buildings, where each of the buildings is capable of taking the role of a heat sink or source,

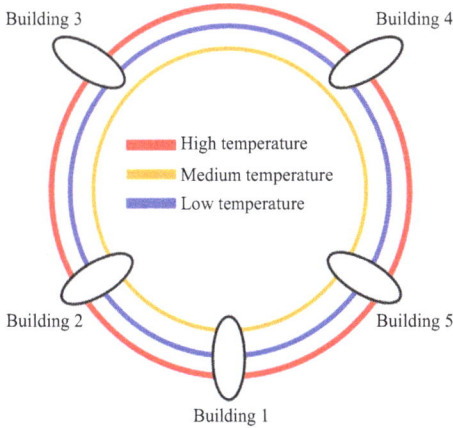

Figure 8.9   Emulated heat grid with two and three temperature level: heat grid configuration (left); grid emulators in the lab (right)

i.e., behaving as a heat energy prosumer. Three buildings are connected by a three-temperature level heat grid that provides flexibility to emulate heating and cooling systems or to supply heating with two different temperature levels.

The base heat grid topology is given in Figure 8.9, with the possibility of reconfiguration to other topologies with limited effort. In order to avoid long pipe systems, special modules are inserted to emulate the dynamic behavior of the arbitrarily long pipes. These modules introduce a controlled delay in the hot water flow, which is accomplished with appropriate cooling or heating of the flowing water (depending on the heat flow change). The central component in the house heat system is the hot water storage, which is heated from different sources (district heating grid, HPs, boilers, solar thermal sources, or CHPs), as shown in Figure 8.10. The clean domestic hot water is obtained through the Domestic Hot Water element (a heat exchanger with or without storage). The amount of consumed hot domestic water is regulated through the valves, after which the water is spilled. The space heating consumption is emulated through the regulated heat exchangers that are cooling hot water to the extent that replicates the desired space temperature. Temperature drop in the heat sink (space heating emulator) is determined through the simulation of the building heat dynamics. The building model is developed using Modelica language in SimulationX and its Green City library.

8.4.3   Monitoring and control system

The equipment described in the previous paragraphs represents the core of the laboratory design. However, the control and monitoring system allows for flexible and efficient utilization, which is essential for a successful research platform. The lab control and monitoring system is based on the National Instruments PXI systems and VeriStand software for building and deploying software-in-the-loop (SIL), model-in-the-loop (MIL), and hardware-in-the-loop (HIL) solutions. The control and

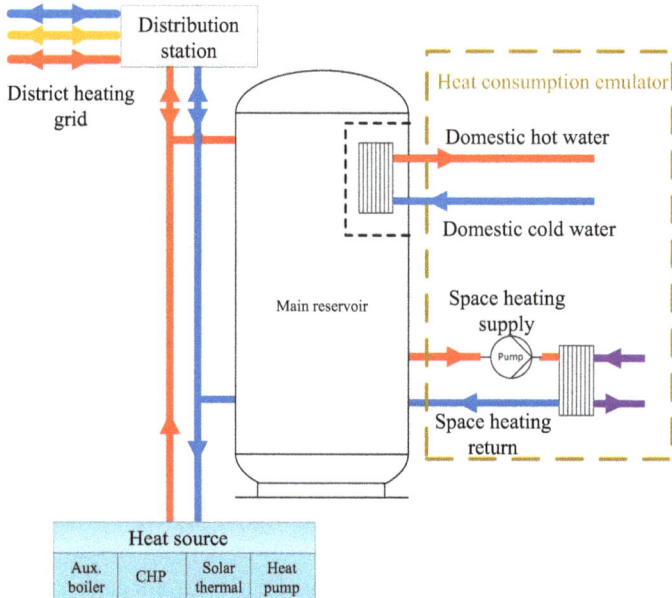

Figure 8.10   Emulated heat system of a building [46]

monitoring architecture is given in Figure 8.11. Three workstations with NI VeriStand software are available and used to program any controller in the system. It is important to note that it is not necessary to run lab experiments on the full system. Instead, a subset of the available controllers (and associated hardware) can be used independently from the remaining system.

Each building has two controllers, one NI PXIe 8880 for the electric system and one industrial controller NI IC 3171. These controllers operate at 10 kHz and 1 kHz rates, respectively. The telecommunication grid consists of a ring that operates at a 1 kHz rate. Reflective memory technology is employed for linking the real-time targets using the GE 5565 reflective memory cards, which create a reflective memory network (RMN). This ensures that all measured data is available to each controller. The industrial controllers communicate with corresponding PXIe systems through Ethernet cable at a 100 Hz rate.

In addition to playing the role of controllers in the experimental MEMG, NI PXIe systems are powerful enough to operate as a real-time simulator, enabling SIL, MIL, and HIL experiments. Once the solution is tested with simulated models, it can be validated in the hardware-in-the-loop or physical test configuration. NI VeriStand software, which implements the V-model of development, enables a seamless transition from the simulation to a full hardware environment. This is accomplished through the standardized NI VeriStand interface (signals) between different models, which replicates the real-life interfaces. Therefore, if the simulation model interfaces are defined to correspond to the hardware interfaces (control inputs, measured

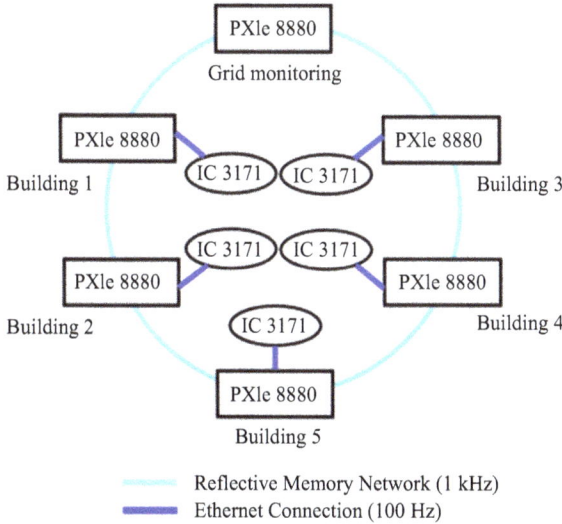

Figure 8.11   Control and monitoring architecture

signals, etc.), the transition to the Physical Test configuration consists of only reassigning the interfaces to the real-life signals in I/O cards. NI VeriStand is able to load any compiled models, which enables the use of different model development tools such as MATLAB®, Simulink®, LabVIEW, and .NET languages. In addition, NI VeriStand can be controlled through the NI VeriStand .NET API.

8.4.4   Experiment: optimal power flow (OPF) in CoSES power hardware in the loop

In order to demonstrate the operation of the laboratory grid, we present a real-time grid optimization by continuously running an Optimal Power Flow (OPF) algorithm that sends the power set points to the generation resources. We use six Egston CSUs—three as generators and three as loads—divided into two clusters, as seen in Figure 8.12. The two clusters can be imagined as two separate MGs, which do not freely share information with each other. Only the electricity price at the ends of the interconnecting node is available to both clusters. The load CSUs receive their setpoints from a time series of an unbalanced three-phase demand profile. The generator CSUs receive their setpoints from the local cluster control and OPF algorithm. The real-time (RT) target (PXI unit) at SFH3 and SFH4 control the power injections in Cluster #1 and Cluster #2, respectively. The RT target for the LV grid measures power at the LV buses and provides PLL references to the Egston CSUs over the RMN network. Two separate Windows PCs are used to deploy the NI VeriStand projects, one each for RT target at SFH3 and SFH4, to represent the two clusters. The PCs run an OPF algorithm, split into two halves, and they communicate with each other over a TCP link. The algorithm receives the measured power at the load buses and directs the generation injection power after an optimization step. We use

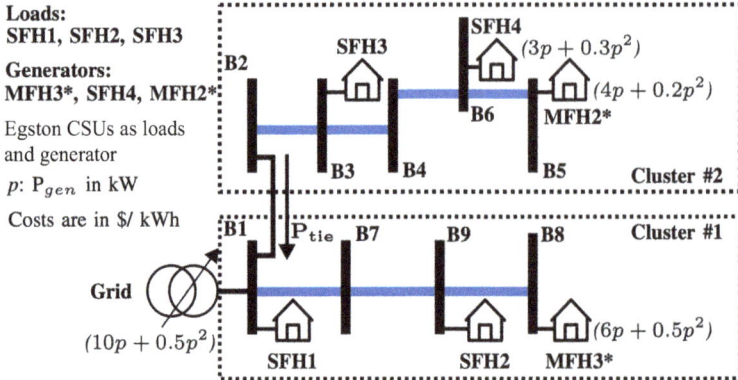

Figure 8.12   PHIL experiment for OPF validation with two clusters. The generation costs are mentioned adjacent to the generators within parentheses.

JSON to exchange the node prices over the TCP link. The VeriStand LabVIEW API is used to transfer data across the static optimization and RT emulation domains. Further explanation of the methodology can be found in [47].

In the top row of Figure 8.13, we show the comparison of the generation mix of phase A for the two cases, Case #1 with OPF and Case #2 without OPF, for the same load profile. The experiment runs for 550s. The generator nodes are initially programmed to provide self-sufficiency for the local cluster as primary control and keep the exchange of power between the two clusters, P_{tie}, at zero. In Case #1, the MFH3* generator completely supplies the required power for Cluster #1, while MFH2* and SFH4 supply power for Cluster #2. Case #2 re-dispatches the three generators to minimize the total cost through online OPF, which reduces the generation from Cluster #1 and increases at Cluster #2, as seen in the bottom left of Figure 8.13. The bottom right of Figure 8.13 shows the power sent from Cluster #2 toward Cluster #1 due to optimized generation costs in the OPF case. The total generation costs are \$0.233 for Case #1 and \$0.198 for Case #2, with an improvement of ≈15%.

8.4.5   Experiment: data-driven heat pump modeling

A challenge in operating MEMGs is their complexity in terms of the number of different components, but even more in the lack of appropriate models that describe interactions between different energy vectors. Namely, the common modeling tools have been developed having in mind only single energy vector applications, while the potential effect on other energy vectors has been neglected. Specifically, modeling heat pumps (HPs) proves to be particularly challenging, primarily due to the black box proprietary control schemes employed. These control systems determine HP dynamic behavior, which is typically not considered in the design of purely heat systems where dynamic behavior holds marginal interests. Consequently, the controllers of HPs on the market have very different characteristics that impact the dynamic

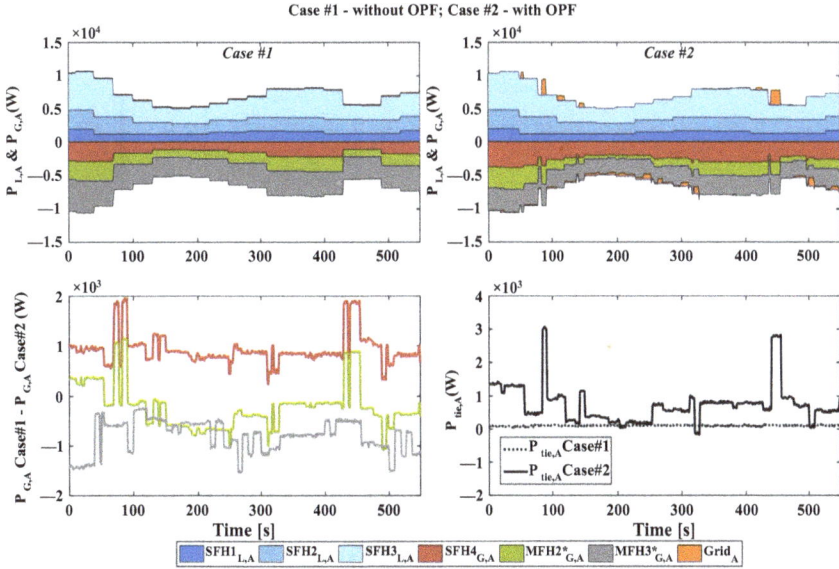

Figure 8.13 *Comparison of generation mix (top), change in $P_{gen,A}$ (bottom left) and power exchanged between clusters (bottom right) with and without OPF*

Figure 8.14 *Air-source HP experimental setup in CoSES laboratory*

behavior of the electric system in unpredictable ways. Therefore, there is a need to develop data-driven models of particular HPs suitable for system-level studies [48], which can be done in laboratory facilities like the one in the CoSES Research Center at the Technical University of Munich.

To demonstrate this use case, a mathematical model of the 4 kW air-source HP (WOLF CHA-10 MONOBLOCK) in the CoSES laboratory has been created using the experimental data. The testing platform (Figure 8.14) includes an HVAC

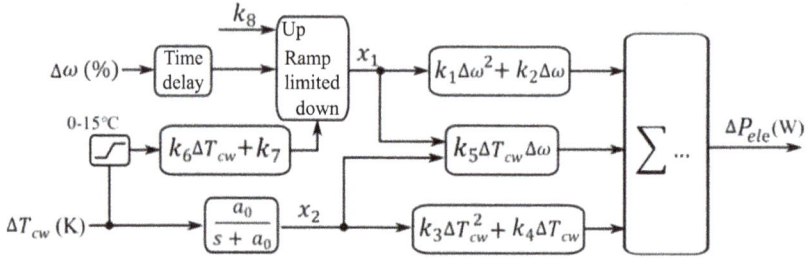

Figure 8.15   HP model structure: (a) a general cascaded Wiener model structure;
(b) model structure identified for the HP in the CoSES laboratory

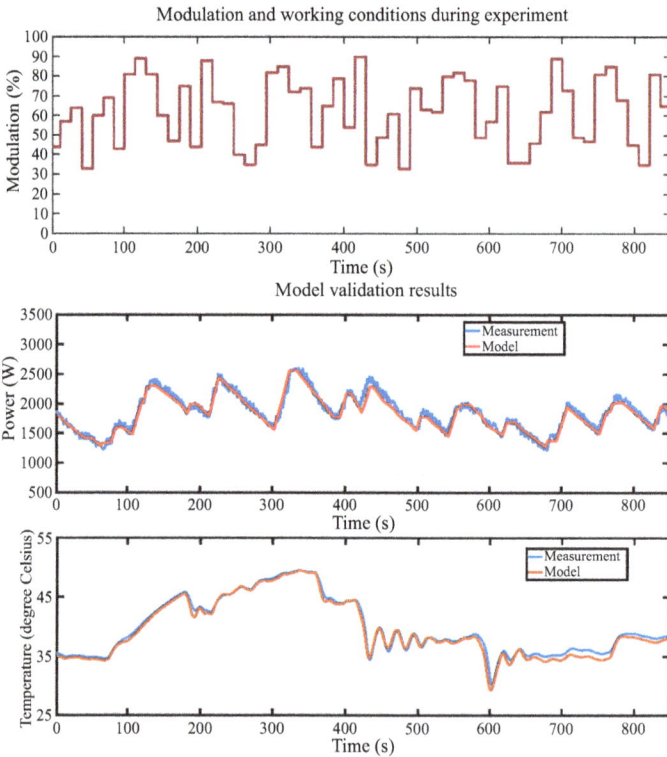

Figure 8.16   Experimental results with CoSES heat pump: (a) reference power
setpoint modulation; (b) electric power output; and (c) output
temperature

system to control the air temperature, a mixing valve to control the water temperature, a hot water tank, and an industrial controller from National Instruments that controls the heat pump. The HP model structure that is found suitable is shown in Figure 8.15. The HP is controlled by changing the power set point (modulation), as shown in Figure 8.16(a) while keeping the constant mass flow rate. The experimental results are shown in Figure 8.16(b) and (c). The results show that the proposed model accurately predicts the HP power consumption and the output water temperature.

8.5 Conclusions

This chapter explores the contribution that a multi-energy microgrid can bring to a flexible and efficient power system operation. It considers both district heating and cooling and electric distribution systems. Energy management schemes of these integrated energy systems are implemented and validated in the CoSES laboratory, assessing their real-world applicability and effectiveness.

References

[1]   M. Tostado-Véliz, P. Arévalo, and F. Jurado, "A comprehensive electrical-gas-hydrogen Microgrid model for energy management applications," *Energy Convers. Manag.*, vol. 228, p. 113726, 2021, doi: 10.1016/j.enconman.2020.113726.

[2]   E. Guelpa, A. Bischi, V. Verda, M. Chertkov, and H. Lund, "Towards future infrastructures for sustainable multi-energy systems: A review," *Energy*, vol. 184, pp. 2–21, 2019, doi: 10.1016/j.energy.2019.05.057.

[3]   P. Mancarella, "MES (multi-energy systems): an overview of concepts and evaluation models," *Energy*, vol. 65, pp. 1–17, 2014, doi: 10.1016/j.energy.2013.10.041.

[4]   V. Kleinschmidt, T. Hamacher, V. Peric, and M. Reza Hesamzadeh, "Unlocking flexibility in multi-energy systems: a literature review," in *2020 17th International Conference on the European Energy Market (EEM)*, 2020, pp. 1–6, doi: 10.1109/EEM49802.2020.9221927.

[5]   M. Ata, A. K. Erenoğlu, İ. Şengör, O. Erdinç, A. Taşcıkaraoğlu, and J. P. S. Catalão, "Optimal operation of a multi-energy system considering renewable energy sources stochasticity and impacts of electric vehicles," *Energy*, vol. 186, p. 115841, 2019, doi: 10.1016/j.energy.2019.07.171.

[6]   P. Liu, T. Ding, Y. He, and T. Chen, "Integrated demand response in multi-energy market based on flexible loads classification," in *2019 IEEE Innovative Smart Grid Technologies – Asia (ISGT Asia)*, 2019, pp. 4346–4350, doi: 10.1109/ISGT-Asia.2019.8881718.

[7]   M. J. Vahid-Pakdel, S. Nojavan, B. Mohammadi-ivatloo, and K. Zare, "Stochastic optimization of energy hub operation with consideration of thermal energy market and demand response," *Energy Convers. Manag.*, vol. 145, pp. 117–128, 2017, doi: 10.1016/j.enconman.2017.04.074.

[8]   S. Stinner, K. Huchtemann, and D. Müller, "Quantifying the operational flexibility of building energy systems with thermal energy storages," *Appl. Energy*, vol. 181, pp. 140–154, 2016, doi: 10.1016/j.apenergy.2016.08.055.

[9]   S. Bahrami and A. Sheikhi, "From demand response in smart grid toward integrated demand response in smart energy hub," *IEEE Trans. Smart Grid*, vol. 7, no. 2, pp. 650–658, 2015, doi: 10.1109/TSG.2015.2464374.

[10]   A. Sheikhi, S. Bahrami, and A. M. Ranjbar, "An autonomous demand response program for electricity and natural gas networks in smart energy hubs," *Energy*, vol. 89, pp. 490–499, 2015, doi: 10.1016/j.energy.2 015.05.109.

[11]   M. Alipour, K. Zare, and M. Abapour, "MINLP probabilistic scheduling model for demand response programs integrated energy hubs," *IEEE Trans. Ind. Informatics*, vol. 14, no. 1, pp. 79–88, 2018, doi: 10.1109/TII.2017.2730 440.

[12]   Y. Zhang, Y. He, M. Yan, C. Guo, and Y. Ding, "Linearized stochastic scheduling of interconnected energy hubs considering integrated demand response and wind uncertainty," *Energies*, vol. 11, no. 9, p. 2448, 2018, doi: 10.3390/en11092448.

[13]   H. Yang, M. Li, Z. Jiang, and P. Zhang, "Multi-time scale optimal scheduling of regional integrated energy systems considering integrated demand response," *IEEE Access*, vol. 8, pp. 5080–5090, 2020, doi: 10.1109/ACCESS.2019. 2963463.

[14]   P. Ge, Q. Hu, Q. Wu, X. Dou, Z. Wu, and Y. Ding, "Increasing operational flexibility of integrated energy systems by introducing power to hydrogen," *IET Renew. Power Gener.*, vol. 14, no. 3, pp. 372–380, 2020, doi: 10.1049/iet-rpg.2019.0663.

[15]   Y. Jia, Z. Mi, W. Zhang, and L. Liu, "Optimal operation of multi-energy sys-tems in distributed energy network considering energy storage," in *2017 IEEE Conference on Energy Internet and Energy System Integration (EI2)*, 2017, pp. 1–6, doi: 10.1109/EI2.2017.8245261.

[16]   F. Brahman, M. Honarmand, and S. Jadid, "Optimal electrical and thermal energy management of a residential energy hub, integrating demand response and energy storage system," *Energy Build.*, vol. 90, pp. 65–75, 2015, doi: 10.1016/j.enbuild.2014.12.039.

[17]   N. Good, L. Zhang, A. Navarro-Espinosa, and P. Mancarella, "Physical mod-eling of electro-thermal domestic heating systems with quantification of economic and environmental costs," in *Eurocon 2013*, 2013, pp. 1164–1171, doi: 10.1109/EUROCON.2013.6625128.

[18]   F. Kienzle, P. Ahcin, and G. Andersson, "Valuing investments in multi-energy conversion, storage, and demand-side management systems under uncer-tainty," *IEEE Trans. Sustain. Energy*, vol. 2, no. 2, pp. 194–202, 2011, doi: 10.1109/TSTE.2011.2106228.

[19]   D. Keihan Asl, A. Hamedi, and A. Reza Seifi, "Planning, operation and flexibility contribution of multi-carrier energy storage systems in integrated energy systems," *IET Renew. Power Gener.*, vol. 14, no. 3, pp. 408–416, 2020, doi: 10.1049/iet-rpg.2019.0128.

[20]   J. Aghaei and M.-I. Alizadeh, "Multi-objective self-scheduling of CHP (combined heat and power)-based microgrids considering demand response

programs and ESSs (energy storage systems)," *Energy*, vol. 55, pp. 1044–1054, 2013, doi: 10.1016/j.energy.2013.04.048.

[21]   N. O'Connell, P. Pinson, H. Madsen, and M. O'Malley, "Benefits and challenges of electrical demand response: A critical review," *Renew. Sustain. Energy Rev.*, vol. 39, pp. 686–699, 2014, doi: 10.1016/j.rser.2014.07.098.

[22]   C. Shao, Y. Ding, J. Wang, and Y. Song, "Modeling and integration of flexible demand in heat and electricity integrated energy system," *IEEE Trans. Sustain. Energy*, vol. 9, no. 1, pp. 361–370, 2018, doi: 10.1109/TSTE.2017.2731786.

[23]   P. Faria and Z. Vale, "Demand response in electrical energy supply: an optimal real time pricing approach," *Energy*, vol. 36, no. 8, pp. 5374–5384, 2011, doi: 10.1016/j.energy.2011.06.049.

[24]   J. M. Lujano-Rojas, C. Monteiro, R. Dufo-López, and J. L. Bernal-Agustín, "Optimum residential load management strategy for real time pricing (RTP) demand response programs," *Energy Policy*, vol. 45, pp. 671–679, 2012, doi: 10.1016/j.enpol.2012.03.019.

[25]   P. Mancarella and G. Chicco, "Real-time demand response from energy shifting in distributed multi-generation," *IEEE Trans. Smart Grid*, vol. 4, no. 4, pp. 1928–1938, 2013, doi: 10.1109/TSG.2013.2258413.

[26]   X. Zhang, L. Che, M. Shahidehpour, A. Alabdulwahab, and A. Abusorrah, "Electricity-natural gas operation planning with hourly demand response for deployment of flexible ramp," *IEEE Trans. Sustain. Energy*, vol. 7, no. 3, pp. 996–1004, 2016, doi: 10.1109/TSTE.2015.2511140.

[27]   L. Wu, "Impact of price-based demand response on market clearing and locational marginal prices," *IET Gener. Transm. Distrib.*, vol. 7, no. 10, pp. 1087–1095, 2013, doi: 10.1049/iet-gtd.2012.0504.

[28]   X. Zhang, M. Shahidehpour, A. Alabdulwahab, and A. Abusorrah, "Hourly electricity demand response in the stochastic day-ahead scheduling of coordinated electricity and natural gas networks," *IEEE Trans. Power Syst.*, vol. 31, no. 1, pp. 592–601, 2016, doi: 10.1109/TPWRS.2015.2390632.

[29]   J. Aghaei and M.-I. Alizadeh, "Demand response in smart electricity grids equipped with renewable energy sources: a review," *Renew. Sustain. Energy Rev.*, vol. 18, pp. 64–72, 2013, doi: 10.1016/j.rser.2012.09.019.

[30]   C. He, X. Zhang, T. Liu, and L. Wu, "Distributionally robust scheduling of integrated gas-electricity systems with demand response," *IEEE Trans. Power Syst.*, vol. 34, no. 5, pp. 3791–3803, 2019, doi: 10.1109/TPWRS.2019.2907170.

[31]   B. Parrish, R. Gross, and P. Heptonstall, "On demand: can demand response live up to expectations in managing electricity systems?," *Energy Res. Soc. Sci.*, vol. 51, pp. 107–118, 2019, doi: 10.1016/j.erss.2018.11.018.

[32]   A. Faruqui, R. Hledik, S. Newell, and J. Pfeifenberger, "The power of five percent: how dynamic pricing can save $35 billion in electricity costs," *Brattle Gr.*, 2007.

[33]   V. Kleinschmidt, S. Troitzsch, T. Hamacher, and V. Peric, "Flexibility in distribution systems — modelling a thermal-electric multi-energy

system in FLEDGE," in *2021 IEEE PES Innovative Smart Grid Technologies Europe (ISGT Europe)*, 2021, pp. 1–5, doi: 10.1109/ISGTEurope52324.2021.9640044.

[34]   V. Kleinschmidt, T. Hamacher, and V. Peric, "Flexibility in active distribution networks – modelling a fully coupled multi-energy system in MESMO," in *2022 IEEE PES Innovative Smart Grid Technologies – Asia (ISGT Asia)*, 2022, pp. 475–479, doi: 10.1109/ISGTAsia54193.2022.10003526.

[35]   H. Cheng, J. Wu, Z. Luo, F. Zhou, X. Liu, and T. Lu, "Optimal planning of multi-energy system considering thermal storage capacity of heating network and heat load," *IEEE Access*, vol. 7, pp. 13364–13372, 2019, doi: 10.1109/ACCESS.2019.2893910.

[36]   L. Kriechbaum, G. Scheiber, and T. Kienberger, "Grid-based multi-energy systems—modelling, assessment, open source modelling frameworks and challenges," *Energy. Sustain. Soc.*, vol. 8, no. 1, p. 35, 2018, doi: 10.1186/s13705-018-0176-x.

[37]   S. Pfenninger and B. Pickering, "Calliope: a multi-scale energy systems modelling framework," *J. Open Source Softw.*, vol. 3, no. 29, p. 825, Sep. 2018, doi: 10.21105/joss.00825.

[38]   S. Hilpert, C. Kaldemeyer, U. Krien, S. Günther, C. Wingenbach, and G. Plessmann, "The open energy modelling framework (oemof) – a new approach to facilitate open science in energy system modelling," *Energy Strateg. Rev.*, vol. 22, pp. 16–25, 2018, doi: 10.1016/j.esr.2018.07.001.

[39]   oemof developer group, "Open energy modeling framework (oemof)." https://github.com/oemof/oemof.

[40]   J. Dorfner and H. Thomas, "urbs: A linear optimization model for distributed energy systems." https://github.com/tum-ens/urbs.

[41]   R. C. Dugan, "Reference guide: the open distribution system simulator (OpenDSS)." https://spinengenharia.com.br/wp-content/uploads/2019/01/OpenDSSManual.pdf.

[42]   S. Troitzsch, "MESMO (FLEDGE) – flexible distribution grid demonstrator." https://mesmo-dev.github.io/mesmo/v0.4.0/index.html#.

[43]   S. Hanif, K. Zhang, C. M. Hackl, M. Barati, H. B. Gooi, and T. Hamacher, "Decomposition and Equilibrium Achieving Distribution Locational Marginal Prices Using Trust-Region Method," *IEEE Trans. Smart Grid*, vol. 10, no. 3, pp. 3269–3281, 2019, doi: 10.1109/TSG.2018.2822766.

[44]   S. Troitzsch, M. Grussmann, K. Zhang, and T. Hamacher, "Distribution locational marginal pricing for combined thermal and electric grid operation," in *2020 IEEE PES Innovative Smart Grid Technologies Europe (ISGT-Europe)*, 2020, pp. 874–878, doi: 10.1109/ISGT-Europe47291.2020.9248832.

[45]   S. Troitzsch and T. Hamacher, "Control-oriented thermal building modelling," in *2020 IEEE Power & Energy Society General Meeting (PESGM)*, 2020, pp. 1–5, doi: 10.1109/PESGM41954.2020.9281503.

[46]   V. S. Perić, T. Hamacher, A. Mohapatra, *et al.*, "CoSES laboratory for combined energy systems at TU Munich," in *2020 IEEE Power & Energy Society*

General Meeting (PESGM), 2020, pp. 1–5, doi: 10.1109/PESGM41954. 2020.9281442.

[47] M. Cornejo, A. Mohapatra, S. Candas, and V. S. Perić, "PHIL implementation of a decentralized online OPF for active distribution grids," in *2022 IEEE Power & Energy Society General Meeting (PESGM)*, 2022, pp. 1–5, doi: 10.1109/PESGM48719.2022.9916705.

[48] R. Song, G. Yon, T. Hamacher, and V. S. Perić, "Data-driven model reduction of the moving boundary heat pump dynamic model," in *2022 IEEE Power & Energy Society General Meeting (PESGM)*, 2022, pp. 1–5, doi: 10.1109/PESGM48719.2022.9916823.

Part V

Case studies

Chapter 9

Electricity market-oriented control of the EUREF Energy Workshop

Michael Rath[1,2] and Alexander Meeder[3]

The EUREF Energy Workshop (EUREF Energiewerkstatt), located in Berlin, is a prominent site of the energy transition. Incorporating Germany's first combined power-to-heat/power-to-cold system, as well as advanced microgrid technology, the energy plant was subject to electricity market-oriented control in several research projects. This chapter reviews the main findings and results of these studies. We detail the development and implementation of a model predictive control framework that effectively interfaces with the day-ahead, intraday, and experimental congestion management markets. Additionally, we discuss the various forecasting algorithms applied, their integration, and potential weaknesses that can be addressed.

9.1 Introduction

In recent years, the expansion of renewable energies in the electricity sector has progressed rapidly. However, the heating sector, with its approximately 56% share of Germany's total final energy consumption, remains one of the greatest challenges of the energy transition—because there will be no energy transition without a heat transition. So far, only a small part of the heat generation is done by renewable sources, whereas oil, gas, and coal still play the most important role for heating homes, offices, and commercial buildings as well as for process heat demand in industry (cf. Figures 9.1 and 9.2 [1]). In order to achieve the Paris climate protection targets, the current government coalition is therefore aiming to generate at least half of municipal heating from climate-neutral sources by 2030 (cf. [2]).

The second phase of the energy transition, which is now beginning, will therefore be about two things: On the one hand, maintaining end-to-end security of supply in the power sector as volatile power sources become the main source of electricity supply, and on the other hand, significantly increasing the share of renewable

[1]Competence Center Integrated Building Energy Systems, Fraunhofer Research Institution for Energy Infrastructures and Geothermal Systems IEG, Germany
[2]Department of Civil and Environmental Engineering, Bochum University of Applied Sciences, Germany
[3]GASAG Solution Plus, Berlin, Germany

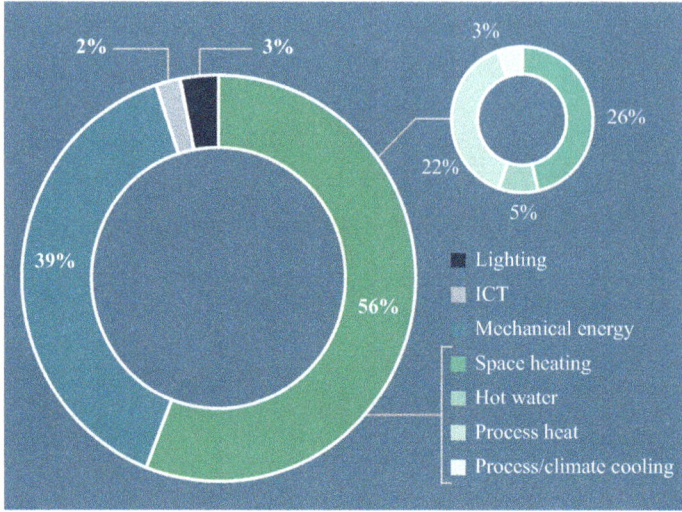

Figure 9.1   *Final energy consumption (2,514 TWh) in 2019 (taken from [1],
© Fraunhofer IEG)*

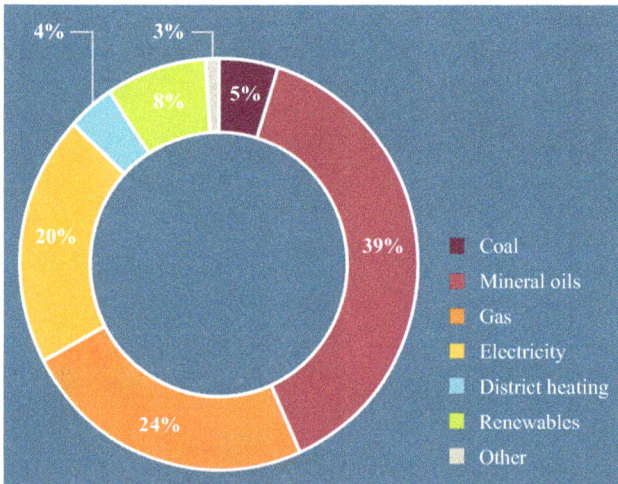

Figure 9.2   *Sources for final energy (2,514 TWh) in 2019 (taken from [1],
© Fraunhofer IEG)*

heat sources in the heating sector from 16% at the moment. Both can be achieved through sector coupling, increased use of geothermal energy and heat pumps in low- to moderate-temperature areas and hydrogen, especially in the high-temperature industrial sector [1].

9.1.1 Sector coupling in the building sector

The concept, according to which power plants follow the momentary demand for electricity and keep the grid frequency stable through flexible generation, is changing. Responsibility is increasingly shifting to consumers, because unlike conventional power plants, wind power or photovoltaic plants can only follow consumers to a limited extent, and their share is growing steadily [3,4]. When people talk about flexibility in the context of the energy transition, they usually mean the ability of elements in the energy system to actively respond to an external signal reflecting the variability of power generation and consumption by changing their output [5]. In order to reduce the load on the power grid and to use electricity above all when it is less CO_2-intensive, it is therefore helpful if consumers and microgrids act on the basis of forecasts and align themselves with the supply. Thermal energy demand in the building sector can also play a crucial role in more flexible and efficient use of renewable wind and solar power—the keyword is sector coupling [6]. Due to the available thermal storage capacity of buildings, generation and demand can be decoupled in time or at least equalized and heat or cold can also be made available at a later point in time [5,7,8].

In order to make optimal use of the efficiency and flexibility potentials of building energy technology, one needs

(a)   a load and/or building model,
(b)   forecasts of the volatile energy sources and sinks and an optimization algorithm that always loads the storage at the most favorable times, and
(c)   the hardware integration of a box at the plants that, while ensuring supply security, keeps the optimized schedules ready for the plant control and always triggers them in the right moment [9,10].

This approach was applied at the EUREF Energy Workshop (cf. Figure 9.3), which will be described in the following.

9.1.2 The EUREF Energy Workshop

On the EUREF campus, a large office quarter in Berlin-Schöneberg, a model predictive control approach was implemented and enabled an existing energy plant—the EUREF Energy Workshop (cf. Figure 9.3)—to react to electricity prices while supplying the campus. First developed within the BENE-Project [11], the technology was implemented in the Federal Ministry of Economic Affairs (Bundesministerium für Wirtschaft, BMWi) funded WindNODE research project [12] in which Germanies first combined power to heat/power-to-cold plant was realized in the EUREF Energy workshop. The technology was further optimized and transferred to other locations and energy plants in the ARCHE project [13], also funded by the BMWi.

Heat is generated by two biomethane combined heat and power (CHP) units, two other CHP units, two low-temperature gas boilers for peak loads, and one electric boiler (cf. Figure 9.3). For the biomethane CHPs in the Energy Workshop, biomethane is taken as balance from the gas grid. The fixed EEG remuneration

Figure 9.3   Visualization of the EUREF Energy Workshop/EUREF
Energiewerkstatt from https://www.energiewende-erleben.de/,
© GASAG Solution GmbH

granted by the German Renewable Energy Act [14] is an important building block for the economic and climate-friendly energy supply (cf. Figure 9.3).

The EUREF Energy Workshop

1. Baseload heat supply with biomethane CHP, providing the base load heat supply for the entire campus with 400 kW electrical power and 431 kW thermal power. Electricity is fed into the public grid with EEG subsidy.
2. Two low-NOx low-temperature gas boilers, activated for peak load heat supply with $2 \times 2,100$ kW thermal power.
3. Four high-efficiency differential pressure-controlled pumps combined as a cascading system.
4. Combined heat and power plant (CHP) providing electricity (50 kW) for the pumps and heat (100 kW) for the campus.
5. Control and Command Center with high-level control and remote maintenance.
6. Two compression chillers providing cooling for the campus throughout the year with a thermal power of $2 \times 1,000$ kW and the option of free cooling.
7. Power-to-heat/power-to-cold system: 2×22 m^3 hybrid buffer storages for intermediate storage of heat (power-to-heat) and/or cold (Power-to-Cold) are supplied by either an electric boiler producing 500 kW heat with 500 kW power or the two compression chillers.
8. Li-ion battery storage: Stabilizes the grid and supports the expansion of renewable energies (Storage capacity: 1.9 MWh, electrical power: 1.25 MW). It can store excess electricity from the grid in second-life traction batteries after their use in test vehicles.

Cooling is supplied by two cooling compression machines powered by green electricity with the option of free cooling, i.e., the integration of cold outside air for more efficient cold supply. A special feature of the plant is the so-called power-to-heat/power-to-cold storage system (P2H/P2C), the first of its kind in Germany (cf. [12]), consisting of two storage tanks with a capacity of 22 m^3 each, which are hydraulically designed in such a way that it is possible to specify for each storage tank individually whether it is to be charged with heat or with cold. For this purpose, renewable electricity is converted into heat or cold as needed by the P2H/P2C, i.e., power to heat or power to cold, respectively) and buffered in the storage system.

The amount of different energy converters and the resulting flexibility gives one the possibility to determine the optimal deployment sequence of the energy converters every 15 min again and again based on market and weather forecasts. The quantities of electricity needed to operate the refrigeration compression machines can be procured in advance on the day-ahead market, and there is then further potential for optimization on the intraday market.

9.2 Process overview

The high level control of the energy system and its infrastructure implementation with all interface coordinations was developed within the research project BENE ([11], see also following sections) by Geo-En Energy Technologies GmbH, which is also a company of the GASAG Group. Within the IT solution Geo-En | EnergyNode, which is now called Nexerion, a digital fingerprint of all consumers was and is created using a self-learning process based on historical measurement and weather data, which then enables a demand forecast using current weather forecasts. Finally, to meet the demand, a stochastic optimization algorithm is used to calculate the most ecologically and economically ideal schedule possible, considering current market data, and this schedule is transferred to the controller (cf. Figure 9.4, [9]).

Process of the high-level control of the EUREF Energy Workshop

1. EUREF Energy Workshop with existing standard Programmable Logic Controller (PLC), classically ensuring security of thermal supply.
2. SCADA (Supervisory Control and Data Acquisition) system, which serves as an interface to the operators, and with which the high-level control/model predictive control (MPC) can be switched of (cf. no. 8).
3. A monitoring database (DB) embedded in a cloud infrastructure including a system for cleaning, collecting, and storing data, serving also to supervise the effectiveness of the MPC.
4. The intelligent core of the MPC—the model: Utilizing the collected load and weather data from the database, this can be used to predict the demands of the EUREF-Campus for the following day with the combination of different machine learning models.

5. Operation plan optimization—with load forecasts, the following day is optimized with respect to operational expenditure or CO_2-emissions.
6. Connected external services such as spot market price signals and weather forecasts.
7. The operation plan is securely transferred to the high-level control.
8. The high-level control receiving the optimized operation plan from the cloud infrastructure delivering it to the classical PLC.

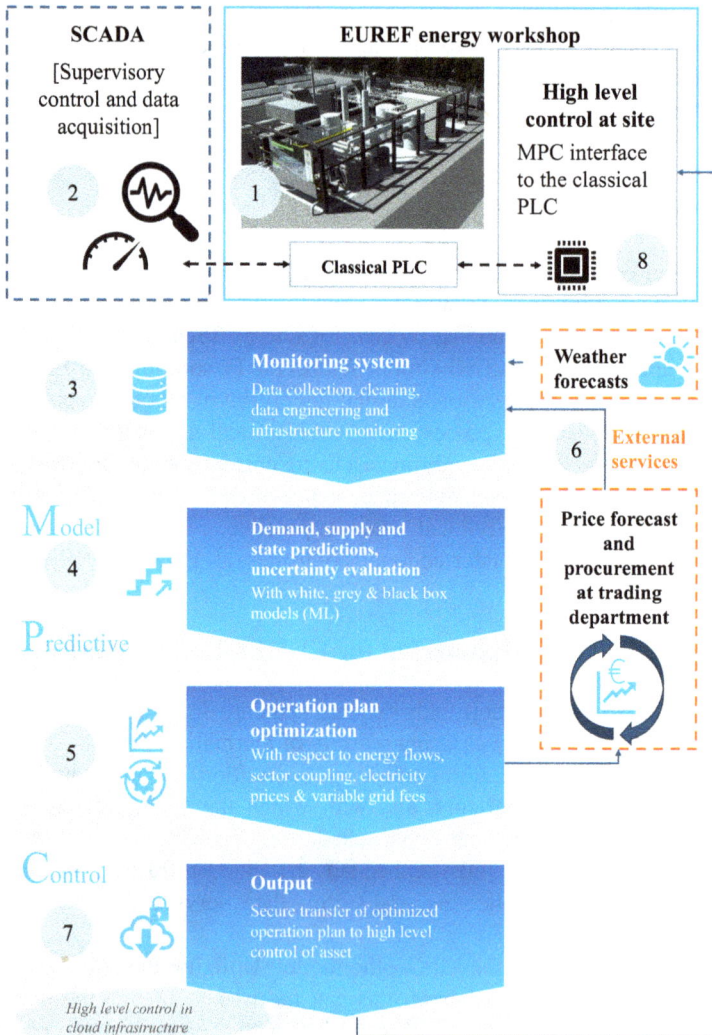

Figure 9.4   Process of the high level control of the EUREF Energy Workshop

9.2.1 Thermal load forecasting with the temperature cluster method

In order to drive a system optimized and foreseeingly, one must have a forecast for the future, particularly heating and cooling load forecasts of the buildings supplied by the system. There are many influencing factors—climate, type of usage, temporal dependencies due to nighttime setback, holidays, etc. and last but not least—the quality and reliable availability of data to feed the forecasting algorithms. The quality of the forecast ultimately determines how optimal the calculated operation plan really was. Physical models for forecasting the energy demand of buildings down to the individual room and user level exist in the literature with varying levels of complexity and effort, appearing, e.g., as parametric models (nRmC resistance–capacitance networks, [15] or non-parametric models (ARX, ARMAX, etc., cf. e.g., [16]). When evaluating the different approaches, the cost-to-benefit ratio needs to be determined in relation to the required depth of information. But even a very good model always remains just a model, and with increasing complexity and depth of information, the effort increases while the benefit stagnates. As shown by [17], the (cost/benefit) over depth of information function already reaches its maximum with comparatively simple models at low cost. So we first also chose to implement a simple forecasting method in the BENE project [11], the so-called temperature cluster method (cf. [17]) for generating load forecasts. The temperature cluster method creates a self-learning matrix for the prediction of an energy demand based on recorded historical temperature and load data. All recorded temperature-load data are logged and written into a matrix of heating energy consumption at given times and outside temperatures (see Figure 9.5).

The values for each outside temperature and time of day were updated with

$$\mu_n = \frac{n-1}{n}\mu_{n-1} + \frac{1}{n}X_n,$$

where

- μ_n: new mean,
- μ_{n-1}: old mean,
- X_n: new measurement value.

The result for another demonstrator from the BENE project with real data for one year is shown in Figure 9.6.

To assess the quality of the prediction, a running standard deviation was additionally computed from the running variance, to obtain a measure of the uncertainty of the data:

$$s_n^2 = \frac{1}{n-1}\left((n-2)s_{n-1}^2 + X_n^2 + (n-1)\mu_{n-1}^2 - n\mu_n^2\right),$$

where:

- s_n^2: new variance,
- s_{n-1}^2: old variance,
- μ_n: new mean,
- μ_{n-1}: old mean,
- X_n: new measurement value.

$$\begin{matrix} T_{amb,1} \\ \vdots \\ T_{amb,n} \end{matrix} \rightarrow \begin{pmatrix} \bar{Q}_{heat\,1,\,00:00} & \cdots & \bar{Q}_{heat\,1,\,23:00} \\ \vdots & \ddots & \vdots \\ \bar{Q}_{heat\,n,\,00:00} & \cdots & \bar{Q}_{heat\,n,\,23:00} \end{pmatrix}$$

Hour of day

Outdoor temperature in °C	00:00	01:00	02:00	03:00	04:00	05:00	06:00	07:00	08:00	09:00	10:00	11:00	12:00	13:00	14:00	15:00	16:00	17:00	18:00	19:00	20:00	21:00	22:00	23:00
20	5	5	5	5	5	5	5	5	5	5	5	5	5	5	5	5	5	5	5	5	5	5	5	5
17,5	10	10	10	10	10	10	10	11	11	11	11	11	11	11	11	11	11	11	11	10	10	10	10	10
15	14	14	14	14	14	14	14	17	17	17	17	17	17	17	17	17	17	17	17	14	14	14	14	14
12,5	19	19	19	19	19	19	19	23	23	23	23	23	23	23	23	23	23	23	23	19	19	19	19	19
10	24	24	24	24	24	24	24	29	29	29	29	29	29	29	29	29	29	29	29	24	24	24	24	24
7,5	28	28	28	28	28	28	28	35	35	35	35	35	35	35	35	35	35	35	35	28	28	28	28	28
5	33	33	33	33	33	33	33	41	41	41	41	41	41	41	41	41	41	41	41	33	33	33	33	33
2,5	38	38	38	38	38	38	38	47	47	47	47	47	47	47	47	47	47	47	47	38	38	38	38	38
0	43	43	43	43	43	43	43	53	53	53	53	53	53	53	53	53	53	53	53	43	43	43	43	43
-2,5	47	47	47	47	47	47	47	58	58	58	58	58	58	58	58	58	58	58	58	47	47	47	47	47
-5	52	52	52	52	52	52	52	64	64	64	64	64	64	64	64	64	64	64	64	52	52	52	52	52
-7,5	57	57	57	57	57	57	57	70	70	70	70	70	70	70	70	70	70	70	70	57	57	57	57	57
-10	61	61	61	61	61	61	61	76	76	76	76	76	76	76	76	76	76	76	76	61	61	61	61	61
-13	66	66	66	66	66	66	66	82	82	82	82	82	82	82	82	82	82	82	82	66	66	66	66	66
-15	71	71	71	71	71	71	71	88	88	88	88	88	88	88	88	88	88	88	88	71	71	71	71	71
-18	75	75	75	75	75	75	75	94	94	94	94	94	94	94	94	94	94	94	94	75	75	75	75	75
-20	80	80	80	80	80	80	80	100	100	100	100	100	100	100	100	100	100	100	100	80	80	80	80	80

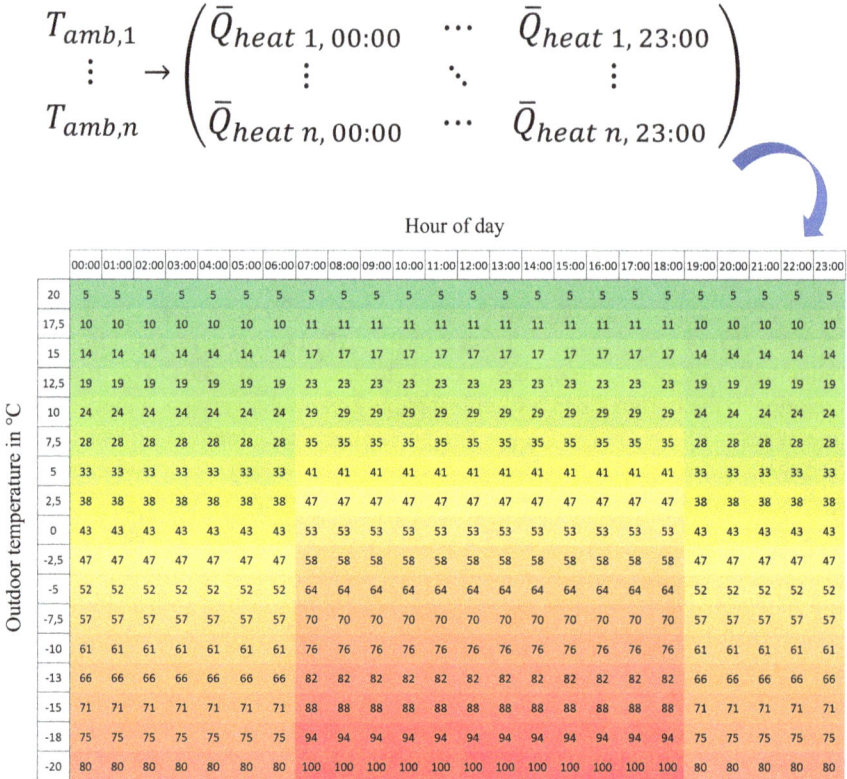

Figure 9.5   Example of temperature cluster method with night setback temperature (taken from [9], cf. [11])

The method delivers a very well-explainable forecast; however, the method works only properly, if there is already existing sufficient data, and the load characteristic is not changing—which was exactly the case for the EUREF-Campus: as a developing office district, we had to cope with increasing and thus changing loads every year, lacking enough time series data to make use of the temperature cluster method. Therefore, we started to look into machine learning methods, which we deepened in the ARCHE project (cf. [13]).

9.2.2   Thermal load forecasting with machine learning

Even if a lot of effort is made—forecasts are never perfect. In the BMWi-funded ARCHE project [13], the electricity market-oriented control was further developed with a focus on forecasts and transferred to further energy plants (cf. [18]). For the EUREF Energy Workshop demonstrator, we focused on the cooling load forecast.

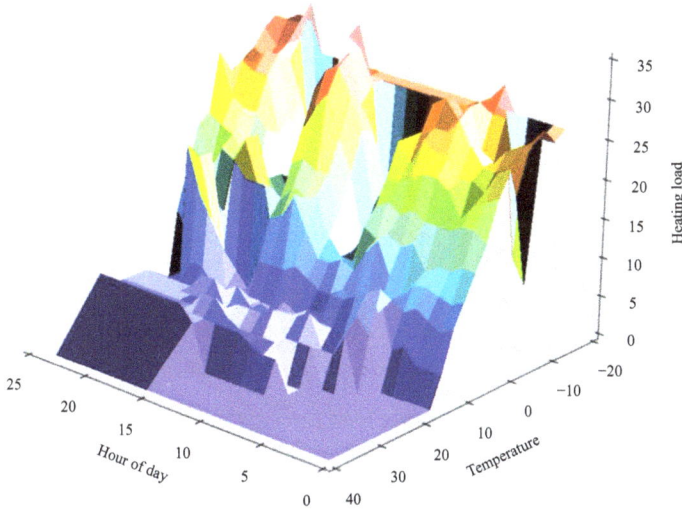

Figure 9.6   Temperature cluster with real data for one year (taken from [9], cf. [11])

9.2.2.1   Data

Our training data had an hourly resolution and contains the following features:

- Weekday
- Differentiation between weekday/weekend
- Load profile (1.10.2020 to 1.9.2021 training data, 1.9.2021 to 11.10.2021 test data
- Outdoor temperatures—we regularly pull weather data and forecasts for the corresponding longitude and latitude of the location in hourly resolution via openweathermap.org

9.2.2.2   Model training

We trained with about one year of data and looked at the predictions we made using Gradient Boost Regression ([19,20], see Figure 9.7), XGBoost ([21], see Figure 9.8), and Random Forest Regression ([22], see Figure 9.9).

As can be seen in the diagrams in Figures 9.7, 9.8, and 9.9, the predictions based on the input variables mentioned are not always optimal. There may be various reasons for this. Among other things, we trained with the cooling load data of an office campus, which was collected during the Corona epidemic.

Irregularities in the use of the buildings and optimizations on the part of the building technology of the individual office buildings, such as a temporal change in the schedules for loading components with cooling, are not considered here due to the lack of availability of information and also due to an inconsistent cost/benefit ratio. A certain degree of uncertainty will always remain in real projects.

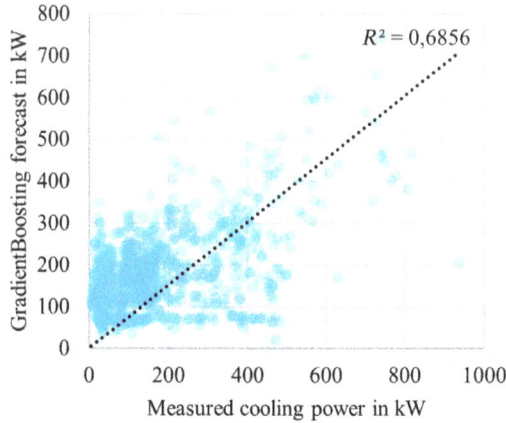

Figure 9.7   Gradient boosting cooling load forecast: predicted versus measured cooling capacities for the period 1.9.21 10 a.m. to 12.10.21 11 a.m. (cf. [18])

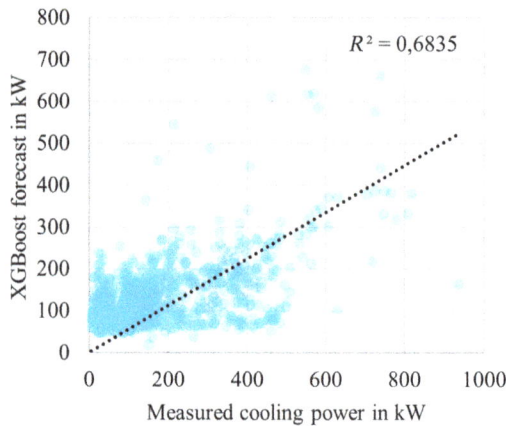

Figure 9.8   XGBoost cooling load forecast: predicted versus measured cooling capacities for the period 1.9.21 10 a.m. to 12.10.21 11 a.m. (cf. [18])

9.2.2.3   Deterministic imprinting of 24-h-ahead load values in the machine learning model

It is a common procedure to include the load value from 24 h ago in order to obtain better forecasts (see, e.g., [23,24]). We originally did this and often obtained good forecasts, as the correlation is generally also good. However, we made an interesting observation: if the machine learning methods are used in a continuous loop during plant operation, it can happen that incorrect measured values from the plant are included in the forecast. When forecasting the heat load of the EUREF-Campus, we observed that the energy meter stood still for a few hours and then delivered the entire amount of energy generated at once (see Figure 9.10). The deflection was not

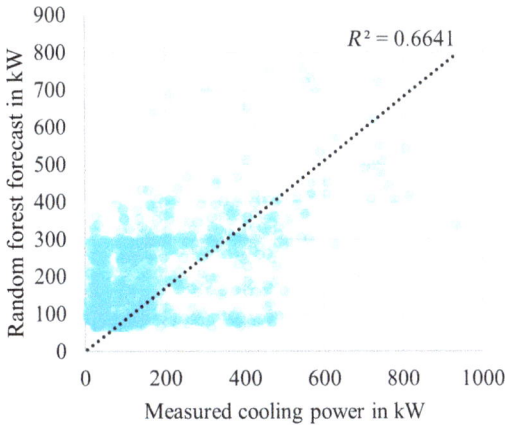

Figure 9.9   RandomForest Cooling Load Forecast: Predicted versus measured cooling capacities for the period 1.9.21 10 a.m. to 12.10.21 11 a.m. (cf. [18])

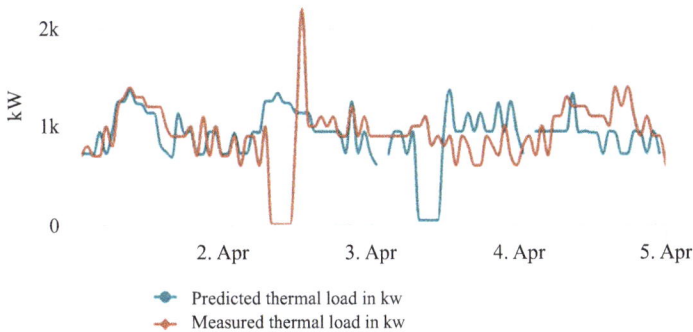

Figure 9.10   Exemplary comparison of the forecast using the Random Forest algorithm with the real measured load values at the beginning of April 2022. The time series motif of the meter stopping for a few hours is unintentionally very strongly imprinted on the following day. The reason for this is the inclusion of the 24-h previous load value in the training (cf. [18]).

yet high enough for plausibility checks to be carried out on the maximum heating load of the system of 4.6 MW.

Machine learning models often show a black box character by lacking interpretability on how the training results and predictions are achieved. But Figure 9.10 clearly shows that here the method gives a very high weighting to the 24-h-ahead load value—a lucky coincidence creating an understanding of the trained model. However, as this is an undesirable effect in such a case, we have decided to leave the 24 h prior load value out of the machine learning training and only ever combine

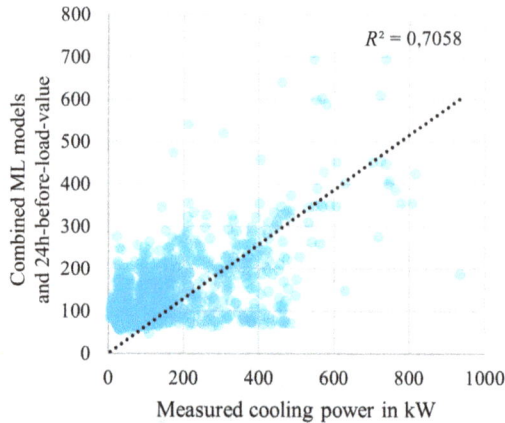

Figure 9.11   Weighted combination of 0.31 gradient boosting, 0.60 XGBoost, 0.05 random forest and 0.04 24-h-ahead load value—the coefficient of determination increases (slightly). Predicted versus measured cooling capacities for the period 1.9.21 10 a.m. to 12.10.21 11 a.m. (cf. [18])

it with the machine learning forecast if it can be expected to increase the quality of the forecast—i.e., especially not in the case of meter jumps.

Figure 9.11 shows the slightly better coefficient of determination when using an optimized combination of the machine learning models and the 24-h-ahead load value for the refrigeration load forecast of the EUREF Energy Workshop.

9.3   Further application in the WindNODE project

During the course of the project, contrary to original assumptions, it turned out that from an optimization point of view, it is advantageous most of the year to load both storage tanks of the P2H/P2C system (cf. Figure 9.12) with the help of the refrigeration compression units with cold, since there is an almost constant demand for cold throughout the year due to a data center on the EUREF campus.

To address congestion management, within the WindNODE Project (cf. [12]), the EUREF-Energy Workshop offered flexibilities at the flexibility platform of the DSO Stromnetz Berlin and the TSO 50Hertz (cf. [4]). We also tested the extent to which uncertainties in forecasts can be quantified in the course of plant and forecast operation and, in turn, incorporated into the management of the storage facilities (cf. [25] and Figure 9.13). A concept was suggested to cope with uncertain load forecasts by reserving a share of the energy storage system for short-term balancing. Depending on the amount of uncertainty in the load forecasts, the energy system is scheduled with a specific reduced storage capacity at the day-ahead market. At the day of delivery, the optimal thresholds are examined when the remaining capacity should be used to balance differences between forecast and reality at the intraday market. It is shown that the energy costs can under certain circumstances be reduced

Figure 9.12   The power-to-heat/power-to-cold system implemented within the WindNode project (cf. [12]) with screen showing historical and live data of the system, © GASAG Solution GmbH

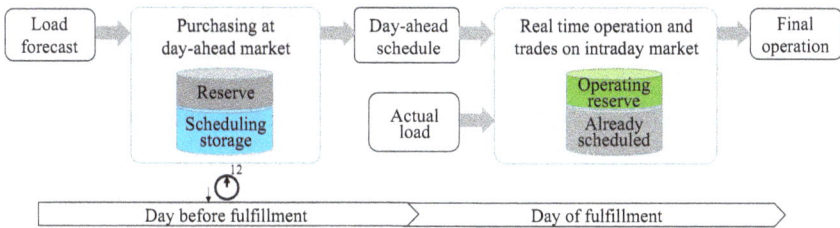

Figure 9.13   Incorporating forecast errors in a two-step, sequential market (taken from [25]).

by up to 10% using the optimal reserve share. The optimal reserve share depends on the forecast quality and the time series of loads and prices.

9.4   Conclusion and Outlook

A key to the EUREF-Energy Workshop project was the mastery of automation technology across the entire process chain—from data origination at the sensor to automation of forecast generation and schedule optimization. Electricity market prices and price volatility will continue to rise due to more renewable energies in the energy grid, which will lead to a further increase in the economic viability of flexible plants of this type and microgrids. The reduction of the high end-user levies that have to be paid when purchasing electricity would thus also indirectly give the economic viability of projects a further necessary boost to CO_2 reduction.

References

[1]   Bracke R, and Huenges E (eds). Roadmap tiefe Geothermie für Deutschland – Handlungsempfehlungen für Politik, Wirtschaft und Wissenschaft für eine erfolgreiche Wärmewende; 2022. [Online]. Available from: https://www.ieg. fraunhofer.de/content/dam/ieg/documents/Roadmap%20Tiefe%20Geotherm ie%20in%20Deutschland%20FhG%20HGF%2002022022.pdf.

[2]   Coalition agreement (2021–2025; SPD, Bündis 90/Die Grünen, FDP); 2021. [Online]. Available from: https://www.spd.de/fileadmin/Dokumente/Koalitio nsvertrag/Koalitionsvertrag_2021-2025.pdf.

[3]   Erdmann G, Graebig M, Rogler N, *et al.* Power, Grid, Flow; 2021. [Online]. Available from: https://bit.ly/2UnngOH.

[4]   Meyer-Braune G, Dreke C, Meinl A, *et al.* DIN SPEC 91410-1:2020-07 – Energy flexibility – Part 1: Flexibility provision for power system congestion management – Requirements for voluntary participation of providers in a flexibility platform; 2020. [Online]. Available from: https://www.beuth.de/ de/technische-regel/din-spec-91410-1/323621795.

[5]   Beuker S, Doderer H, Funke A, *et al.* Synthesis Report Flexibility, Market and Regulation; 2021. [Online]. Available from: https://www.windnode.de/fi leadmin/Daten/Downloads/FMR.pdf.

[6]   Ramsebner J, Haas R, Ajanovic A, Wietschel M. The sector coupling concept: A critical review. *WIREs Energy Environ.* 2021;10:e396. Available from: htt ps://doi.org/10.1002/wene.396.

[7]   Rath M, Neumann J, and Wilhelm G. CO2-neutral neighborhood supply in the 21st century. In: *The Smart Energy Showcase (2017–2020)*; 2021. p. 144–147. Available from: https://www.windnode.de/fileadmin/Daten/Downloads/ Jahrbuch/WindNODE_Jahrbuch_2020_Web_150dpi.pdf.

[8]   George KN, Rath M, and Bracke R. Quantifying demand flexibilities of buildings for an optimal design and operation of integrated district energy systems. In: *Proceedings of ECOS 2023 – 36th International Conference on Efficiency, Cost, Optimization, Simulation and Environmental Impact of Energy Systems.* Las Palmas de Gran Canaria, Spain; 2023.

[9]   Meeder A, Rath M, Schuldt-Gruner M, *et al.* Project final report "Control room of hybrid energy plants"; 2019. [Online]. Available from: https://www. berlin.de/sen/uvk/_assets/umwelt/foerderprogramme/berliner-programm-fu er-nachhaltige-entwicklung-bene/bene-projekte/bene-1137_abschlussberich t_pbl.pdf.

[10]   Meeder A, Rath M, Koch M, *et al.* Control system for intelligent cross-sector coupling in an urban quarter. In: *ICREN 2019*; 2019. p. 109. Available from: https://premc.org/doc/ICREN2019/ICREN2019_Book_Of_Abstracts.pdf.

[11]   BENE 1137-B5-O: Entwicklung und Test einer Leitstandtechnologie zum zentralen Monitoring und zur effizienten und vorausschauenden Lenkung hybrider Energieanlagen innerstädtischer Gebäude. Research project within 'Berliner Programm für Nachhaltige Entwicklung (BENE), gefördert aus Mitteln des Europäischen Fonds für Regionale Entwicklung und des Landes

Berlin' funding code 1137-B5-O; 2016–2019. [Online]. Available from: http s://www.berlin.de/sen/uvk/_assets/umwelt/foerderprogramme/berliner-progr amm-fuer-nachhaltige-entwicklung-bene/bene-projekte/plakat_klimaschutz partner_2019_ansicht_v7.pdf.

[12]   WindNODE: The showcase for smart energy from the northeast of Germany. Subproject: Construction, integration and operation of a power to heat/cold plant at the EUREF-Quartier with showcase. Joint research project, funding code 03SIN515; 2016–2021. [Online]. Available from: https://www.enargus.de/pub/bscw.cgi/?op=enargus.eps2&q=%220117 1449/1%22&v=10&p=3&id=548523.

[13]   EnOB: ARCHE – Architekturen und Entwurfsmethodik für selbstopti- mierende Regelverfahren in verteilten Energiesystemen; Teilvorhaben: sEMS-Architekturen und Demonstratoren 2. Joint research project, funding code 03ET1567D; 2018–2022. [Online]. Available from: https://www.enarg us.de/pub/bscw.cgi/?op=enargus.eps2&q=GASAG&m=2&v=10&s=3&y=1 &id=388165.

[14]   Erneuerbare Energien Gesetz (EEG) 2023; 2023. [Online]. Available from: https://www.gesetze-im-internet.de/eeg_2014/EEG_2023.pdf.

[15]   Oldewurtel F, Parisio A, Jones CN, *et al.* Use of model predictive control and weather forecasts for energy efficient building climate control. *Energy and Buildings*. 2012;45:15–27. Available from: https://www.sciencedirect.com/s cience/article/pii/S0378778811004105.

[16]   Ljung L. In: Procházka A, Uhlíř J, Rayner PWJ, *et al.*, editors. *System Identi- fication*. Boston, MA: Birkhäuser Boston; 1998. p. 163–173. Available from: https://doi.org/10.1007/978-1-4612-1768-8_11.

[17]   Fassnacht T, Öestreicher H, and Wagner A. Gebäudemodelle für modell- basierte Regler und Energiemanagementsysteme. In: *Proceedings of the 5th German–Austrian Conference of the International Building Perfor- mance Simulation Association, RWTH Aachen University*; 2014. Available from: https://publications.ibpsa.org/proceedings/bausim/2014/papers/bausim 2014_1191.pdf.

[18]   Rath M, Ray H, van Treek M, *et al.* Untersuchung verschiedener Lastprog- noseverfahren für die prognosebasierten Steuerung dezentraler Energiean- lagen. In: *Proceedings of BauSim Conference 2022: 9th Conference of IBPSA-Germany and Austria*; 2022.

[19]   Hastie T, Tibshirani R, and Friedman J. *The Elements of Statistical Learning*. New York: Springer; 2009.

[20]   Pedregosa F, Varoquaux G, Gramfort A, *et al.* Scikit-learn: machine learning in Python. *Journal of Machine Learning Research*. 2011;12:2825–2830.

[21]   Chen T, and Guestrin C. XGBoost: a scalable tree boosting system. In: *Proceedings of the 22nd ACM SIGKDD International Conference on Knowl- edge Discovery and Data Mining*. New York, NY, USA: Association for Computing Machinery; 2016. p. 785–794.

[22]   Breiman L. Random forests. *Machine Learning*. 2001;45:5–32.

[23]   Tritschler M, and Trischtler M. Monitoring und Betriebsoptimierung – Vergleich der Prognose des Energieverbrauchs mit neuronalen Netzen und linearen Modellen. *GI – Gebäudetechnik in Wissenschaft & Praxis*. 2017;138: 294–303.

[24]   Dziubany M, Schneider J, Schmeink A, *et al.* Prognose von Wärmeverbräuchen. In: *Workshop der INFORMATIK 2018. Lecture Notes in Informatics (LNI)*. Bonn: Gesellschaft für Informatik; 2018.

[25]   Koch J, Bensmann A, Eckert C, *et al.* Planning of reserve storage to compensate for forecast errors. *Energies*. 2024;17(3):720. Available from: https://www.mdpi.com/1996-1073/17/3/720.

Chapter 10

A fog computing-based architecture for the decentralized energy management of microgrids

Alessandro Armando[1], Daniel Fernandez Valderrama[1], Giulio Ferro[1], Giacomo Longo[1], Alessandro Orlich[1], Michela Robba[1], Mansueto Rossi[2] and Enrico Russo[1]

In this chapter, a novel approach is presented for the day-ahead optimal power scheduling of polygenerative microgrids. The innovative method simplifies power scheduling optimization by relying on local computations by agents within the system and secure communication among these agents. This eliminates the need for a centralized scheduling unit.

Furthermore, this study introduces a cloud-fog-based framework as a convenient and cost-effective infrastructure to support a decentralized method. To assess its effectiveness, the proposed approach is tested using the test-bed facilities at the University of Genova Savona Campus, which include a Smart Polygeneration Microgrid (SPM) and a Sustainable Energy Building (SEB) connected to the microgrid. Simulation results showcase the overall performance of the proposed framework in the real case study utilizing CISCO devices.

Acronyms

EMS Energy Management System
CEMS Centralized EMS
DEMS Decentralized EMS
ICT Information and Communication Technology
SCADA Supervisory Control and Data Acquisition Systems
DSL Digital Subscriber Line
PLC Power Line Communication
MG Microgrid
SPM Smart Polygeneration Microgrid
SEB Sustainable Energy Building

[1]Department of Informatics, Bioengineering, Robotics and Systems Engineering, University of Genoa, Italy
[2]Department of Naval, Electrical, Electronic and Telecommunication Engineering – DITEN, Italy

ADMM alternating direction method of multipliers
HVAC heating, ventilating, and air conditioning

10.1   Introduction and related work

A microgrid is an innovative and localized power grid situated in a specific site, such as a university campus, hospital, or neighborhood. It enables the integration of various distributed energy sources, such as solar panels and wind turbines, to supply power to nearby loads. The primary objective of a microgrid is to provide energy self-sufficiency and resilience, allowing it to operate independently from the main electrical network in islanding mode if necessary. In addition to generating electricity from renewables, microgrids can also cater to the thermal energy requirements of the site, further enhancing their efficiency and sustainability.

To effectively manage the complex operation of a microgrid and coordinate the behavior of its diverse energy sources, an Energy Management System (EMS) is essential [1]. EMS acts as central intelligence, orchestrating control algorithms and minimizing costs and emissions while optimizing power management within the microgrid. There are two predominant classes of EMS for microgrids: centralized EMS (CEMS) and decentralized EMS (DEMS) [2].

CEMS adopts a hierarchical approach, where a central controller collects data from the various devices within the microgrid and sends commands to implement the control algorithm. This centralized control structure offers several advantages, including comprehensive system-wide data gathering and the absence of conflicts in decision-making. Traditionally, microgrids have utilized Supervisory Control and Data Acquisition Systems (SCADA) for centralized control and communication technology.

However, DEMS aims to address the limitations of CEMS. It decomposes the problem into a network of autonomous local controllers, resulting in reduced computational requirements and increased flexibility. With DEMS, any modifications or additions to the microgrid only necessitate updating the relevant network nodes, minimizing the impact on the overall system [3].

From an Information and Communication Technology (ICT) perspective, the literature often describes a microgrid as a multilayer structure comprising three fundamental layers: control, communication, and power as depicted in Figure 10.1. The power layer encompasses the physical devices that constitute the microgrid, such as transformers, loads, renewable sources, and other distributed generators. The communication layer serves as the bridge connecting all these devices, forming the backbone of the microgrid architecture. Finally, the control layer assumes responsibility for the operation and management of the microgrid. With a high-performing communication layer, the control layer can promptly respond to contingencies occurring in the power layer, ensuring efficient and reliable microgrid operation.

Overall, microgrids represent a transformative approach to energy distribution, offering localized and sustainable power solutions with the capability to adapt to evolving energy needs.

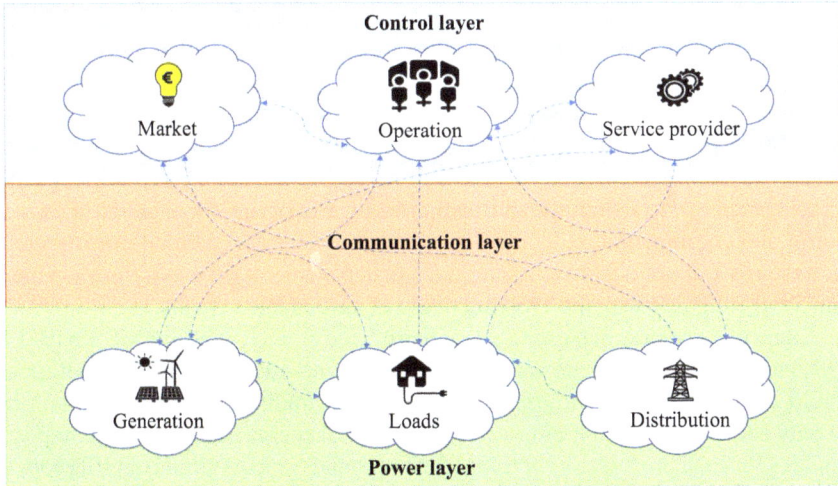

Figure 10.1   Layers of the MG architecture

The operation of a microgrid relies heavily on the exchange of data, making it crucial for the communication infrastructure to meet stringent requirements. One key consideration is the communication latency, which refers to the delay in transmitting data between devices. Certain functions of the microgrid control system necessitate a high refresh rate for the variables involved, while others are less sensitive to latency. Therefore, the choice of the communication technology should take into account these diverse requirements. For instance, wide-area situational awareness systems may require data rates exceeding 600 kbps, whereas simple distributed generator monitoring can suffice with rates as low as 10–56 kbps [4].

Moreover, other requirements such as interoperability, reliability, security, and scalability differ across available communication technologies. Wired options encompass optical networks, Digital Subscriber Line, Power Line Communication, and Ethernet. Wireless alternatives include 3G/4G/5G, Wi-Fi, ZigBee, and satellite communications. A comprehensive review of the standards and protocols developed for these different technologies can be found in [5]. For further information on the ICT requirements for various applications in microgrids, [6] provides valuable insights.

The motivation behind this work relies on recent technological advancements that have facilitated the integration of advanced EMS into small, energy-efficient computing devices. These systems adopt decentralized architectures, distributing computing capabilities across a network of nodes. This approach not only enhances scalability but also ensures the robustness and resilience of the overall system. The use of low computational units enables real-time monitoring, analysis, and optimization of energy consumption, production, and storage at a highly detailed level [7]. By empowering individual devices and components to make cooperative decisions, these decentralized systems contribute to a more adaptive and responsive energy grid.

To address the previously mentioned issue, the development of a Fog computing architecture represents a possible solution [8]. Fog computing plays a key role in revolutionizing microgrid EMS, offering experts a dynamic and efficient solution to address the complexities of decentralized energy systems. By facilitating real-time data processing at the edge of the network, fog computing minimizes latency and enhances the responsiveness of microgrid EMS. Furthermore, fog computing enhances the reliability and resilience of microgrids by enabling localized control and coordination, reducing dependence on centralized cloud infrastructure. The result is a more adaptive, secure, and scalable microgrid energy management framework, aligning with the evolving needs of sustainable systems.

Some recent works that propose a decentralized architecture for EMS are [9–14]. In [9] authors suggest an innovative peer-to-peer transactive trading system utilizing a multi-actor algorithm, coupled with a peer-to-peer trading platform. This platform not only encourages prosumers to participate in local energy trading but also imposes penalties for each prosumer's contribution to rebound peaks. Authors of [10] propose a framework for day-ahead microgrid energy management using multiple agents. The goal is to minimize energy loss and operation costs for various agents, including traditional distributed generators, wind turbines, photovoltaics, demands, battery storage systems, and the microgrid aggregator agent. To protect agents' information privacy, the alternating direction method of multipliers (ADMM) is employed to find the optimal operating point of the microgrid in a distributed manner. The work in [11] proposes a consensus-based distributed privacy-preserving energy management strategy with an event-triggered scheme to achieve privacy preservation and network resource-saving simultaneously by means of a consensus-based privacy-preserving algorithm that employs differential privacy. While [12] describes two kinds of decentralized economic dispatch framework for the coordinated operation of multi-microgrids in a distribution network using a decentralized framework that can respect the independence and privacy of different operators, and be beneficial to solve technical and economic challenges brought by centralized optimization. In [13], a decentralized EMS consensus-based algorithm that focuses on tackling the problem of random packet drops. A novel consensus based algorithm is proposed, which tracks and exchanges the accumulated value of the power mismatch estimation so that the information loss can be recovered. Finally, Li *et al.* [14] address the issue of privacy and scalability to decentralized EMS in microgrids. The absence of a central controller brings significant challenges to promoting trusted collaboration and avoiding possible collusion. To address these issues, a blockchain-based framework is proposed, which adopts a consensus-based algorithm with a collusion prevention mechanism.

The organization of this chapter is as follows. Section 10.2 describes the fog-based architecture developed for the decentralized operation of the microgrid. Section 10.3 introduces the optimization problem and models of each component of the microgrid. Section 10.4 describes the overall decentralized algorithm. Section 10.5 presents the case study of the EMS on the SPM, including the architecture and implementation details on CISCO devices. Section 10.6 provides some concluding remarks.

10.2   Fog computing for decentralized microgrid management

Fog computing is a standard architecture [15] that extends cloud computing capabilities to the network edge. It refers to a scenario in which ubiquitous and decentralized devices, e.g., routers, gateways, and IoT devices, collaborate, and communicate with each other and the network [16]. The above devices, namely *fog devices*, are designed and optimized for fog computing, enabling them to execute storage, and processing tasks directly, bypassing the need for external intermediaries. These tasks encompass essential network functions, innovative services, and applications operating within a secure, isolated environment. By bringing computation closer to the data source, fog computing minimizes latency, optimizes bandwidth usage, improves privacy, and enables real-time data analysis and decision-making.

A typical fog-based architecture comprises three hierarchical layers. The lowermost layer, namely the *Edge layer*, includes edge components responsible for managing hardware units such as microturbines, storage systems, or geothermal heat pumps. In the middle, the *Fog layer* hosts fog devices that perform immediate and localized data processing, analysis, and storage. The *Cloud layer* is a remote data center providing the fog layer with robust and long-term computational and storage capabilities, effectively addressing the inherent limitations of the capacities of fog devices.

Furthermore, the fog layer can be classified into *tiers*, primarily based on the degree to which the functions performed by the devices are oriented toward either the Cloud or Edge layers.

Figure 10.2 depicts the fog-based architecture we consider for running the decentralized optimization. A decentralized optimization scheme is one in which there exist decision-making capabilities at both the local device level and a centralized

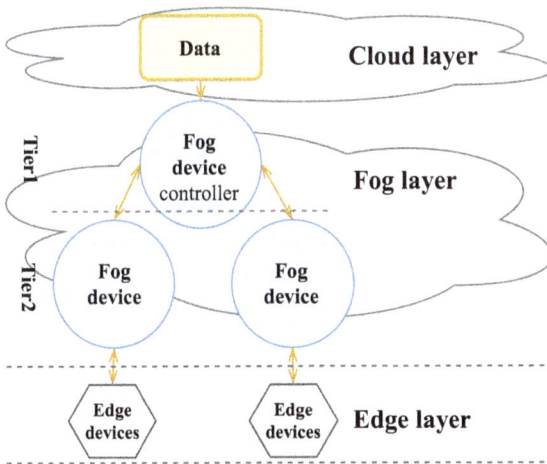

Figure 10.2   Fog-based architecture

coordinator. This definition makes a distinction between local and decentralized optimization schemes, where local schemes have purely agent-level decision-making and no communication between agents or a central coordinator [1].

The fog-based architecture comprises the above three hierarchical layers [17]. In particular, the fog layer hosts the network devices that enable the execution of different fog applications in charge of running the code of the decentralized optimization model and the required support services, e.g., the middleware supporting the network protocols for interfacing with the edge devices. Moreover, fog devices are structured in a 2-tier arrangement. In the lowest tier, namely *tier2*, each device runs a sub-task of the optimization algorithm related to a source or equipment assigned. This role is enabled by their ability to communicate with the edge components, gathering inputs and issuing commands as necessary. In the upper *tier1*, a leader device, namely the *Central Controller* (CC), coordinates the iterations of the *tier2* devices during the execution of the decentralized algorithm and interfaces with the Cloud layer. At the end of the execution of the optimization algorithm, it translates the results into action by sending the corresponding setpoints to *tier2* devices for activation in the controllable generation plants.

Finally, the Cloud layer is a remote data center providing the Fog layer with data to be integrated into the optimization process, that is, preliminary computations, and forecasting data about the availability of renewable energy, and user demand. Once implemented, this architecture can exhibit several desirable properties suitable for EMS for microgrids, including:

- **Scalability**. It enables adding or removing fog nodes as needed. Consequently, the optimization algorithm can seamlessly adapt to changes in demand and accommodate microgrid growth without performance degradation.
- **Decentralization**. It facilitates a decentralized architecture by distributing computing resources and decision-making capabilities across multiple fog nodes. Moreover, any device can assume the controller role. This property improves reliability and autonomy.
- **Resilience**. It enhances uninterrupted functionality by enabling the optimization process to sustain local operations in case of a fog device failure or during network or cloud connectivity problems. Resilience and decentralization synergize with the capability of microgrids to operate in islanded mode, where they can function autonomously and independently from having outsourced facilities available.
- **Low latency**. It achieves low-latency data processing and decision-making, enabling fast responses to changing conditions and optimizing energy distribution within the grid.

10.3   Microgrid mathematical modeling

The sets of decision variables of the optimization problem for the EMS of a microgrid are listed below:

- N is the set of power grid nodes;
- $H_{F,i} = \{1, ..., h_{f,i}\}$ is the set of controllable generation plants at node $i \in N$;

- $H_{R,i} = \{1, ..., h_{r,i}\}$ is the set of renewable generators at node $i \in N$;
- $S_i = \{1, ..., k_{s,i}\}$ is the set of storage systems at node $i \in N$.
- $T = \{0, ..., T_f\}$ is the set of time instants.

10.3.1   Power balance

The model consists of a unique bus bar in which each component of the microgrid is connected, therefore the active power balance is given by the sum of plants and consumers at each node:

$$\sum_{i \in N} \left(\sum_{h \in H_{F,i}} P^E_{h,i,t} + \sum_{l \in H_{R,i}} P^R_{l,i,t} - \sum_{k \in S_i} P^S_{k,i,t} - P^D_{i,t} + P^G_{i,t} - P^H_{i,t} \right) = 0, \quad t \in T \quad (10.1)$$

where $P^E_{h,i,t}$ [kW] is the active power produced by the controllable generator h; $P^R_{l,i,t}$ [kW] is the active power by the power plant l using renewables; $P^S_{k,i,t}$ [kW] is the active power exchanged with the storage k; $P^D_{i,t}$ [kW] is the overall power load; $P^G_{i,t}$ [kW] is the power exchanged with the main grid; and $P^H_{i,t}$ [kW] is the power consumed by heat pumps. The power exchanged with the distribution network is characterized by the constraints related to the limitation of the substation, i.e., maximum and minimum powers, respectively, $P^{G,Max}_i$ and $P^{G,Min}_i$, as it is reflected in (10.2):

$$\underline{P^G_i} \le P^G_{i,t} \le \overline{P^G_i}, \quad i \in N \quad t \in T \tag{10.2}$$

10.3.2   Storage system

The following discrete-time equation models the state of charge of the storage systems:

$$x_{k,i,t+1} = a_{k,i,t} x_{k,i,t} + P^S_{k,i,t} \Delta \quad k \in S_i \quad i \in N \quad t \in T \tag{10.3}$$

$$x^{Min}_{k,i,t} \le x_{k,i,t} \le \overline{x_{k,i,t}} \quad k \in S_i \quad i \in N \quad t \in T \tag{10.4}$$

$$\underline{P^S_{k,i}} \le P^S_{i,k,t} \le \overline{P^S_{k,i}}, \quad k \in S_i \; i \in N \; t \in T \tag{10.5}$$

where $x_{k,i,t}$ is the energy stored in the storage system k of node i at time t; $a_{k,i,t}$ is a loss coefficient; Δ is the time step.

10.3.3   Controllable generation plants

Microturbines are modeled by the following expressions:

$$P^{PE}_{h,i,t} = \mu_{h,t} P^E_{h,i,t} \quad i \in N \quad h \in H_{F,i} \quad t \in T \tag{10.6}$$

$$P^{TH}_{h,i,t} = \bar{\mu}_{h,t} P^E_{h,i,t} \quad i \in N \quad h \in H_{F,i} \quad t \in T \tag{10.7}$$

$$\underline{P^E_{h,i}} \le P^E_{h,i,t} \le \overline{P^E_{h,i}}, \quad i \in N \quad h \in H_{F,i} \quad t \in T \tag{10.8}$$

where $P^{PE}_{h,i,t}$ is the chemical power of the fuel; $P^E_{h,i,t}$ is the electrical power generated by the microturbine; $P^{TH}_{h,i,t}$ is the thermal power generated by the microturbine; $\mu_{h,t}$ and

$\bar{\mu}_{h,t}$ are parameters that depend on the time-varying external air temperature. More-over, constraints regarding the tax discounts on high-efficiency CHP power plants are to be considered:

$$Q_{h,i,t}^{UTX} = (1 - f\!f)\beta P_{h,i,t}^{E} \quad i \in N \quad h \in H_{F,i} \quad t \in T \tag{10.9}$$

$$Q_{h,i,t}^{TX} = Q_{h,i,t} - Q_{h,i,t}^{UTX} \quad i \in N \quad h \in H_{F,i} \quad t \in T \tag{10.10}$$

$$P_{h,i,t}^{PE} = Q_{h,i,t}LHV, \quad i \in N \quad h \in H_{F,i} \quad t \in T \tag{10.11}$$

where $Q_{h,i,t}^{UTX}$ [m^3/y] is the untaxed share (due to Italian legislation on high-efficiency CHP systems) of natural gas $Q_{h,i,t}$ [m^3/y] used to feed the microturbines; $Q_{h,i,t}^{TX}$ [m^3/y] is the taxed quantity of natural gas used in the microturbines; $f\!f$ and β are parameters related to taxation policies for the Italian regulation and *LHV* is the low heating value of the fuel.

10.3.4   Thermal units and demand satisfaction

Heat demand can be satisfied by microturbines and boilers, while cool demand can be met through chillers, supplied by electricity and microturbines respectively. First, boiler thermal expression is given by (10.12):

$$P_{i,t}^{PE,B}\eta_i^B = P_{i,t}^{TH,B}, \quad i \in N \quad t \in T \tag{10.12}$$

where $P_{i,t}^{PE,B}$ is the power related to the primary energy of the boiler of node i at time t; and η_i^B is the boiler's efficiency. Similar to the electric power balance in (10.1), the thermal balance is expressed by (10.13).

$$\underline{a}D^H \le \sum_{i \in N}\sum_{h \in H_{F,i}} P_{h,i,t}^{TH} + P_t^{TH,B} \le \bar{a}D^H \quad t \in T \tag{10.13}$$

$$P_{i,t}^{CHI}COP_i = P_{i,t}^{TH,CHI}, \quad i \in N \quad t \in T \tag{10.14}$$

where D^H is the thermal demand for heat; $\underline{a}$ and $\bar{a}$ are bound parameters; $P_{i,t}^{TH,CHI}$ is the thermal power generated from chillers; COP_i is a known coefficient of performance.

10.3.5   Smart energy building

The SEB is heated and cooled by a geothermal heat pump integrated with thermal energy storage, which provides thermal power to the different spaces in the building through fan coils. The electric power from the microgrid feeds the heat pump to produce this thermal power $P_{SEB,t}^{TH,HP}$, which expression is (10.15).

$$P_{SEB,t}^{TH,HP} = COP_{SEB}, P_{SEB,t}^{E,HP} \quad t \in T \tag{10.15}$$

where COP_{SEB} is the heat pump coefficient of performance.

The thermal energy storage system responds to a dynamic behavior, which is formulated as follows:

$$T_{t+1}^S = \frac{P_{SEB,t}^{TH,HP} - P_{S,out,t}^{TH}}{C_S} \Delta + T_t^S \quad t \in T \tag{10.16}$$

where T_t^S [K] is the temperature of the thermal storage; [kWh/K] is the thermal capacity of thermal storage; and $P_{S,out,t}^{TH}$ [kW] is the power exiting the thermal energy storage system. Such stored power is related to the thermal power of the ventilation system $P_{FC,t}^{TH}$ [kW] through (10.17).

$$P_{S,out,t}^{TH} = \frac{P_{FC,t}^{TH}}{\eta_{FC}}, \quad t \in T \tag{10.17}$$

where η_{FC} is the efficiency of the ventilation system.
The entire SEB is modeled as a single equivalent room, and the expression that defines the temperature T_t^B is given by

$$T_{t+1}^B = \frac{\left(\frac{T_{ext,t} - T_t^B}{R_b} + \alpha_{FC} P_{FC,t}^{TH}\right)}{C_b} \Delta + T_t^B, \quad t \in T \tag{10.18}$$

where R_b and C_b [kWh/K] are the thermal resistance and capacity of the building that defines the heat exchange with the outside, respectively; α_{FC} is the efficiency of the fan coil. In summer, when the outside temperature is higher than the building temperature, the expression $\frac{T_{ext,t} - T_t^B}{R_b}$ is positive, $P_{FC,t}^{TH}$ is negative to provide cooling, and $P_{HP,SEB,t}^{TH}$ is also negative. Moreover, constraints on $P_{FC,t}^{TH}$, T_t^S, and T_t^B must be considered:

$$0 \leq P_{FC,t}^{TH} \leq \eta_{FC} m_{air} c_{p,air} \left(T_t^S - T_t^B\right) \quad t \in T \tag{10.19}$$

$$\underline{T_S} \leq T_t^S \leq \overline{T_S} \quad t \in T \tag{10.20}$$

$$\underline{T_B} \leq T_t^B \leq \overline{T_B}, \quad t \in T \tag{10.21}$$

where m_{air} [kg/h] is the air mass flow rate passing through the fan coils; $c_{p,air}$ [kWh/(kg K)] is the specific heat of air at constant pressure.

10.3.6 Objective function

The objective function of the developed EMS is represented by the minimization of the weighted sum of economic costs, the costs of CO_2 emissions costs and the monitoring of the desired temperatures for the SEB. In particular, the specifications considered in the objective function for costs and CO_2 emissions are:

- The cost for the electrical energy absorbed from the main grid.
- The costs for the purchase of natural gas that is used in microturbines and boilers.
- CO_2 emissions generated by the natural gas supplying boilers and microturbines.
- CO_2 emissions from the power coming from the main grid.

The objective function comprises three terms, which are related to SEB, storage and generators, and given by (22), (23), and (24), respectively. These terms will be used by the decentralization of the algorithm.

$$J_B = \Delta \sum_{t \in T} (T_{S,t} - T_S^*)^2 + (T_t^B - T_B^*)^2 \tag{10.22}$$

$$J_S = \Delta \sum_{t \in T} \left\{ \sum_{k=0}^{S_i} \left\{ -(C_t + f_e C_{CO_2}) P_{k,i,t}^S \right\} \right\} \tag{10.23}$$

$$J_G = \Delta \sum_{t \in T} (C_t + f_e C_{CO_2}) P_{HP,i,t} + C_{CO_2} P_{PE,B,t} \tilde{E}_{f-ng} + P_{th,B,t} TES_{pp} +$$
$$+ \sum_{i=1}^{N} \sum_{h=1}^{H_{F,i}} Q_{utx,h,i,t} NG_{ppwf} + Q_{tx,h,i,t} NG_{pp} + C_{CO_2} P_{PE,h,i,t} E_{f-ng}, \tag{10.24}$$

where C_t [€/kWh] is a unit cost to be paid by the microgrid to buy electrical energy from the grid; f_e [kgCO2/m3] is the emission factor related to the electrical energy taken from the grid; C_{CO_2} [€/ton] is the cost associated to the emission of 1 ton of CO_2; TESpp [€/kWhth] is the purchasing price of the primary energy used in the boilers; NG_{ppwf} [€/m3] is the untaxed gas price; NG_{pp} [€/m3] is the full price of gas; E_{f-ng} and $\tilde{E}_{f-ng}$ [tCO2/m3] are the emission factors of the fuel; finally T_B^* and T_S^* are temperature set points for the building and the SEB tank. The objective function for the entire optimization problem is given by the sum of the terms (10.22), (10.23), and (10.24), as shown in (10.25).

$$J = J_B + J_S + J_G \tag{10.25}$$

10.4   Decentralized optimization algorithm

A key contribution of this chapter is the decentralized optimization algorithm to create a decentralized EMS. To introduce the algorithm, we first consider a general convex optimization problem of the form:

$$\min_{x \in \mathcal{X}} \; f(x)$$
$$s.t \; Ax = b, \tag{10.26}$$

where A and b are the constraint matrix and vector, respectively, $\mathcal{X}$ is the feasibility set of the variables, and $f(x)$ is a general convex objective function. The corresponding Lagrangian function of (10.26) is:

$$\mathcal{L}(x, \nu) = f(x) + \nu^T (Ax - b), \tag{10.27}$$

where the constraint has been dualized with dual variable ν. The proposed method is based on the Augmented Lagrangian approach that modifies the classical Lagrangian including a regularization term, which ensures better convergence properties [18]. For this specific application, a proximal augmentation term has been chosen. A general form of the proposed algorithm is given as follows:

$$\hat{x}\left[\tau+1\right] = \underset{x}{\mathrm{argmin}}\left\{\mathcal{L}\left(x,\nu\left[\tau\right]\right) + \frac{1}{2\rho}\|x - \hat{x}\left[\tau\right]\|_2^2\right\} \tag{10.28}$$

$$x\left[\tau+1\right] = \underset{x\in\mathcal{X}}{\mathrm{argmin}}\left\{\|x - \hat{x}\left[\tau+1\right]\|_2^2\right\} \tag{10.29}$$

$$\nu\left[\tau+1\right] = \nu\left[\tau\right] + \rho\left(Ax\left[\tau+1\right] - b\right), \tag{10.30}$$

where ρ is the step size. The overall algorithm is based on three steps: the first one (10.28) is creating an unconstrained minimization of the augmented Lagrangian by fixing dual variables at the previous iterations; the second step is the projection of the previous variables into the feasibility set $\mathcal{X}$; finally, the last step is a classical gradient dual update based on the actual value of decision variables.

Problem (10.1)–(10.24) can be rewritten in standard quadratic programming (QP) form:

$$\min_{y} \frac{1}{2}y^T Q y + p^T y$$
$$\text{s.t. } \hat{A}y \le b, \tag{10.31}$$

where y is the vector of decision variables, and Q, p, $\hat{A}$ and b are suitable vectors and matrices. In order to apply algorithm (10.28)–(10.30), it is necessary to rewrite the above QP problem in the form of (10.26) by using positive slack variables z. The resulting optimization problem will be of the form:

$$\min_{y,z} \frac{1}{2}y^T Q y + p^T y$$
$$\text{s.t } \begin{bmatrix} \hat{A} & I \end{bmatrix}\begin{bmatrix} y \\ z \end{bmatrix} = b.$$
$$z \ge 0 \tag{10.32}$$

It is important to note that, for this problem, $\mathcal{X} = \{z : z \ge 0\}$.

At this point, it is important to analyze the overall optimization problem to understand a possible decomposition. We can note that the only variables included in the objective function (10.25) are related to storage units, thermal units and the SEB; furthermore, the only coupling constraint (i.e., the relation that involves more than one variable) is the limit on the power exchange with the main grid (1), which physically corresponds to the point of common coupling (PCC) of the microgrid. Therefore, it is possible to split the variables $y = \begin{bmatrix} y_s & y_t & y_b \end{bmatrix}$ corresponding to each collection of variables related to the different units (storage, thermal, and SEB). Thus, it is possible to rewrite Q, p, $\hat{A}$ and b as follows:

$$Q = \begin{bmatrix} Q_s & 0 & 0 \\ 0 & Q_t & 0 \\ 0 & 0 & Q_b \end{bmatrix} \tag{10.33}$$

$$p = \begin{bmatrix} p_s \\ p_t \\ p_b \end{bmatrix} \qquad (10.34)$$

$$\hat{A} = \begin{bmatrix} A_s & 0 & 0 \\ 0 & A_t & 0 \\ 0 & 0 & A_b \\ I & I & I \end{bmatrix} \qquad (10.35)$$

$$b = \begin{bmatrix} b_s \\ b_t \\ b_b \\ b_g \end{bmatrix} \qquad (10.36)$$

Based on these considerations the overall primal update of the algorithm can be replaced with three parallel updates as follows:

$$\hat{x}_{s,k+1} = \arg\min_{x_s} \left\{ \frac{1}{2} x_s{}^T Q_s x_s + p_s^T x_s + \lambda_k^T (Ax) + \frac{1}{2\rho} \|x_s - x_{s,k}\|_2^2 \right\} = \qquad (10.37)$$
$$- \left(Q_s + \frac{1}{\rho} \right)^{-1} \left(p_s^T + (\lambda_k^T A)_s - \frac{x_{s,k}}{\rho} \right)$$

$$\hat{x}_{t,k+1} = \arg\min_{x_t} \left\{ \frac{1}{2} x_t{}^T Q_t x_t + p_t^T x_t + \lambda_k^T (Ax) + \frac{1}{2\rho} \|x_t - x_{t,k}\|_2^2 \right\} = \qquad (10.38)$$
$$- \left(Q_t + \frac{1}{\rho} \right)^{-1} \left(p_t^T + (\lambda_k^T A)_t - \frac{x_{t,k}}{\rho} \right)$$

$$\hat{x}_{b,k+1} = \arg\min_{x_b} \left\{ \frac{1}{2} x_b{}^T Q_b x_b + p_b^T x_b + \lambda_k^T (Ax) + \frac{1}{2\rho} \|x_b - x_{b,k}\|_2^2 \right\} = \qquad (10.39)$$
$$- \left(Q_b + \frac{1}{\rho} \right)^{-1} \left(p_b^T + (\lambda_k^T A)_b - \frac{x_{b,k}}{\rho} \right).$$

10.5   Implementation and results

In this section, we present the case study and the implementation setup that serves as the basis for validating the proposed EMS. In particular, we use experimental data from the SPM at the Savona Campus of the University of Genoa (SPM) [19].

10.5.1   *Case study*

The SPM is a 3-phase low voltage (400 V line-to-line) distribution system, coupled with a thermal network, composed of electrical/thermal loads and generation units.

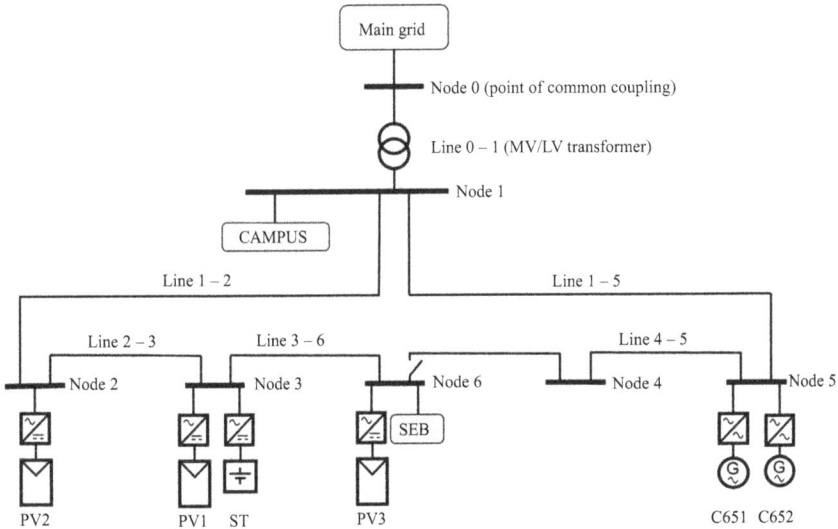

Figure 10.3   Simplified one-line diagram of the SPM at Savona Campus

The SPM electrical grid topology, illustrated in Figure 10.3, follows a ring configuration with six switchboards: QEG is the main one, connected to the Campus' MV/LV substation, while Q01, Q02, Q03, and Q04 are the connection points for the different equipment installed in the SPM. A fifth switchboard, Q05, is integrated into the SEB, realized about three years after the SPM.

To connect the SPM to the 15 kV distribution network, a dedicated MV/LV transformer is employed. This transformer is linked to the MV busbar of the Savona Campus MV/LV substation. While the buildings, excluding the SEB, are not directly linked to the SPM, the power generated by its installed sources can still supply the buildings through the PCC with the public medium voltage distribution network.

The following sources and equipment are connected to the SPM:

* an electrical storage (ES) system (SoNick sodium/nickel chloride technology, rated power: about 30 kW, rated energy: about 140 kWh), Figure 10.4;
* two absorption chillers, Figure 10.5;
* two photovoltaic (PV) fields (rated power: 80 kWp and 15 kWp), plus a third one installed on the SEB (21 kWp);
* two cogeneration microturbines (Capstone C65) fed by natural gas (rated electrical power: 65 kW, rated thermal power: 112 kWth), Figure 10.6;
* two electric vehicle (EV) charging stations.

Due to the cogeneration microturbines and the chillers, the SPM is a coupled electrical and thermal system.

The SEB (Figure 10.7) is an environmentally sustainable building connected to the SPM as a "prosumer" and characterized by energy efficiency solutions (such as high-performance thermal insulation materials and ventilated facades). The building

Figure 10.4   Overview of the storage systems of the SPM

Figure 10.5   Absorption chiller

Figure 10.6   Cogenerative microturbines

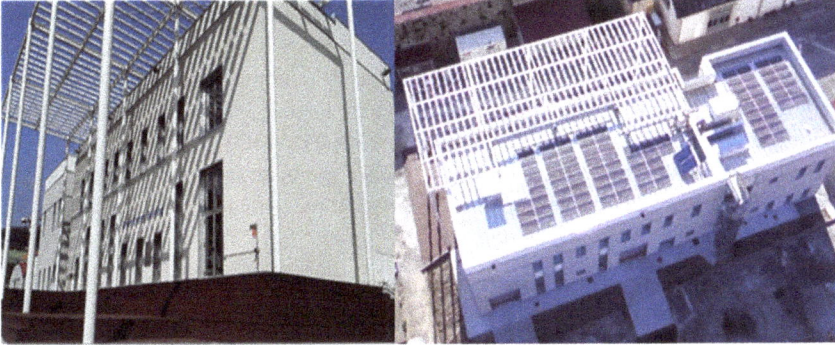

Figure 10.7   Two views of the Savona Campus SEB

has two floors, and it is 10 m high. It is made of reinforced concrete with predellas prestressed slabs, and the ground floor elevation has been adopted to prevent water entering the building in the event of floodings, which may occur due to a nearby river. High performance thermal insulation materials for building applications (ventilated facades and claddings) and acoustic insulation systems are also applied. The building is certified as A4, which is the most efficient "energy class" in accordance with Italian Classification of Building Energy Efficiency. The SEB hosts three laboratories, a gym, two classrooms, and some offices. In addition, the building is equipped with a Geothermal Heat Pump (GHP), which provides both winter heating and summer cooling, low-consumption LED lamps, and a rainwater collection system.

From the software development point of view, currently, different EMS have been developed and tested at the Savona Campus, but they are centralized with a classical ICT architecture. In [20] and [21] a multi-objective EMS for polygenerative microgrids is presented. The proposed tool has been developed within the LIVING GRID project, and it is characterized by a detailed representation of generation units and flexible loads, as well as electric/thermal networks and storage systems, which can be present in microgrids and sustainable districts. The optimization model includes an objective function related to the minimization of costs, the maximization of the overall exergy efficiency of the system, and the minimization of stress indexes.

10.5.2   Implementation

Our Proof-Of-Concept implementation relies on the containerization tool Podman [22]. Using Podman, we establish two separate virtual networks: one called Enterprise and the other named Edge. A "Pod" refers to a cluster of one or more containers that share storage and network resources.

In our setup, one Pod hosts a container responsible for simulating the SEB, and this container is exclusively connected to the Edge Network. Additionally, we have three instances of another Pod, which serve as Fog devices. These instances are represented as routers and are connected to both the Edge Network and the Enterprise Network. To emulate these routers, we employ a container running QEMU [23], specifically loaded with a Cisco Cloud Services Router 1000V (CSR1000V) ISO. These routers are designed to be compatible with Cisco IOx, an application framework that facilitates the deployment of Docker applications at the network edge. In practical terms, each of these routers hosts three containers packaged as IOx applications:

• A FOSS message broker for efficient communication between devices.
• A node participating in the decentralized optimization.
• An edge gateway that translates messages from/to the device controller communication protocol and the protocol of the message broker.

For the microservice inter-communication, we chose NATS [24], a CNCF incubating project that enables both the pub/sub and remote procedure call (RPC) communication patterns and also offers a key-value database. Each Fog Device runs a NATS server instance, configured to be in a cluster with the others.

The optimizer microservice is implemented in Python. At startup, one of the nodes is elected as leader via the RAFT algorithm. Each node is responsible for one among the SEB, the cogenerative units or the storage. At each iteration step, the leader communicates the corresponding part of dual variable ν to the other two nodes, asking for their set of primal variables, based on which unit (storage, thermal, and SEB); meanwhile, it computes its own part. With the three results, it can compute the dummy and tau values for the next iteration.

In the case of the SEB, one of the edge gateways transfers the temperatures read from Modbus registries to the NATS KV store and converts to Modbus the setpoints obtained by the optimization.

Thanks to the publish/subscribe pattern, an event can trigger the start of the optimization process.

10.5.3   Results and discussion

From an ICT point of view, the SPM is characterized by a local controller for each plant. As to other nameplate values of each component (e.g., efficiencies, coefficient of performances) of the SPM can be found in [21].

The EMS developed in this chapter has been tested in a winter scenario (with high thermal load) with the EMS re-scheduling every 15-min for a 24-h period. The optimal electric and thermal dispatch schedule is obtained by solving the optimization problem (10.1)–(10.25) with the decentralized algorithm. The dispatch results are presented in Figures 10.8 and 10.9. The microgrid can supply 85% of electrical load using local resources, relying heavily on cogenerative units and PV during the day (6 am to 7 pm). Notably, the storage unit is able to charge during hours of higher PV production (1 pm to 4 pm) and discharge overnight to limit the amount of power drawn from the external grid. Further, the microturbines can supply the thermal and electrical loads without exceeding thermal balance constraints. The total operational costs for this scenario is 405.48$.

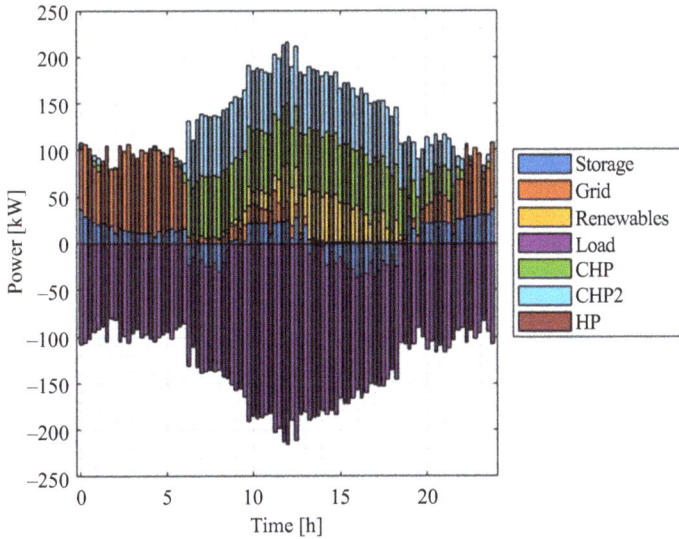

Figure 10.8   Optimal electrical plant dispatch

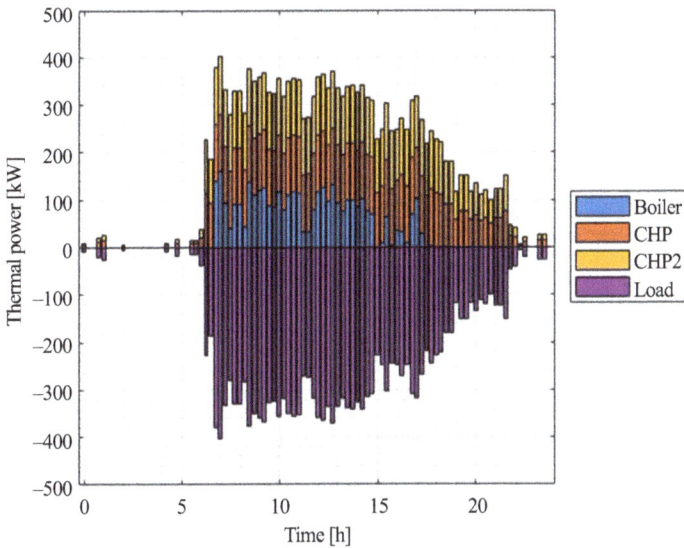

Figure 10.9   Optimal thermal plant dispatch

10.6   Conclusions

This chapter introduces a novel fog-based architecture designed for a decentralized EMS. The architecture was crafted to fulfill several key objectives, including scalability, decentralization, resiliency, interoperability, efficiency, responsiveness, and security. These goals were realized by capitalizing on the capabilities of emerging fog-enabled devices. A tailored decentralized optimization algorithm was employed to harness the computing resources of these devices without overtaxing their computational power or saturating the communication network. Moreover, the framework enhances security and privacy by minimizing external data transmission, and it improves resiliency, ensuring continued operation in the event of fog device failure or connectivity issues.

Validation of the architecture was conducted using a real test bed, the SPM of the Savona Campus, featuring a diverse array of components such as renewable and gas-fired generation units, storage devices, and HVAC equipment in buildings, including heat pumps and boilers. The results demonstrated the satisfactory performance of the proposed architecture in achieving the intended goals. Furthermore, the developed algorithm exhibited the ability to converge within a reasonable number of iterations, supporting its practical applicability.

Future efforts will focus on refining the algorithm's convergence speed and testing it in more intricate scenarios. This includes addressing power network constraints and exploring possibilities such as providing services to the distribution network.

Acknowledgments

This work was supported by the Silicon Valley Community Foundation (SVCF) for the project "Energy management systems: Distributed optimization and control on Edge nodes for smart grids." This work was also partially supported by the RAISE PNRR project, the SEED project P2PEM from the University of Genoa, the "Network 4 Energy Sustainable Transition—NEST" project (MUR project code PE000021, Concession Decree No. 1561 of 11.10.2022), in the framework of the NextGenerationEu PNRR plan (CUP D33C22001330002).

References

[1] Vuddanti S, and Salkuti SR. Review of energy management system approaches in microgrids. *Energies*. 2021;14(17):5459.
[2] Karavas CS, Arvanitis K, and Papadakis G. A game theory approach to multi-agent decentralized energy management of autonomous polygeneration microgrids. *Energies*. 2017;10(11):1756.
[3] Sharma P, Mathur HD, Mishra P, *et al.* A critical and comparative review of energy management strategies for microgrids. *Applied Energy*. 2022;327:120028.
[4] Lezama F, Palominos J, Rodríguez-González AY, *et al.* Agent-based microgrid scheduling: an ICT perspective. *Mobile Networks and Applications*. 2019;24:1682–1698.
[5] Gungor VC, Sahin D, Kocak T, *et al.* Smart grid technologies: communication technologies and standards. *IEEE Transactions on Industrial Informatics*. 2011;7(4):529–539.

[6]   Kuzlu M, Pipattanasomporn M, and Rahman S. Communication network requirements for major smart grid applications in HAN, NAN and WAN. *Computer Networks*. 2014;67:74–88.

[7]   Sitharthan R, Vimal S, Verma A, *et al.* Smart microgrid with the internet of things for adequate energy management and analysis. *Computers and Electrical Engineering*. 2023;106:108556.

[8]   Karthik SS, and Kavithamani A. Fog computing-based deep learning model for optimization of microgrid-connected WSN with load balancing. *Wireless Networks*. 2021;27:2719–2727.

[9]   Ye Y, Tang Y, Wang H, *et al.* A scalable privacy-preserving multi-agent deep reinforcement learning approach for large-scale peer-to-peer transactive energy trading. *IEEE Transactions on Smart Grid*. 2021;12(6):5185–5200.

[10] Afrasiabi M, Mohammadi M, Rastegar M, *et al.* Multi-agent microgrid energy management based on deep learning forecaster. *Energy*. 2019;186:115873.

[11] Yan L, Chen X, and Chen Y. A consensus-based privacy-preserving energy management strategy for microgrids with event-triggered scheme. *International Journal of Electrical Power & Energy Systems*. 2022;141:108198.

[12] Zhou X, Ai Q, and Yousif M. Two kinds of decentralized robust economic dispatch framework combined distribution network and multi-microgrids. *Applied Energy*. 2019;253:113588.

[13] Li H, Hui H, and Zhang H. Consensus-based energy management of microgrid with random packet drops. *IEEE Transactions on Smart Grid*. 2023;14(5):3600–3613.

[14] Li H, Hui H, and Zhang H. Decentralized energy management of microgrid based on blockchain-empowered consensus algorithm with collusion prevention. *IEEE Transactions on Sustainable Energy*. 2023;14(4):2260–2273.

[15] IEEE Standard for Adoption of Openfog Reference Architecture for Fog Computing; 2018. Standard No. 1934–2018. https://standards.ieee.org/standard/1934-2018.html.

[16] Yi S, Hao Z, Qin Z, *et al.* Fog computing: platform and applications. In: *2015 Third IEEE Workshop on Hot Topics in Web Systems and Technologies (HotWeb)*. IEEE; 2015. Available from: https://doi.org/10.1109/hotweb.2015.22.

[17] Srivastava P, Haider R, Nair VJ, *et al.* Voltage regulation in distribution grids: a survey. *Annual Reviews in Control*. 2023;55:165–181. Available from: https://www.sciencedirect.com/science/article/pii/S1367578823000123.

[18] Bertsekas DP. *Constrained Optimization and Lagrange Multiplier Methods*. Academic Press; 2014.

[19] Bracco S, Delfino F, Pampararo F, *et al.* A mathematical model for the optimal operation of the University of Genoa Smart Polygeneration Microgrid: evaluation of technical, economic and environmental performance indicators. *Energy*. 2014;64:912–922.

[20] Caliano M, Delfino F, Di Somma M, *et al.* An energy management system for microgrids including costs, exergy, and stress indexes. *Sustainable Energy, Grids and Networks*. 2022;32:100915.

[21]   Delfino F, Ferro G, Robba M, *et al.* An energy management platform for the optimal control of active and reactive powers in sustainable microgrids. *IEEE Transactions on Industry Applications*. 2019;55(6):7146–7156.
[22]   What is Podman? https://docs.podman.io/en/latest/ (2 October 2023).
[23]   qemu org. About QEMU. https://www.qemu.org/docs/master/about/index.html (accessed 2 October 2023).
[24]   NATS. https://nats.io/.

Index

www.ingramcontent.com/pod-product-compliance
Lightning Source LLC
Chambersburg PA
CBHW050511190326
41458CB00005B/1502